Fractional Slot Concentrated Windings

IEEE Press

445 Hoes Lane

Piscataway, NJ 08854

Fractional Slot Concentrated Windings

Design, Analysis, and Optimization

Gurakuq Dajaku

FEAAM GmbH
Research Center for Electrical Drives and
Actuators Munich, Neubiberg, Germany

IEEE*Press*
WILEY

Published by John Wiley & Sons, Inc., Hoboken, New Jersey.

For general information on our other products and services or for technical support, please contact our Customer Care Department within the United States at (800) 762-2974, outside the United States at (317) 572-3993 or fax (317) 572-4002.

Wiley also publishes its books in a variety of electronic formats. Some content that appears in print may not be available in electronic formats. For more information about Wiley products, visit our web site at www.wiley.com.

Library of Congress Cataloging-in-Publication Data:

Names: Dajaku, Gurakuq, author
Title: Fractional slot concentrated windings : design, analysis, and optimization / Gurakuq Dajaku.
Description: Hoboken, New Jersey : Wiley-IEEE Press, [2026] | Includes bibliographical references and index.
Identifiers: LCCN 2026020300 | ISBN 9781394423927 hardback | ISBN 9781394423941 adobe pdf | ISBN 9781394423934 epub | ISBN 9781394423958 obook
Subjects: LCSH: Electric generators—Windings
Classification: LCC TK2474 .D35 2026
LC record available at https://lccn.loc.gov/2026020300

Cover Design: Wiley
Cover Image: Ebru Avci

To my teacher,
who sacrificed much
for our education

Contents

About the Author *xi*
Preface and Acknowledgments *xiii*
Abbreviations and Symbols *xv*

1 **Introduction** *1*
1.1 Stator Winding Types *2*
1.2 Winding Realization *4*
1.2.1 Number of Layers *4*
1.2.2 Coil Polarity *4*
1.2.3 Phase Zone *5*
1.2.4 Base Winding *6*
1.3 Winding Topologies *6*
1.4 Fundamental Winding Specifications *7*
1.4.1 Slots per Pole and Phase *7*
1.4.2 Winding Periodicity *8*
1.4.3 Winding Feasibility *8*
1.4.4 Number of Slots *8*
1.4.5 Coil-pitch *8*
1.4.6 Turns per Coil and Phase *8*
1.5 MMF Harmonic Definition *9*
1.6 Manufacturing Methods *9*
1.7 Applications *10*

2 **Standard Fractional Slot Concentrated Windings** *13*
2.1 Single-coil MMF Formulation *13*
2.1.1 Definition of MMF Parameters *14*
2.1.2 Coil MMF Function *17*
2.2 MMF Analysis of Standard FSCWs *18*
2.2.1 Single-layer FSCWs *18*
2.2.2 Double-layer FSCWs *24*
2.3 Winding Factors *28*
2.3.1 Maximum Winding Factors *29*
2.3.2 Selection of Working Harmonic *30*
2.3.3 Winding Factors for Different SL-FSCWs *31*
2.3.4 Winding Factors for Different DL-FSCWs *32*

2.4 Three-phase MMF Distribution 33
2.5 MMF Characteristics of Various Three-phase Standard FSCWs 35
2.5.1 SL-FSCWs with $z = 1$ 35
2.5.2 SL-FSCWs with $z = 2$ 37
2.5.3 DL-FSCWs with $z = 1$ 39
2.5.4 DL-FSCWs with $z = 2$ 41
2.5.5 Comparison of Winding Performances 43

3 Fractional Slot Concentrated Windings with Unequal Zones 45
3.1 Winding Configurations 45
3.2 MMF Analysis 47
3.2.1 Single-layer FSCWs with Unequal Zones 48
3.2.2 Double-layer FSCWs with Unequal Zones 53
3.2.3 FSCWs with Asymmetric Zone Distribution 58
3.3 Winding Factors 66
3.3.1 Maximum Winding Factors 74
3.3.2 MMF Working Harmonics 78
3.4 MMF Characteristics of Various Three-phase FSCWs with Unequal Zones 79
3.4.1 SL-FSCWs with Unequal Zones 79
3.4.2 DL-FSCWs with Unequal Zones 81
3.4.3 DL-FSCWs with Unequal Zone-groups 83
3.4.4 DL-FSCWs with Asymmetric Zone-groups 84
3.4.5 Concluding Remarks 85

4 Fractional Slot Concentrated Windings with Specific Coil Configuration 87
4.1 FSCWs with Asymmetric Coils 87
4.1.1 Coil MMF Function 88
4.1.2 FSCWs with $z = 1$, $n_c > 1$, and Asymmetric Coils 91
4.1.3 FSCWs with $z = 2$, $n_c > 1$, and Asymmetric Coils 93
4.2 FSCWs with Unequal Turns per Coil 95
4.2.1 FSCWs with $z = 1$, $n_c > 1$, and Unequal Turns per Coil 96
4.2.2 FSCWs with $z = 2$, $n_c > 1$, and Unequal Turns per Coil 99
4.3 FSCWs with Unequal Coil-pitch 101
4.3.1 MMF Approach 102
4.3.2 Coil-pitch Effect 103
4.4 Exemplary Windings with Special Coil Configurations 104
4.4.1 Summary of the Results 106

5 Multiphase Fractional Slot Concentrated Windings 107
5.1 Multiphase FSCW Topologies 107
5.1.1 Conversion of a Three-phase FSCW into a Six-phase Winding 108
5.1.2 Other Multiphase FSCW Configurations 110
5.2 Multi-three-phase FSCWs 112
5.2.1 Asymmetric Six-phase FSCW 112
5.2.2 Asymmetric Nine-phase FSCW 114
5.2.3 Concluding Remarks 118

5.3 FSCWs with an Arbitrary Number of Phases *118*
5.3.1 MMF Function Analysis *120*
5.3.2 Operation Modes *122*
5.3.3 Multiphase FSCWs with Various Slot/Pole Combination *124*
5.4 Representative Multiphase FSCWs *125*
5.4.1 Multiphase Double-layer FSCWs *126*
5.4.2 Multiphase Single-layer FSCWs *127*
5.4.3 Concluding Remarks *131*

6 Fractional Slot Concentrated Windings with Flux-barrier Stator *133*
6.1 Stator Flux-barrier Techniques *135*
6.1.1 Flux-barrier Effect *136*
6.1.2 Realization *136*
6.2 FSCWs with Stator-yoke Flux-barriers *138*
6.2.1 Equivalent MMF Function *141*
6.2.2 SL-FSCWs with Stator-yoke Flux-barrier *143*
6.2.3 DL-FSCWs with Stator-yoke Flux-barrier *149*
6.2.4 Key Features and Potentials *153*
6.3 FSCWs with Stator-teeth Flux-barriers *155*
6.3.1 SL-FSCWs with Stator-teeth Flux-barriers *158*
6.3.2 SL-FSCWs with Unequal Zones and Stator-teeth Flux-barriers *163*
6.4 Multiphase SL-FSCWs with Flux-barrier Stator *165*
6.4.1 Multiphase FSCWs with Stator-yoke Flux-barriers *165*
6.4.2 Multiphase FSCWs with Stator-teeth Flux-barriers *169*

7 Two-tooth Fractional Slot Winding *179*
7.1 Implementation and Operating Principles *180*
7.2 Single-layer Two-tooth/Three-phase FSWs *184*
7.2.1 Single-layer FSWs with One Zone/Phase *184*
7.2.2 Single-layer FSWs with Two Zones/Phase *186*
7.3 Double-layer Two-tooth/Three-phase FSWs *188*
7.3.1 Double-layer FSWs with One Zone/Phase *188*
7.3.2 Double-layer FSWs with Two Zone/Phase *191*
7.3.3 Double-layer FSWs with Unequal-zones *193*
7.4 Multiphase Two-tooth FSWs *194*
7.4.1 Single-layer Multiphase Two-tooth FSWs *195*
7.4.2 Double-layer Multiphase Two-tooth FSWs *199*
7.4.3 Concluding Remarks *202*
7.5 Two-tooth FSWs with Specific Coil Configurations *203*
7.5.1 Two-tooth FSWs with Asymmetric Coils *204*
7.5.2 Two-tooth FSWs with Unequal Coils per Zone *205*
7.5.3 Two-tooth FSWs Employing an Unequal Tooth-width Stator Core *207*

8 Fractional Slot Concentrated Windings with Star-delta Connection *211*
8.1 Star-delta Windings *212*
8.1.1 Star-delta Phase Currents *213*
8.1.2 Winding MMF Function *214*

8.2 Star-delta FSCW Topologies *216*
8.2.1 Symmetric Star-delta FSCWs *216*
8.2.2 Asymmetric Star-delta FSCWs *219*
8.3 Winding Analysis *221*
8.3.1 Balanced Star-delta FSCWs *222*
8.3.2 Unbalanced Star-delta FSCWs *228*
8.3.3 Star-delta FSCWs with Sinusoidal MMF Distribution *233*
8.3.4 Concluding Remarks *234*
8.4 FSCWs with Sequential Star-delta Coil Arrangements *234*
8.4.1 MMF Winding Function *236*
8.4.2 FSCW Examples with Sequential Star-delta Coil Arrangements *239*
8.5 Star-delta FSCWs with a Dual Slot-layer Stator *242*
8.6 Exemplary Star-delta FSCW Machines *244*

9 Fractional Slot Concentrated Windings for Axial Flux Machines *249*
9.1 AFMs – Construction and Realization *249*
9.2 AFMs – Stator Winding MMF Harmonics *251*
9.3 Stator-shifting Technique for AFMs with Concentrated Windings *254*
9.3.1 AFMs with Double-layer FSCWs *255*
9.3.2 AFMs with Single-layer FSCWs *259*
9.4 AFMs with Asymmetric Stator Shifting *261*
9.5 AFMs with Unequal Turns per Coil *264*
9.6 AFMs with Flux-barrier Stator *264*
9.7 AFMs Topologies *265*

References *267*
Index *275*

About the Author

Gurakuq Dajaku received the diploma degree in electrical engineering from the University of Prishtina, Kosovo, in 1997 and the PhD degree from the Universität der Bundeswehr München, Munich, Germany, in 2006. Since 2007, he has been working for the FEAAM company (Research Center for Electrical Drives and Actuators), where he has served as CEO since 2018. Since 2008, he has been a lecturer at the Universität der Bundeswehr München, Munich, Germany. From 2010 to 2013, he was an invited lecturer at the University of Prishtina, Kosovo. Since 2013, he has been a lecturer at the "Haus der Technik" offering three specialized seminars in the field of electric machines. Since 2025, he has been an Associate Editor for *IET Electric Power Applications* (EPA). His research interests are in the field of electrical machines and drives. He has published numerous technical papers in different IEEE journals and conferences and has several international patents and patent-pending applications.

Preface and Acknowledgments

Over the last two decades, fractional slot concentrated windings (FSCWs) play a crucial role in modern electrical machines. Although they were invented more than a century ago, this type of winding has become popular since the 2000s, coinciding with the electrification of various transportation sectors, including road, water, and air vehicles. This surge in use has led to more and more attention being paid to their analysis and development. Initially, most of the research focused mainly on winding design, analyzing, and identifying the key issues caused by the magnetomotive force (MMF) winding harmonics, such as eddy-current losses in permanent magnets, increased core losses, and acoustic noise. However, a decade later, research has shifted to winding optimization and performing more in-depth analysis to overcome these challenges. Despite their simple structure with several turns wound around a single stator tooth, recent studies have revealed a surprising range of optimization possibilities using simple techniques. For example, a simple change in the coil structure, such as using unequal coils or different numbers of turns per coil-side, can remarkably reduce the subharmonics of the winding. Alternatively, a reduction in subharmonics can also be achieved by leaving the winding unchanged and modifying the stator core instead, e.g. by implementing flux-barriers in different stator regions. If both the subharmonics and the higher critical harmonics need to be reduced, the stator-shifting technique proves to be an effective solution. Another straightforward optimization approach is the use of a star-delta combination in FSCWs. Last but not least, the study of multiphase configurations, particularly FSCWs with an arbitrary number of phases, offer variable MMFs and various modes of operation.

Research on the analysis and optimization of FSCWs has grown rapidly over the past decades, leading to numerous publications in research papers and textbook chapters. The main motivation behind this work is to compile existing FSCWs and their optimization methods into a single comprehensive volume that provides a thorough analysis of all winding types. At the beginning of this work, I did not anticipate the vast number of FSCW configurations and the resulting new winding groups that I would come across at the end of my research. After a brief introduction to FSCWs and their practical implementation in Chapter 1, Chapter 2 deals with an in-depth analysis of the winding function, MMF, and the winding factors for all standard FSCWs based on the winding function theory. Beyond these well-known windings, seven additional winding groups are identified, classified, and examined in detail. Chapter 3, for example, investigates a particular class of FSCWs characterized by unequal zones and asymmetric zone distributions, thereby extending the range of standard concentrated winding topologies. The discussion of optimization methods begins in Chapter 4, which introduces FSCWs with specific coil configuration. Chapter 5 addresses the analysis of standard multiphase FSCWs, as well as windings with an arbitrary number of phases. In Chapter 6, the flux-barrier technique is introduced, including the implementation of flux-barriers in both the stator yoke and stator teeth regions. Chapter 7 deals with the winding category with two-tooth coil-pitch structure, commonly referred to as two-tooth fractional slot winding (FSWs). Chapter 8 examines various star-delta combination concepts applicable to FSCWs, highlighting their advantages and limitations. Finally, Chapter 9 presents a range of winding topologies and associated optimization methods specifically developed for axial flux machines. Overall, this work provides both the theoretical foundations and practical considerations of FSCWs, offering detailed insights into their benefits, challenges, and solutions. By covering winding configuration principles, performance analysis, and advanced optimization techniques, this book serves as a comprehensive reference for engineers, researchers, and practitioners involved in the design and optimization of modern electrical machines employing concentrated windings.

This work was carried out during my research at FEAAM – Research Center for Electric Drives and Actuators in Munich – Germany. Over the last two decades, FEAAM has made significant contributions to the theory and development of FSCWs. Several innovative methods, such as coil configuration with unequal turns per coil-side, the flux-barrier technique, two-tooth FSWs, FSCWs with variable modes of operation, as well as various star-delta FSCW concepts were initially developed at FEAAM. Therefore, I would like to express my sincere gratitude to all team members for their contributions, exceptional work, and brilliant ideas. My sincere thanks also go to all colleagues who critically reviewed the original manuscript of this book and provided valuable suggestions. I am also deeply grateful to Professor Dieter Gerling, head of the Institute for Electrical Drives (EAA) at the Universität der Bundeswehr Munich, Germany, for his tireless support and for the excellent cooperation between FEAAM and EAA teams, which has resulted in the two institutions becoming a renowned partnership in the field of electrical drives.

Last but certainly not least, I would like to express my deepest gratitude to my family, especially my wife and our children, for their understanding and unwavering support. My writing journey has often required sacrifices of precious time with them, and I am deeply grateful for their patience and encouragement throughout.

Munich, February 2026 *Gurakuq Dajaku*

Abbreviations and Symbols

A_i	auxiliary magnetomotive force function, magnitude ($i = 1, 2, 10, 20$), [A]
AC	alternating current
AFM	axial flux machine
ASM	asynchronous machine
B	magnetic flux density, [T]
C	constant variable (integer number)
DC	direct-current
DL	double-layer
DW	distributed winding
EV	electric vehicles
F	magnetomotive force function, [A]
F_{coil}	magnetomotive force function for one coil, [A]
$F_{\text{conv},j}$	magnetomotive force function for one conventional coil, phase-j, [A]
$F_{\text{go\_in},j}$	magnetomotive force function for one conductor, phase-j, [A]
F_I	auxiliary magnetomotive force function, [A]
F_j	magnetomotive force function for phase-j, [A]
$F_{j,\text{fb}}$	magnetomotive force function for phase-j and flux-barrier stator, [A]
$F_{j,z}$	magnetomotive force function of a phase-zone, [A]
$F_{j,\text{zg}}$	magnetomotive force function of a zone-group, [A]
$F_{3\text{ph}}$	magnetomotive force function of a three-phase winding system, [A]
F_{Δ}	magnetomotive force function of delta winding, [A]
F_Y	magnetomotive force function of star winding, [A]
F_0^*	coil function magnitude of a region including flux-barrier effect, [A]
F_{11}^*	coil function magnitude of a region including flux-barrier effect, [A]
F_{12}^*	coil function magnitude of a region including flux-barrier effect, [A]
F_0	negative magnitude of magnetomotive force distribution, [A]
F_1	positive magnitude of magnetomotive force distribution, [A]
FB	flux-barrier
fb_w	flux-barrier region width

FEA	finite element analysis
FSCW	fractional slot concentrated winding
FSW	fractional slot winding
GCD	greatest common divisor
g_0	air-gap length, [m]
H	magnetic field strength, [A/m]
h	positive integer
h_C	magnetomotive force harmonic content
h_{fb}	flux-barrier region depth
h_{yoke}	stator yoke thickness
i	current, [A]
i_j	phase-j current, [A]
k	integer number
k_{max}	maximum number of injected current harmonics
$k_{\alpha,\tau}$	coil-pitch variable
k_{ws}	number of m-phase winding systems
L	integration path
l	length [m], closed line
MEA	more electric aircraft
MMF	magnetomotive force, [A]
m	number of phases
N	north pole
N_{ph}	number of turns per phase
N_w	number of turns per coil
N_{w11}	number of turns, first coil-side
N_{w12}	number of turns, second coil-side
$N_{w,\mathrm{eq}}$	equivalent number of turns per coil
$N_{w,z}$	total number of turns per zone
$N_{w\Delta}$	number of turns in delta winding
$N_{w\mathrm{Y}}$	number of turns in star winding
n_c	number of coils per zone
$n_{c\Delta}$	number of coils in delta winding
$n_{c\mathrm{Y}}$	number of coils in star winding
n_{fb}	flux-barrier coefficient
n_l	number of layers
n_z	number of zones per zone group
P_{path}	number of parallel paths
PM	permanent magnet

PMSM	permanent magnet synchronous motor
p	number of pole pairs
Q_S	number of stator slots
$Q_{S\Delta}$	number of stator slots for delta winding
Q_{SY}	number of stator slots for star winding
Q_b	number of stator slots for the base winding
q	number of slots per pole per phase
r	air-gap radius
RFM	radial flux machine
S	unit vector of the surface, south pole
SESM	self-exciting synchronous machine
SL	single-layer
t	time, [s]
THD	total harmonic distortion
U	depiction of a phase
V	depiction of a phase
ν	harmonic order
ν_0	injected magnetomotive force harmonic
ν_{cr}	critical magnetomotive force harmonic
ν_p	working magnetomotive force harmonic
W	depiction of a phase
X	depiction of a phase
Y	depiction of a phase
Z	depiction of a phase
z	number of zones per phase
α_{fb}	flux-barrier angle, [rad]
α_S	stator peripheral angle, [rad]
α_τ	coil-pitch angle, [rad]
$\alpha_{z1,z2}$	phase shift between zone groups, [rad]
$\alpha_{\Delta Y}$	deviation spatial angle between delta and star winding sets, [rad]
β	integration angle, [rad]
Γ_{fby}	auxiliary function for flux-barrier effect
δ	current load angle, [rad]

$\wp$	number of winding periodicity
ξ_p	coil-pitch factor
$\xi_{p,\mathrm{eq}}$	equivalent coil-pitch factor
$\xi_{p,\mathrm{fb}t}$	coil-pitch factor including flux-barrier effect
$\xi_{p,\mathrm{fb}y}$	coil-pitch factor including flux-barrier effect
ξ_d	distribution factor
$\xi_{d,n_{\wp 0}}$	distribution factor for individual winding sets
$\xi_{d\Delta}$	distribution factor of delta winding
$\xi_{d\mathrm{Y}}$	distribution factor of star winding
ξ_z	phase-zone factor
ξ_{z12}	distribution zone factor
$\xi_{zg,\mathrm{unsm}}$	zone factor for asymmetric zone distribution
ξ_w	winding factor
$\xi_{w,\mathrm{fb}y}$	equivalent winding factor
$\xi_{w,\mathrm{fb}t}$	equivalent winding factor
$\xi_{w\Delta}$	winding factor of delta winding
$\xi_{w\mathrm{Y}}$	winding factor of star winding
φ_{coil}	spatial location of the coil, [rad]
φ_K	spatial angle between coils, [rad]
φ_m	spatial angle between winding sets, [rad]
$\varphi_{z,\mathrm{unsm}}$	unsymmetric spatial angle between zones, [rad]
$\varphi_{z1,z2}$	spatial angle between zone-groups, [rad]
$\varphi_{\Delta\mathrm{Y}}$	spatial angle between delta and star winding sets, [rad]
ω	electric angular frequency, [rad/s]
ω_ν	angular speed of νth harmonic, [rad/s]
Δ	difference
Δn_w	turn difference between coil-side

Subscripts

0	region
1	region
2	region
d	distribution
FB	flux-barrier
fb	flux-barrier
j	phase number
k	coil number
p	coil-pitch
r	radial

ν	harmonic order
Y	star
Δ	delta

Superscripts

$\hat{}$	peak/maximum value, amplitude
$\grave{}$	referred to two two-tooth stator, star-delta winding
*	flux-barrier effect, p.u. number

1

Introduction

The stator winding is the most crucial component in an electric machine, as it directly determines key performance metrics, including electromagnetic torque, power density, torque ripple, losses, acoustic noise, and field weakening capability. Therefore, an in-depth understanding of winding theory is essential for the efficient design and analysis of electric machines. This theory includes the principles and methods used in the winding design, performance calculation, and optimization.

The development of the winding theory of electric machines has been the subject of intensive research for over a century. Its fundamental principles were established in the early 20th century, mainly through the contributions of Arnold, Richter, Kauders, Heller, and Sequenz. A significant early work in this field was Arnold's book, "Die Ankerwicklungen der Gleichstrom-dynamomaschinen," published in March 1891, which introduced the "circuit rules" to determine the common and specific characteristics of the windings known at that time [1]. A few years later, in 1904, Arnold [2] developed the first method for determining the winding factors for both alternating current (AC) and direct current (DC) machines, based on the so-called "*star of slots*" (the original German term was "*Nutenstern*"). Richter [3] later improved this method, particularly for AC machines. Further advancements in winding theory were made by Kauders [4], Heller and Kauders [5], and Liwschitz [6]. They developed comprehensive formulas for calculating the winding factors for both two- and three-phase windings, including single-layer and double-layer configurations. These contributions marked a significant step forward in this field. In the following years, the theoretical foundations of winding design were further consolidated by the publication of several valuable textbooks, e.g. Veinott [7], Liwschitz [8], Sequenz [9], Richter [10], and Heiles [11]. After this period, the "star of slots" method has been used systematically and appears in many textbooks and research papers, e.g. Say [12], Heller and Hamata [13], Bianchi and Dai Pre [14], and Müller et al. [15].

Another widely used approach for analyzing the winding characteristics, developed around the same period as the "star of slots," is the "*winding function*" theory. This method describes the spatial distribution of conductors along the machine air-gap circumference and the magnitudes and polarities of the currents carried by the conductors. The resulting winding function is often referred to as current-linkage distribution or magnetomotive force (MMF) distribution. The foundations of this approach were laid as early as 1917, when Hague [16] provided a detailed analysis of the MMF distribution for a winding composed of full-pitch coils, deriving equations for the coil-pitch factor and distribution factor. A decade later, in 1927, Graham [17] investigated the MMF waveform for particular windings, identifying harmonic components with wavelength greater than twice the pole-pitch, referred to as sub-synchronous harmonics. A few years later, Calvert [18] developed general formulas for computing the distribution factor and coil-pitch factor for any MMF harmonics in three-phase fractional slot windings (FSWs). Since then, the "winding function" method, together with the "star of slots," has become a standard tool for the analysis and design of electric machines and has been extensively reported in numerous studies and textbooks, including works of Schuisky [19], Krause [20], Toliyat et al. [21], Sadarangani [22], Lipo [23], Ouyang et al. [24], Binder [25], Pyrhönen et al. [26], Gerling [27], and others.

The star of slots and the winding function methods were initially developed for integral-slot and fractional slot distributed windings (DWs). These methods, however, are equally applicable to fractional slot concentrated windings (FSCWs), which were discovered at the same time, and have become increasingly popular since about 2002 with the rise of vehicle electrification. Therefore, it can be said that this period marked the beginning of a second phase in electrical machine windings research, focusing primarily on the design, analysis, and optimization of FSCWs. In 2002, Cros and Viarouge [28] explored the realization of permanent magnet (PM) machines using various slot/pole concentrated winding combinations. They demonstrated methodologies to identify winding arrangements with high winding factors. In the same year, Koch and Binder [29] compared two PM motors with different FSCWs for traction applications. In 2003, Magnussen and Sadarangani [30] introduced a theory for calculating winding factors based on the star of slot and winding function methods. Their work included detailed calculations of winding factors for both single-layer and double-layer windings, covering combinations up to sixteen pole-pairs. Subsequently, in 2004, Libert and Soulard [31] conducted a detailed and comprehensive study of various slot/pole combinations of concentrated windings. They systematically presented the winding factors for all standard FSCW configurations known to date. Ishak et al. [32–34] introduced a comparison study of different FSCW PM machines, examining both single-layer and double-layer windings, as well as the configurations with equal and unequal stator teeth widths. A significant milestone was achieved by Bianchi et al. [35, 36], who extended and completed the theory of the star of slots method specifically for FSCWs. One of the key contributions of their work was the definition of design rules for concentrated windings based on this method. In 2007, El-Refaie et al. [37] proposed a general synthesis method for high-performance multi-phase PM machines using FSCWs, supported by tables of key performance indicators (winding factors, cogging torque, and net radial force) for optimal slot/pole selection. Later, El-Refaie [38] presented a comprehensive overview of FSCW PM synchronous machines, discussing design trade-offs, performance optimization, winding configurations, losses, fault tolerance, and comparisons across different PM topologies. Since then, research in the field of FSCWs has intensified, resulting in an extensive volume of literature. Therefore, I extend my sincere apologies to all authors whose work I have not been able to cite here. Your contributions to the advancement of machine theory are deeply appreciated and highly acknowledged.

1.1 Stator Winding Types

Electric machines have been of great importance since the beginning of industrial development and since then their role has consistently increased. Today, they play a crucial role in nearly every aspect of modern life, from household appliances and robotics to the automotive and aerospace industries. Compared with combustion engines, electric machines offer higher efficiency, lower costs, and simpler construction, making them a preferred choice for many applications. An electric machine consists of two main components:

- the stator (stationary part) and
- the rotor (rotating part).

Both the stator and rotor consist of a laminated iron core made of high permeability magnetic materials. Between these two components there is a small region known as the air-gap. The stator adjacent to the air-gap is equipped with several slots, which are distributed circumferentially and into which the winding coils are inserted. On the other hand, the rotor can be configured in different ways depending on the type of machine, e.g. as a PM rotor, wound rotor, squirrel cage rotor, or reluctance rotor.

In electric machines, two standard winding configurations are commonly used:

1. Distributed overlap winding: In this configuration, the windings are distributed across several slots, which provide a more uniform sinusoidal MMF distribution.
2. Tooth concentrated winding: In this design, the windings are wound around individual stator teeth, creating a more compact structure with significantly shorter end windings. However, unlike DWs, the resulting MMF distribution deviates substantially from a sinusoidal waveform, which is the main disadvantage of this configuration.

Figure 1.1 illustrates the stator structures for both winding configurations. The winding layout of the concentrated winding is much simpler and more compact, with significantly shorter end windings compared to the DW. In addition, the end windings do not overlap, and the respective Ohmic losses are lower. These properties make concentrated windings particularly advantageous for applications in which installation space, manufacturing, and machine efficiency are crucial.

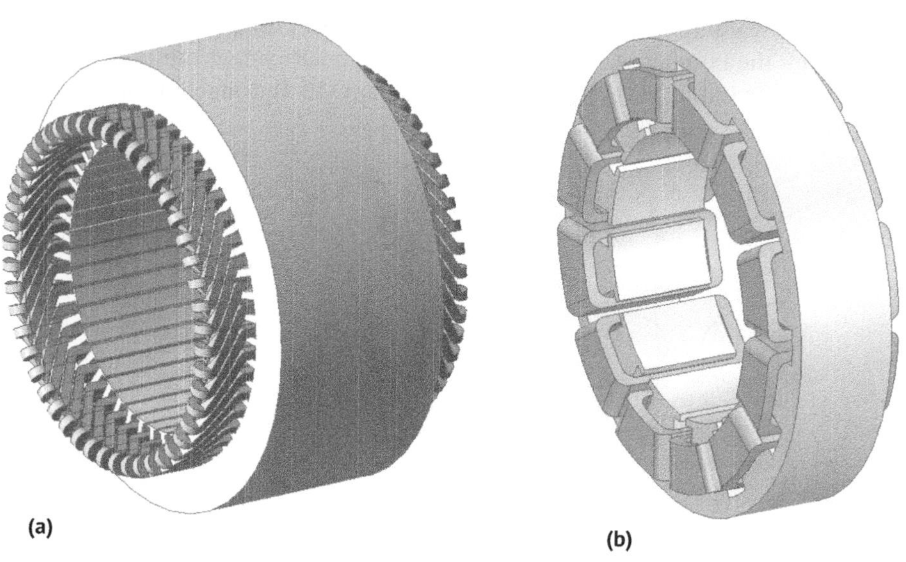

Figure 1.1 Stator construction of an electric machine. (a) Distributed winding. (b) Tooth concentrated winding.

1.2 Winding Realization

A stator winding consists of several key design parameters, such as turns, coils, coil polarity, coil groups, and winding layers. Depending on the coil arrangement, connection, and their combinations, a wide range of winding topologies can be achieved to meet different design requirements. Although these windings may appear similar and simple in their external structure, each topology is uniquely suited to a particular application due to its internal characteristics, such as MMF distribution and harmonic content.

1.2.1 Number of Layers

FSCWs can be designed as either single-layer or double-layer windings, as illustrated in Figure 1.2a,b. In the case of single-layer windings, the coils are wound on alternate teeth only, whereas in the double-layer configuration, each tooth carries its own coil. These winding types are often referred to in literature as one-layer and two-layer windings, respectively. Each configuration offers distinct advantages and challenges, both in terms of manufacturing process and machine performance, as discussed by Ishak et al. [32–34], El-Refaie and Jahns [39], and Meier and Soulard [40].

The selection of the number of layers is not limited to two. In order to enhance the MMF characteristics, several multilayer FSCWs have been proposed and studied in several papers, for instance, Ito et al. [41], Cistelecan et al. [42], Reddy et al. [43], Alberti and Bianchi [44], and Bianchi et al. [45]. Although these winding configurations demonstrate improved MMF performance, they are characterized with a reduced copper slot fill factor and increased manufacturing complexity.

1.2.2 Coil Polarity

In an FSCW, each coil is wound N_w time in a "*positive*" or "*negative*" direction around a single tooth, where N_w denotes the number of turns per coil. The coil polarity determines the current flow direction when the winding is excited with the current. In the machine illustrations

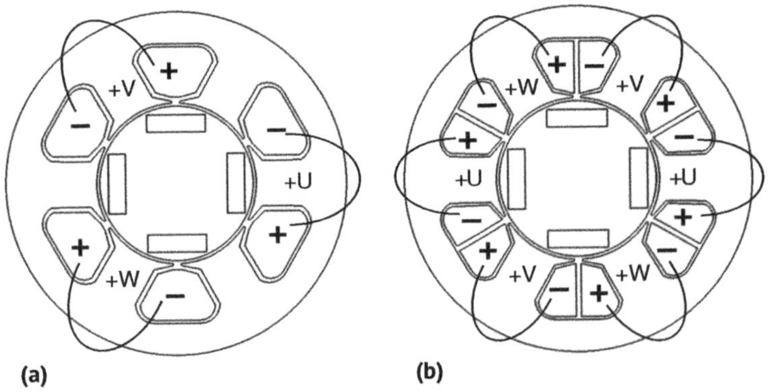

(a) **(b)**

Figure 1.2 Stator winding layouts for a four-pole PM machine. (a) Single-layer winding. (b) Double-layer winding.

presented in the following, "+" indicates that the assumed positive current direction is along the stator length (into the paper), and "−" indicates a current direction out of the paper.

1.2.3 Phase Zone

Another important term in the winding configuration is the coil-group or phase zone. This refers to a group of adjacent coils that belong to the same phase. Each phase zone consists of n_c single coils connected in series. The number of coils per zone, as well as the polarity of the coils, significantly influences the MMF characteristics of the winding. Figure 1.3 illustrates various one-zone single-layer and double-layer windings. The symbols "+" and "−" on the winding coil notation indicate the coil polarity, where "+U" represents a positive coil polarity and "−U" denotes a negative coil polarity.

As will be discussed in later chapters, coil arrangements with the same winding polarity, such as +U+U, lead to a higher winding factor for single-layer windings. However, the opposite is true for double-layer windings. Although certain winding groups may exhibit less favorable MMF properties, they are not ignored during the initial theoretical investigation in this book. Instead, they are thoroughly analyzed and studied along with other winding configurations to

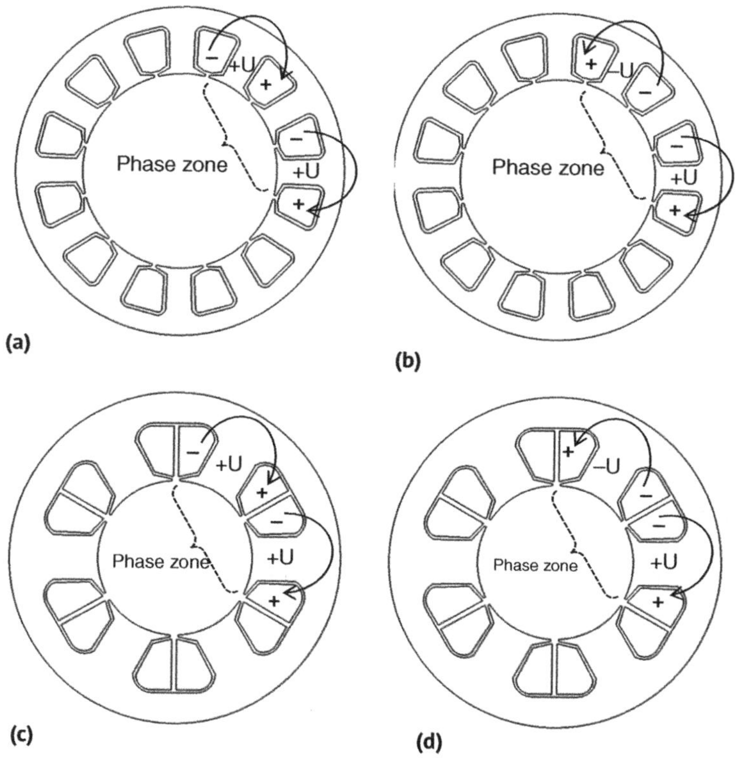

Figure 1.3 Coil arrangement in one phase zone. (a) Single-layer winding with same-polarity coils. (b) Single-layer winding with opposite polarity coils. (c) Double-layer winding with same-polarity coils. (d) Double-layer winding with opposite polarity coils.

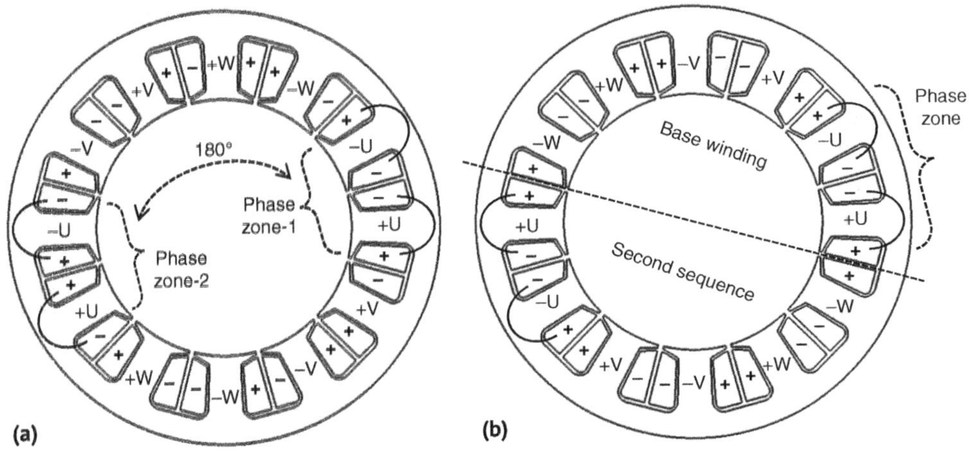

Figure 1.4 Definition of "base winding" and "phase zone." (a) FSCW with one base winding and two zones per phase. (b) FSCW with the base winding repeated twice and only one zone per base winding.

validate their characteristics and understand their potential applications. This comprehensive approach ensures a complete evaluation of all possible winding topologies.

1.2.4 Base Winding

A conventional stator winding may include several identical sequences of coil groups. The shortest such sequence can be considered the minimal units of the winding, referred to as the "base winding" layout, as illustrated in Figure 1.4. Since each winding sequence generates identical spatial MMF distribution in the air-gap, it is sufficient to focus only on the base winding arrangement when performing the winding analysis.

The base winding may comprise either one or two phase zones. In the case of two zones per phase, they are spaced 180 electrical degrees apart and have opposite polarities, as depicted in Figure 1.4a.

1.3 Winding Topologies

Common FSCWs can be realized in different configurations depending on the number of coils per zone, number of zones per phase, and polarity of the coils:

- One-zone windings
  1. $z = 1, n_c = 1,$
  2. $z = 1, n_c > 1,$ and same polarity coils,
  3. $z = 1, n_c > 1,$ and opposite polarity coils.
- Two-zones windings
  4. $z = 2, n_c = 1,$
  5. $z = 2, n_c > 1,$ and same polarity coils,
  6. $z = 2, n_c > 1,$ and opposite polarity coils.

1.4 Fundamental Winding Specifications

In order to classify stator windings uniquely, the following useful winding parameters are defined:

N_w – no. of turns per coil
n_c – no. of coils per zone
n_l – no. of coil layers
m – no. of phases
P_{path} – no. of parallel paths (series/parallel)
p – no. of rotor pole pairs
Q_S – no. of stator slots
q – no. of slots per pole per phase
z – no. of coil-groups (zones)
$\wp$ – winding periodicity

To demonstrate the winding parameters, Figure 1.5 shows the winding layout for an example winding (one-phase) with $n_c = 2$, $z = 2$, and two different parallel path combinations. The stator consists of $Q_S = 12$ slots, and the winding is viewed from the inner circumference of the stator, so only the tooth-tip surfaces are visible.

1.4.1 Slots per Pole and Phase

The term "fractional slot winding" refers to the number of slots per pole and phase, and is defined as:

$$q = \frac{Q_S}{2 \cdot p \cdot m} = \frac{q_n}{q_d} \tag{1.1}$$

where q_d and q_n are the numerator and denominator of q reduced to the lowest term. If q is not an integer, the winding is an FSW. However, in FSCWs the q value is smaller than unity $(q < 1)$.

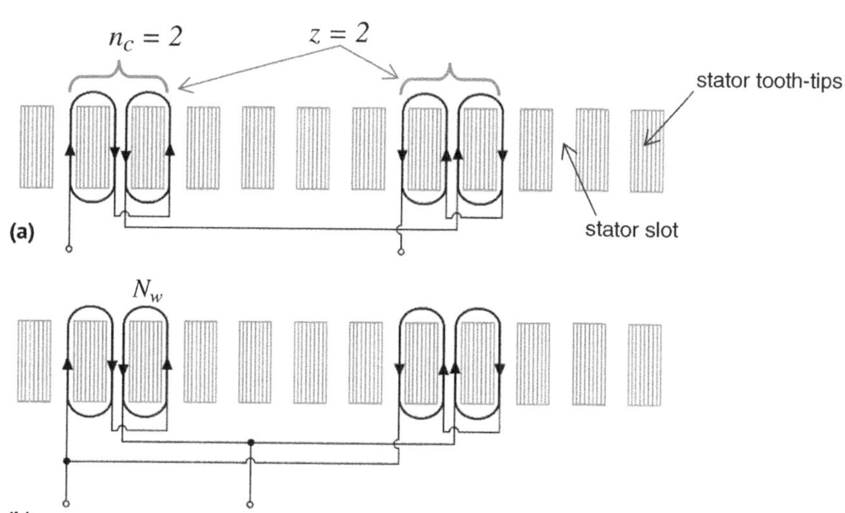

Figure 1.5 Winding layout for an example double-layer concentrated winding. (a) Zones in series $(P_{\text{path}} = 1)$. (b) Zones in parallel $(P_{\text{path}} = 2)$.

1.4.2 Winding Periodicity

The second winding property used in the classification scheme is the winding periodicity $\wp$ parameter determined as the greatest common divisor (GCD) between the number of slots Qs and the number of pole pairs p, i.e.

$$\wp = \text{GCD}\{Q_S, p\} \tag{1.2}$$

For an FSCW stator with the same q but a higher number of slots and poles, the winding layout is determined by repeating the base winding layout $\wp$ times. Each of the $\wp$ basic windings has Q_b slots and p_b poles, therefore:

$$Q_b = \frac{Q_S}{\wp} \text{ and } p_b = \frac{p}{\wp} \tag{1.3}$$

Consequently, for each base winding:

$$\wp = \text{GCD}\{Q_b, p_b\} = 1 \tag{1.4}$$

1.4.3 Winding Feasibility

For a m-phase winding, a symmetric FSCW is feasible if [46]:

$$\frac{Q_S}{\text{GCD}\{Q_S, p\}} = mC \tag{1.5}$$

where C is a positive integer.

1.4.4 Number of Slots

If n_c denotes the number of adjacent coils per zone and z the number of zones per base winding, an m-phase FSCW with n_c and z winding parameters can be constructed in a stator with Q_b slots:

$$Q_b = \begin{cases} 2mzn_c, & \text{for single-layer FSCW} \\ mzn_c, & \text{for double-layer FSCW} \end{cases} \tag{1.6}$$

The total number of stator slots when $\wp > 1$ can be determined from (1.3).

1.4.5 Coil-pitch

The coil-pitch is defined as the circumferential angle between the coil-sides. It is convenient to express the coil-pitch in terms of the number of slots:

$$\alpha_\tau = \frac{2\pi}{Q_S} \tag{1.7}$$

1.4.6 Turns per Coil and Phase

A winding coil consists of N_w turns wound around a stator tooth, while the phase winding, on the other hand, is realized by combining several coils and zones together. When the number of coils n_c and zones z are greater than one, the n_c-coils within a zone are connected in series, whereas the phase zones can be connected either in series or in parallel. Additionally, when the winding is constructed by repeating the base winding layout $\wp$ times, the total number of turns per phase is defined as:

$$N_{\mathrm{ph}} = \wp \frac{z n_c N_w}{P_{\mathrm{path}}} \tag{1.8}$$

where P_{path} is number of parallel paths:

$$P_{\mathrm{path}} = \begin{cases} 1, & \text{all zones in series} \\ z, & \text{zones of base winding in parallel} \\ z \cdot \wp, & \text{all zones in parallel} \end{cases} \tag{1.9}$$

1.5 MMF Harmonic Definition

The MMF distribution of FSCWs does not have a sinusoidal waveform and therefore contains many harmonics. The MMF harmonics, denoted by ν, are defined as shown in Equation (1.10). All harmonics are related to the 2π bore of the machine, meaning the *fundamental harmonic* corresponds to $\nu = 1$ and forms one pole pair. The harmonic that interacts with the rotor's $2p$ poles has order $\nu_p = p$ and is called the *working harmonic*. This component is often also referred to as the *operating harmonic* or *operating wave*.

In addition to the working harmonic, the MMF distribution of many FSCWs exhibits lower-order harmonics, so-called *subharmonics*, and a significant presence of higher-order harmonics. Components whose order is close to ν_p and that usually have the same winding factor are called critical harmonics.

$$\nu^{\mathrm{th}} = \begin{cases} \nu_p = p, & \text{working harmonic} \\ \nu^{\mathrm{th}} < p, & \text{sub-harmonics} \\ \nu^{\mathrm{th}} > p, & \text{higher harmonic} \\ \nu_{\mathrm{cr}} = \begin{cases} \nu_p \pm 1 \\ \nu_p \pm 2 \end{cases}, & \text{critical harmonic} \end{cases} \tag{1.10}$$

1.6 Manufacturing Methods

Concentrated windings are known for their structural simplicity and low manufacturing costs. Several well-established techniques can be used to produce this type of winding, including [47, 48]:

- linear winding,
- flyer winding,
- needle winding,
- insertion winding.

In linear winding, the tooth or the coil rotates while the wire feeder moves only linearly to distribute the wire evenly, as illustrated in Figure 1.6a. In contrast, in flyer winding, a rotating flyer guides the wire around a stator tooth. Both techniques enable manufacturing of FSCW stators in a modular manner while achieving a high copper fill factor.

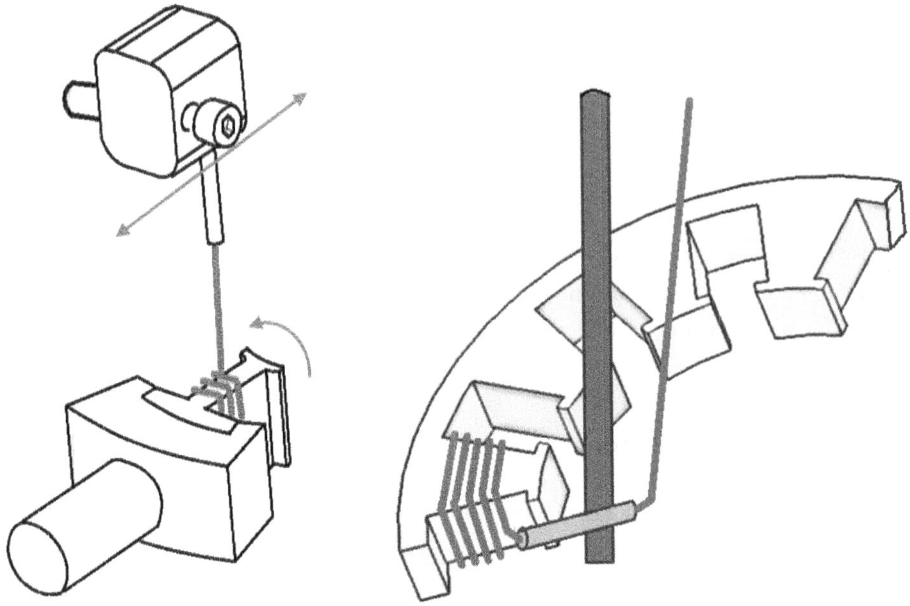

(a) (b)

Figure 1.6 Different manufacturing techniques for an FSCW stator. (a) Linear winding. (b) Needle winding.

Another practical technique is the needle winding method, which allows a complete stator to be wound without separating it into individual teeth. As shown in Figure 1.6b, needle winding method uses a movable needle that is inserted into the slot openings of the stator. The needle is then moved around the tooth to place the wire accurately at the designated position. Although this technique offers several advantages, it also involves certain compromises. In particular, needle motion requires space within the slot that must remain free, reducing the achievable copper fill factor compared with other methods.

1.7 Applications

Concentrated windings have gained popularity in various electric machine applications due to their unique advantages over traditional DWs. As mentioned earlier, the use of concentrated coils around the teeth offers clear benefits in terms of manufacturing simplicity and more efficient utilization of available space. Other advantages include higher power and torque density, lower Ohmic losses, improved fault tolerance, nonoverlapping coils, shorter end windings, and lower production costs [28, 30, 34, 40]. Over the past two decades, this winding type has been used in a variety of high-performance applications, ranging from low-power devices to those with several megawatts of power:

- Traction motors,
- Aerospace and more electric aircraft (MEA),
- Industrial drives and automation,

- Renewable energy systems,
- Household appliances,
- Robotics,
- Servo motors,
- Auxiliary motors and actuators.

Despite its many advantages, an FSCW is not always the perfect solution. The discrete distribution of coils leads to a non-sinusoidal MMF distribution along the air-gap, which contains considerable low- and high-order spatial harmonics. In PM machines, these harmonics contribute to significant rotor magnet losses. In asynchronous machines, they cause increased rotor bar losses and generate parasitic torque components that degrade the average torque of the machine. In synchronous reluctance machines, they lead to high torque ripple and a low power factor. Furthermore, spatial MMF harmonics cause local magnetic saturation in the magnetic core, additional stator and rotor iron losses, as well as noise and vibrations. Therefore, the accurate determination and analysis of the winding's MMF characteristics, including harmonic contents and winding factors, are crucial for the efficient design and optimization of FSCW machines.

2

Standard Fractional Slot Concentrated Windings

As discussed in Chapter 1, a concentrated winding comprises several turns connected in series and wound around a stator tooth, also referred to as a winding coil. These coils can be grouped into one or more adjacent coils to create a coil-group (also called a zone). The entire phase winding is then made from one or two zones. When the winding configuration has common coils and equal, symmetrically distributed zones, the winding is referred to as a standard fractional slot concentrated windings (FSCWs) in the following.

Figure 2.1 illustrates the generalized structure of a standard FSCW using two examples, where the different colors represent the zones corresponding to the individual phases (U, V, W). Within a zone, the coil polarity can be the same or different, and the windings can be arranged in either single-layer (SL) or double-layer (DL) form. Combining key parameters, such as number of coils per zone, coil polarity, number of layers, and also the choice between one-zone or two-zone configuration, a wide variety of FSCW topologies can be achieved. This flexibility allows numerous configurations to address diverse application requirements. One of the key advantages of FSCWs is therefore the large number of possible slot and pole combinations, which offer a wide variety of winding solutions. Figure 2.2 illustrates three common FSCW examples. The main difference between these topologies lies in the number of coils per zone and the number of zones per phase.

The first example in Figure 2.2a has one coil per zone ($n_c = 1$), one zone per phase ($z = 1$), and winding periodicity of two ($\wp = 2$). The base winding for this configuration is the 3-slot/2-pole winding with $q = 0.5$. The second example in Figure 2.2b has $z = 1$, $n_c = 3$, and $\wp = 1$. This design is referred to as 9-slot/8-pole winding with $q = 0.375$. Finally, the example in Figure 2.2c is the well-known 12-slot/10-pole FSCW with $z = 1$, $n_c = 2$, and $\wp = 1$. The q parameter for this winding type is $q = 0.4$.

2.1 Single-coil MMF Formulation

The MMF of a stator winding in an electric machine is a measure of the magnetic field potential or "forcing function" that generates a magnetic flux within the machine's magnetic circuit. This parameter directly determines the magnetic field distribution in the air-gap and consequently, the overall machine performance. Accurate modeling of the MMF distribution is therefore essential for predicting and understanding key machine characteristics, including electromagnetic torque, flux-linkage, back-EMF, inductance, and core losses.

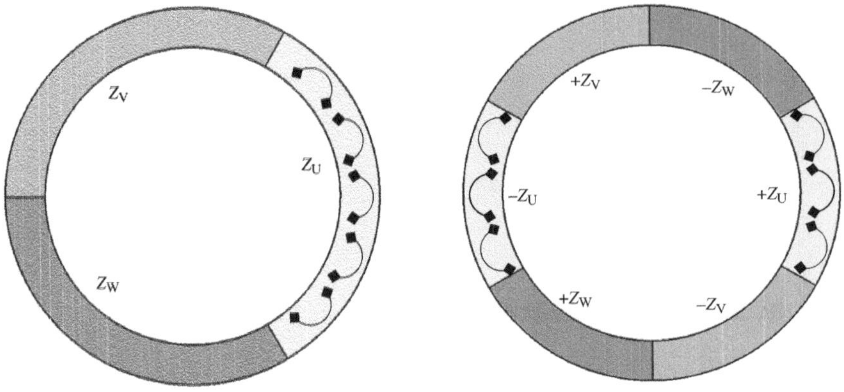

Figure 2.1 Standard FSCWs. (a) One zone per phase with $n_c = 5$ coils per zone. (b) Two zones per phase with $n_c = 3$ coils per zone.

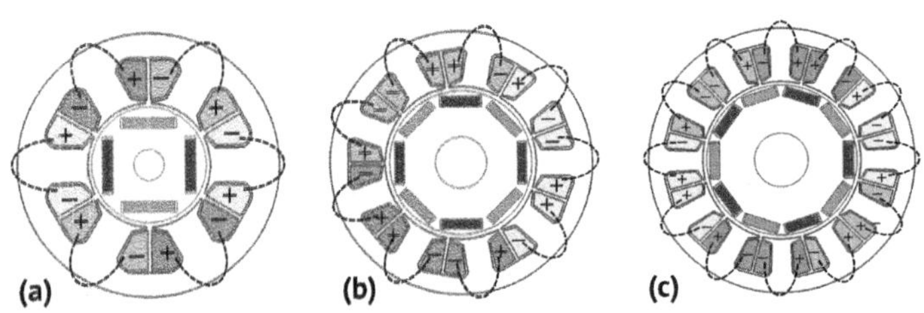

Figure 2.2 Permanent magnet synchronous motors (PMSMs) with standard concentrated windings. (a) FSCW with $z = 1, n_c = 1$, and $\wp = 2$. (b) FSCW with $z = 1, n_c = 3$, and $\wp = 1$. (c) FSCW with $z = 1, n_c = 2$, and $\wp = 1$.

In this study, winding function theory is used to derive the mathematical expressions of the MMF for different winding configurations. The analysis begins with the formulation of the MMF for a single coil based on Ampère's law and flux-conservation principle (Gauss's law). For windings consisting of multiple coils, the total MMF is obtained by superimposing the contributions of all individual coils. To enable a simple analysis of the air-gap MMF, the following assumptions are used:

- The magnetic permeability of the iron μ_{Fe} is considered infinite, resulting in negligible MMF across any flux path within the ferromagnetic circuit.
- The magnetic circuit is assumed to be linear, enabling the application of linear system principles, particularly the superposition theorem.
- The slotting effect is neglected by assuming a uniform air-gap length (g_0).

2.1.1 Definition of MMF Parameters

To discuss the air-gap MMF, a single-coil stator model shown in Figure 2.3a is considered. When the coil is excited with a current, the resulting flux density distribution is depicted in Figure 2.3b. For simplicity, the stator is modeled with only two slots, each containing a coil-side, and the rotor is assumed to be cylindrical.

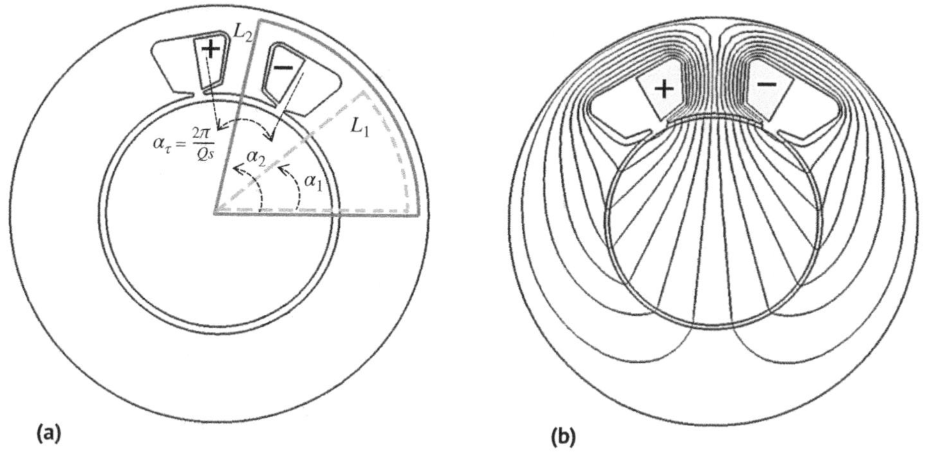

Figure 2.3 (a) Single-coil stator. (b) Flux density distribution due to a single-coil.

From Ampère's law, the line integral of the magnetic field intensity H along a closed Amperian path is equal to the ampere-turns linked by this path. In Figure 2.3a, two Amperian paths labeled with L_1 and L_2 are considered. Taking the path defined as loop L_1, the line integral does not enclose any current-carrying conductor. Thus,

$$\oint_{L_1} \vec{H} \cdot d\vec{l} = 0 \tag{2.1}$$

Assuming that the field intensity within the stator and rotor core is negligible (due to the high permeability of the steel), we can simplify and evaluate Equation (2.1) as follows:

$$H_r(\alpha_1)\, g_0 - H_r(0)\, g_0 = 0 \tag{2.2}$$

Here, $H_r(0)$ and $H_r(\alpha_1)$ denote the radial component of magnetic field intensity in the air-gap corresponding to positions 0 and α_1, respectively. Equation (2.2) indicates that the field intensity is uniformly constant in the region before any current is enclosed. Since the term $H_r(\alpha_s)\, g_0$ represents the magnetic potential between the stator and rotor surfaces at any angular position α_s along the air-gap, it is beneficial to define the MMF as $F(\alpha_s) = H_r(\alpha_s)\, g_0$. In terms of MMF, Equation (2.2) can be written as:

$$F(\alpha_1) = F(0) \tag{2.3}$$

Now, let us consider the path L_2. Applying Ampere's law to this path yields:

$$\oint_{L_2} \vec{H} \cdot d\vec{l} = N_w i \Rightarrow F(\alpha_2) - F(0) = N_w i \tag{2.4}$$

Assuming $F(0) = 0$, Equations (2.3) and (2.4) indicate that MMF in the air-gap region outside the coil is zero, and within the coil region, the MMF distribution exhibits a step change with a magnitude of $N_w i$, as approximated in Figure 2.4a. Under this assumption, Ampere's law results in a unidirectional MMF distribution for the coil, which would produce a magnetic flux density in only one direction across the air-gap.

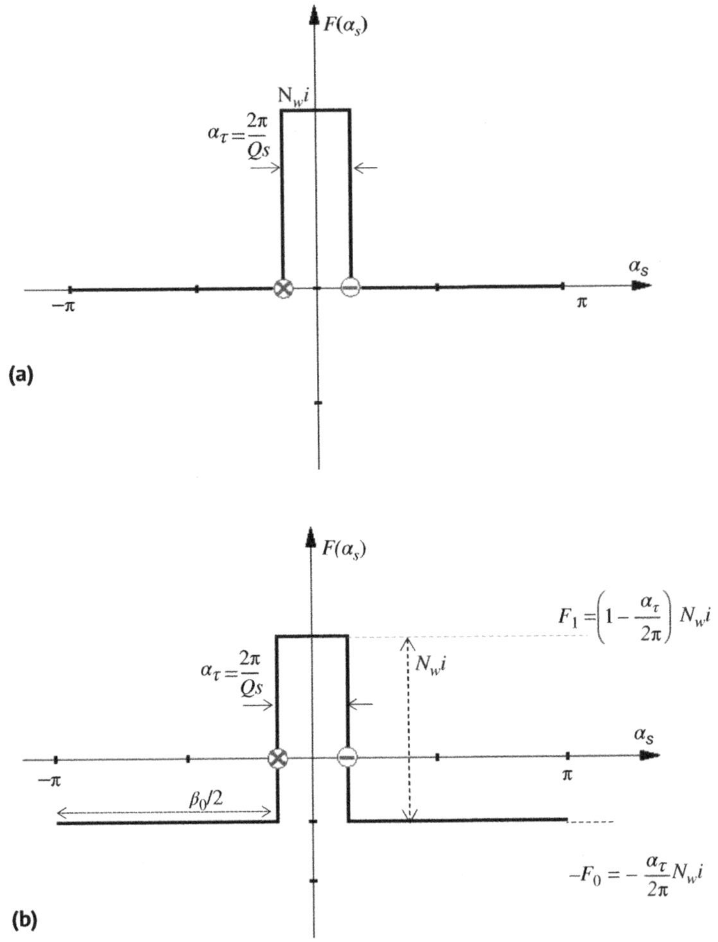

Figure 2.4 (a) Air-gap MMF for a single coil assuming $F(\alpha_s) = 0$ outside the coil region. (b) Modified air-gap MMF satisfying the Gauss's law. The angular origin along the air-gap periphery is defined at the center of the coil.

However, Gauss's law of magnetism must also be satisfied:

$$\oint_S \vec{B} \cdot d\vec{S} = 0 \tag{2.5}$$

or, for the given example ($rld\alpha_S$ is the incremental area of the cylindrical air-gap surface):

$$\int_0^{2\pi} B_r(\alpha_S) rld\alpha_S = 0 \tag{2.6}$$

where l is the axial length of the machine, r the air-gap radius, and α_S the stator peripheral angle. Equation (2.6) means that no net flux may travel across the air-gap. Therefore, to satisfy the Gauss's law, the air-gap MMF distribution must have a zero average value. This condition can be understood by examining the flux density generated by a single coil, as depicted in Figure 2.3b.

Here, the flux crossing the air-gap within the coil-pitch angle α_τ returns at an angle $\beta_0 = 2\pi - \alpha_\tau$. Based on this, the MMF of the coil can be approximated using the modified curve shown in Figure 2.4b. In this scenario, the positive and negative MMF values must satisfy the following relationship:

$$\int_0^{\alpha_\tau} F(\alpha_S)d\alpha_S = \int_{2\pi-\alpha_\tau}^{2\pi} F(\alpha_S - \alpha_\tau)d\alpha_S \tag{2.7}$$

If F_1 and F_0 are defined as the magnitudes of the MMF distribution under the coil and outside the coil region within the air-gap, the terms in Equation (2.7) can be solved as follows:

$$\begin{aligned} &\int_0^{\alpha_\tau} F(\alpha_S)d\alpha_S = F_1 \cdot \alpha_\tau \\ &\int_{\alpha_\tau}^{2\pi-\alpha_\tau} F(\alpha_S)d\alpha_S = F_0 \cdot (2\pi - \alpha_\tau) \end{aligned} \tag{2.8}$$

By combining Equations (2.7) and (2.8) leads to:

$$\alpha_\tau F_1 = (2\pi - \alpha_\tau)\, F_0 \tag{2.9}$$

Furthermore, based on the coil MMF distribution shown in Figure 2.4b, i.e. $F(\alpha_2) = F_1$ and $F(0) = -F_0$, Equation (2.4) may be expressed as:

$$F_1 + F_0 = N_w i \tag{2.10}$$

Finally, solving the system of Equations (2.9) and (2.10) yields the MMF function parameters:

$$\begin{aligned} F_0 &= \frac{\alpha_\tau}{2\pi} N_w i \\ F_1 &= \left(1 - \frac{\alpha_\tau}{2\pi}\right) N_w i \end{aligned} \tag{2.11}$$

2.1.2 Coil MMF Function

In a Q_S-slot FSCW stator, each coil is wound around a single tooth, resulting in a square pulse spatial MMF distribution with a width of $2\pi/Q_S$ mechanical radians, as presented in Figure 2.4b. According to relation (2.11), the MMF waveform of a single coil can be defined as:

$$F_{\text{Coil},k}(\alpha_S) = \begin{cases} F_1 = \left(1 - \dfrac{\alpha_\tau}{2\pi}\right) N_w i, & \alpha_k < \alpha_S < \alpha_k + \dfrac{2\pi}{Q_S} \\ F_0 = \dfrac{\alpha_\tau}{2\pi} N_w i, & \text{elsewhere} \end{cases} \tag{2.12}$$

The Fourier series describing this function is:

$$F_{\text{Coil}}(t, \alpha_S) = \frac{2(F_1 + F_0)}{\pi} i(t) \sum_\nu \frac{1}{\nu} \sin\left(\nu \frac{\alpha_\tau}{2}\right) \cdot \cos(\nu \alpha_S) \tag{2.13}$$

From Equations (2.10) and (2.13), the MMF function for a single coil becomes:

$$F_{\text{Coil}}(t,\alpha_S) = \frac{2N_w i(t)}{\pi} \sum_{\nu} \frac{{}^{\nu}\xi_p}{\nu} \cdot \cos(\nu \alpha_S) \tag{2.14}$$

Here, ν represents the MMF winding harmonics, and ${}^{\nu}\xi_p$ the coil-pitch factor, defined as:

$${}^{\nu}\xi_p = \sin\left(\nu \frac{\alpha_\tau}{2}\right) = \sin\left(\nu \frac{\pi}{Q_S}\right) \tag{2.15}$$

2.2 MMF Analysis of Standard FSCWs

2.2.1 Single-layer FSCWs

In a single-layer FSCW (SL-FSCW), coils are wound around every second tooth, with each coil occupying two adjacent slots. Consequently, the number of stator teeth is twice the number of winding coils. As outlined in Chapter 1, this winding topology allows numerous coil combination options, each of which has a significant impact on the resulting spatial MMF distribution. In the analytical approach developed below, the focus is restricted to the base winding layout and the corresponding single-phase MMF waveform.

2.2.1.1 SL-FSCW with $z = 1$ and $n_c = 1$

The base winding layout comprises one coil per zone ($n_c = 1$), one zone per phase ($z = 1$), and a stator with 6 slots ($Q_s = 6$). The winding arrangement and the corresponding spatial MMF distribution are illustrated in Figure 2.5a,b. For the case of a single coil per phase, the MMF distribution is identical to the coil MMF function, i.e.,

$$F_j(t,\alpha_S) = F_{\text{Coil}}(t,\alpha_S) \tag{2.16}$$

Accordingly, the phase-j MMF distribution becomes:

$$F_j(t,\alpha_S) = \frac{2N_w i_j(t)}{\pi} \sum_{\nu} \frac{{}^{\nu}\xi_p}{\nu} \cos(\nu(\alpha_S - \varphi_{\text{coil}})) \tag{2.17}$$

where φ_{coil} denotes the spatial location of the coil relative to the phase-j axis. In the MMF representation shown in Figure 2.5b, the winding coil is fixed on the zero axis, so $\varphi_{\text{coil}} = 0°$.

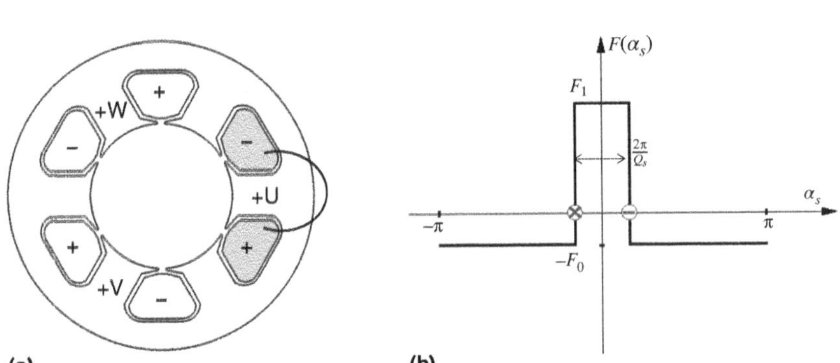

Figure 2.5 (a) Three-phase SL-FSCW with $z = 1$ and $n_c = 1$. (b) MMF distribution of phase-j.

According to Equation (2.15), the coil-pitch factor for the studied winding, i.e. for $Q_S = 6$, is

$$ {}^{\nu}\xi_p\Big|_{Q_S=6} = \sin\left(\nu\frac{\pi}{6}\right) \tag{2.18} $$

2.2.1.2 SL-FSCW with $z = 1$ and $n_c > 1$

In the base winding layout, a zone refers to a group of adjacent coils, with each coil displaced from the next by a mechanical angle φ_κ. In an SL configuration, this parameter is defined as

$$ \varphi_\kappa = 2\alpha_\tau = \frac{4\pi}{Q_S} \tag{2.19} $$

When a zone contains more than one coil (i.e. $n_c > 1$), two winding groups can be realized:

- Same polarity adjacent coils: +U +U +U,..., +U +U
- Opposite polarity adjacent coils: +U −U +U,..., +U −U.

Figures 2.6–2.9 illustrate the winding configurations for $n_c = 2$ and $n_c = 3$ in combination with various coil arrangements. The coil polarities within a winding zone have a strong impact on the

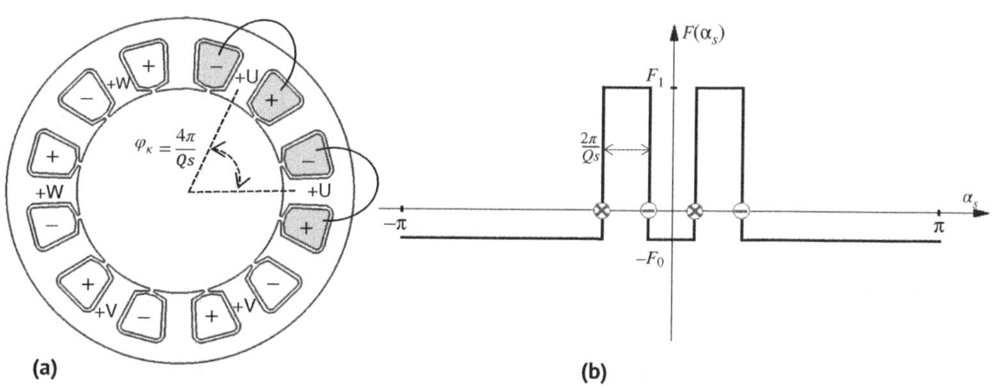

Figure 2.6 (a) Three-phase SL-FSCW with $z = 1$, $n_c = 2$, and same polarity coils. (b) MMF distribution of phase-j.

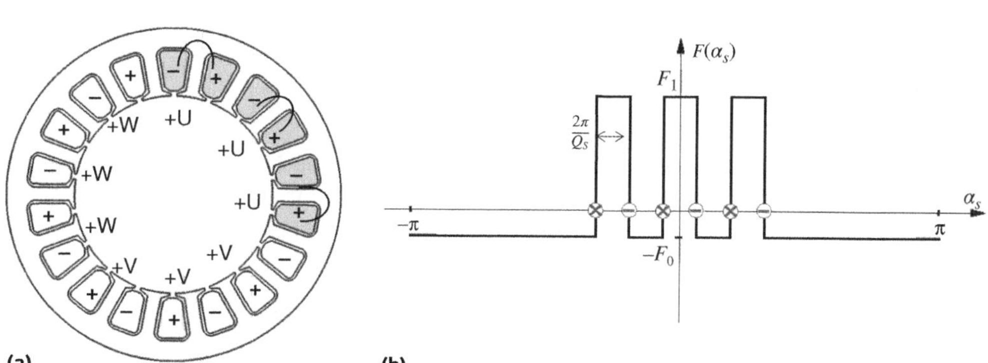

Figure 2.7 (a) Three-phase SL-FSCW with $z = 1$, $n_c = 3$, and same polarity coils. (b) MMF distribution of phase-j.

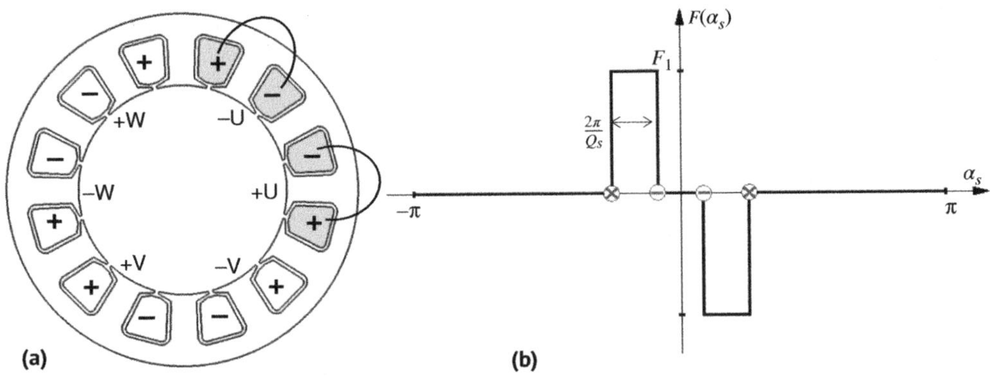

Figure 2.8 (a) Three-phase SL-FSCW with $z = 1$, $n_c = 2$, and opposite polarity coils. (b) MMF distribution of phase-j.

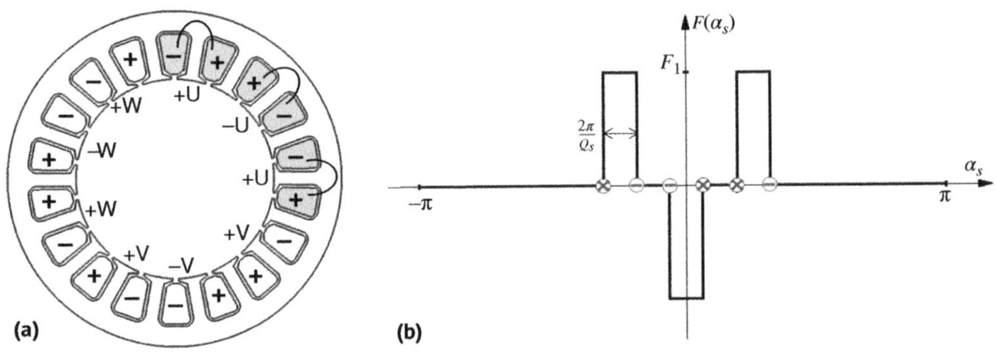

Figure 2.9 (a) Three-phase SL-FSCW with $z = 1$, $n_c = 3$, and opposite polarity coils. (b) MMF distribution of phase-j.

resulting MMF waveform, as the following examples show. Figures 2.6 and 2.7 present the winding layouts and the corresponding spatial MMF distributions (phase-j) for the case with same polarity coils. Figures 2.8 and 2.9, on the other hand, examine the corresponding topologies featuring coils with opposite polarity.

Case 1: Same polarity adjacent coils

When the number of coils per zone is greater than one, the winding MMF distribution can be obtained by summing the contributions of all coils, accounting for both their polarity and spatial displacement. Referring to Equation (2.19), the spatial displacement between adjacent coils in an SL-FSCW is $4\pi/Q_S$. In the "same polarity" coil configuration, each coil produces an identical MMF waveform, with each subsequent coil shifted spatially by φ_κ radians. As a result, the MMF distribution of phase-j is given by:

$$F_j(t, \alpha_S) = F_{j,1}(t, \alpha_S) + F_{j,2}(t, \alpha_S - \varphi_\kappa) + F_{j,3}(t, \alpha_S - 2\varphi_\kappa) + \dots \tag{2.20}$$

Substituting Equation (2.17) into (2.20) yields:

$$
\begin{aligned}
F_j(t,\alpha_S) &= \frac{2N_w i_j(t)}{\pi} \sum_{\nu} \frac{{}^{\nu}\xi_p}{\nu} \cdot \\
&\cdot \left\{ \cos(\nu\alpha_S) + \cos\left(\nu\left(\alpha_S - 1\frac{4\pi}{Q_S}\right)\right) + \cos\left(\nu\left(\alpha_S - 2\frac{4\pi}{Q_S}\right)\right) + \ldots \right\} \\
&= \frac{2N_w i_j(t)}{\pi} \sum_{\nu} \frac{{}^{\nu}\xi_p}{\nu} \cdot \sum_{\kappa=0}^{n_c-1} \cos\left(\nu\left(\alpha_S - \kappa\frac{4\pi}{Q_S}\right)\right)
\end{aligned}
\tag{2.21}
$$

According to trigonometric function theory, the sum of cosine functions can be expressed as:

$$
\begin{aligned}
\sum_{\kappa=0}^{n_c-1} \cos\left(\nu\left(\alpha_S - \kappa\frac{4\pi}{Q_S}\right)\right) &= \sum_{\kappa=0}^{n_c-1} e^{j\nu\left(\alpha_S - \kappa\frac{4\pi}{Q_S}\right)} = e^{j(\nu\alpha_S)} \sum_{\kappa=0}^{n_c-1} e^{j\kappa\left(-\nu\frac{4\pi}{Q_S}\right)} \\
&= e^{j(\nu\alpha_S)} \frac{\sin\left((n_c - 1 + 1)\frac{1}{2}\left(\nu\frac{4\pi}{Q_S}\right)\right)}{\sin\left(\frac{1}{2}\left(\nu\frac{4\pi}{Q_S}\right)\right)} e^{j\frac{n_c-1}{2}\left(-\nu\frac{4\pi}{Q_S}\right)} \\
&= \frac{\sin\left(\nu n_c\frac{2\pi}{Q_S}\right)}{\sin\left(\nu\frac{2\pi}{Q_S}\right)} e^{j\left(\nu\left(\alpha_S - (n_c-1)\frac{2\pi}{Q_S}\right)\right)}
\end{aligned}
\tag{2.22}
$$

Substituting Equation (2.22) into (2.21), the MMF distribution for the SL-FSCW with $z = 1$, $n_c > 1$, and the same polarity coil arrangement is expressed in Equation (2.23), while the resulting distribution factor ${}^{\nu}\xi_d$ is defined in Equation (2.24).

$$
F_j(t,\alpha_S) = \frac{2n_c N_w i_j(t)}{\pi} \sum_{\nu} \frac{{}^{\nu}\xi_p\, {}^{\nu}\xi_d}{\nu} \cdot \cos\left(\nu\left(\alpha_S - (n_c - 1)\frac{2\pi}{Q_S}\right)\right)
\tag{2.23}
$$

$$
{}^{\nu}\xi_d = \frac{\sin\left(\nu n_c\frac{2\pi}{Q_S}\right)}{n_c \cdot \sin\left(\nu\frac{2\pi}{Q_S}\right)}
\tag{2.24}
$$

Case 2: Opposite polarity adjacent coils

The phase winding MMF distribution for the coil configuration with "opposite polarity" is derived in a similar manner, with the main difference being that the term "$\kappa\pi$" is included in the respective MMF functions of the coils to account for the reverse polarity. Thus, based on Equation (2.21), we obtain:

$$
\begin{aligned}
F_j(t,\alpha_S) &= \frac{2N_w i_j(t)}{\pi} \sum_{\nu} \frac{{}^{\nu}\xi_p}{\nu} \\
&\cdot \left\{ \cos(\nu\alpha_S) - \cos(\nu(\alpha_S - \varphi_\kappa)) + \cos(\nu(\alpha_S - 2\varphi_\kappa)) - \ldots \right\} \\
&= \frac{2N_w i_j(t)}{\pi} \sum_{\nu} \frac{{}^{\nu}\xi_p}{\nu} \cdot \sum_{\kappa=0}^{n_c-1} \cos\left(\nu\left(\alpha_S - \kappa\frac{4\pi}{Q_S}\right) - \kappa\pi\right) \\
&= \frac{2N_w i_j(t)}{\pi} \sum_{\nu} \frac{{}^{\nu}\xi_p}{\nu} \cdot \sum_{\kappa=0}^{n_c-1} \cos\left(\nu\alpha_S - \kappa\left(\nu\frac{4\pi}{Q_S} + \pi\right)\right)
\end{aligned}
\tag{2.25}
$$

Using

$$\sum_{\kappa=0}^{n_c-1} \cos\left(\nu\alpha_S - \kappa\left(\nu\frac{4\pi}{Q_S} + \pi\right)\right) = \frac{\sin\left(\nu n_c \frac{2\pi}{Q_S} + \frac{n_c}{2}\pi\right)}{\sin\left(\nu\frac{2\pi}{Q_S} + \frac{\pi}{2}\right)} e^{j\left(\nu\left(\alpha_S - (n_c-1)\frac{2\pi}{Q_S}\right) - (n_c-1)\frac{\pi}{2}\right)} \tag{2.26}$$

The final expression for the MMF distribution for the SL-FSCW with $z = 1$, $n_c > 1$, and opposite polarity coils is

$$F_j(t, \alpha_S) = \frac{2n_c N_w i_j(t)}{\pi} \sum_\nu \frac{{}^\nu\xi_p\, {}^\nu\xi_d}{\nu} \cdot \cos\left(\nu\left(\alpha_S - (n_c - 1)\frac{2\pi}{Q_S}\right) - (n_c - 1)\frac{\pi}{2}\right) \tag{2.27}$$

The quantity,

$${}^\nu\xi_d = \frac{\sin\left(\nu n_c \frac{2\pi}{Q_S} + \frac{n_c}{2}\pi\right)}{n_c \cdot \sin\left(\nu\frac{2\pi}{Q_S} + \frac{\pi}{2}\right)} \tag{2.28}$$

is defined as the distribution factor for this winding group.

2.2.1.3 SL-FSCW with $z = 2$, $n_c = 1$, and $n_c > 1$

In an FSCW with two zones per phase ($z = 2$), the zones in the base winding are shifted by 180 electrical degrees and have opposite polarity. Apart from this phase shift, the internal winding structure, such as the coil polarity and the coil combination within each zone, remains identical to that of the $z = 1$ case. Several representative topologies within this category are presented and analyzed below.

Case 1: FSCW with $z = 2$ and $n_c = 1$

The winding layout and the corresponding MMF waveform for phase-j are shown in Figure 2.10. In this configuration, each phase consists of two zones (with one coil per zone), a positive phase zone and a negative phase zone, spatially shifted by 180 electrical degrees. By accounting for their polarities and spatial offsets, the MMF distribution is obtained by summing the contributions from the corresponding zone functions $F_{j,z1}$ and $F_{j,z2}$:

$$F_j(t, \alpha_S) = F_{j,z1}(t, \alpha_S) + F_{j,z2}(t, \alpha_S - \pi) \tag{2.29}$$

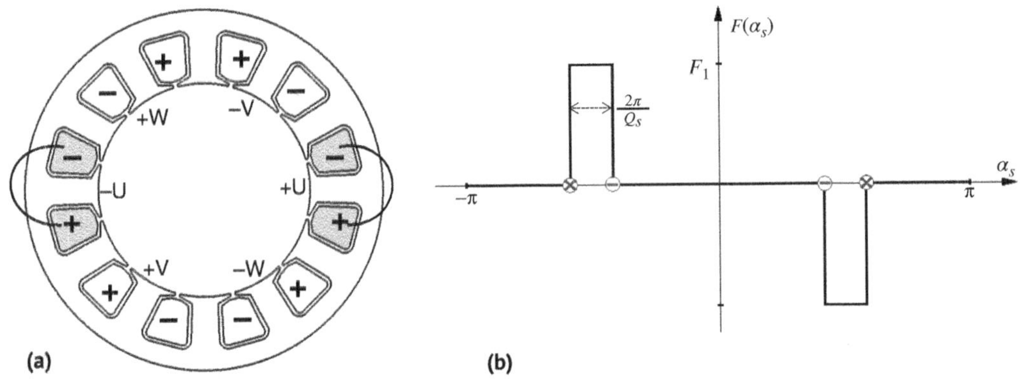

Figure 2.10 (a) Three-phase SL-FSCW with $z = 2$ and $n_c = 1$. (b) MMF distribution of phase-j.

Substituting Equation (2.17) into (2.29), the MMF distribution for the phase-j is:

$$\begin{aligned} F_j(t,\alpha_S) &= \frac{2N_w i_j(t)}{\pi}\sum_{\nu}\frac{{}^{\nu}\xi_p}{\nu}\{\cos(\nu\alpha_S)-\cos(\nu(\alpha_S-\pi))\} \\ &= \frac{4N_w i_j(t)}{\pi}\sum_{\nu}\frac{{}^{\nu}\xi_p}{\nu}\sin\left(\nu\frac{\pi}{2}\right)\sin\left(\nu\left(\alpha_S-\frac{\pi}{2}\right)\right) \\ &= \frac{2zN_w i_j(t)}{\pi}\sum_{\nu}\frac{{}^{\nu}\xi_p\,{}^{\nu}\xi_z}{\nu}\sin\left(\nu\left(\alpha_S-\frac{\pi}{2}\right)\right) \end{aligned} \tag{2.30}$$

In the above relation, the zone factor is defined as:

$${}^{\nu}\xi_z = \sin\left(\nu\frac{\pi}{2}\right) \tag{2.31}$$

Case 2: FSCW with $z = 2$ and $n_c > 1$

For this winding group, two representative examples with $z = 2$, $n_c = 2$, and different coil polarity combinations are shown in Figures 2.11 and 2.12. Analogous to the previous case, the MMF distribution for phase-j is determined by applying the respective zone MMF functions, while accounting

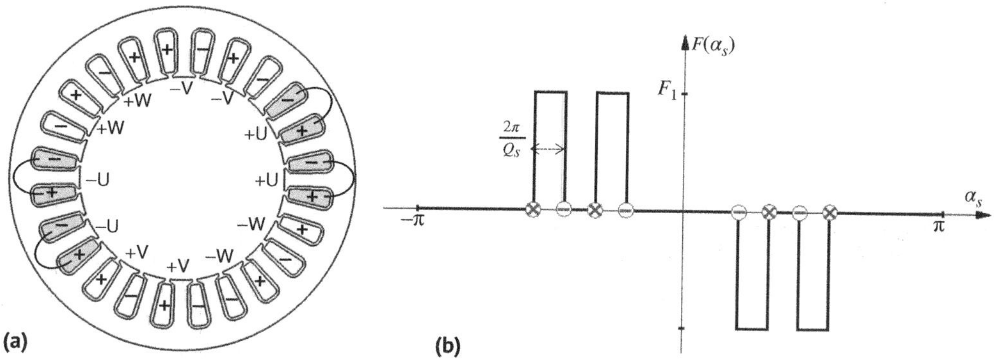

Figure 2.11 (a) Three-phase SL-FSCW with $z = 2$, $n_c = 2$, and same polarity coils. (b) MMF distribution of phase-j.

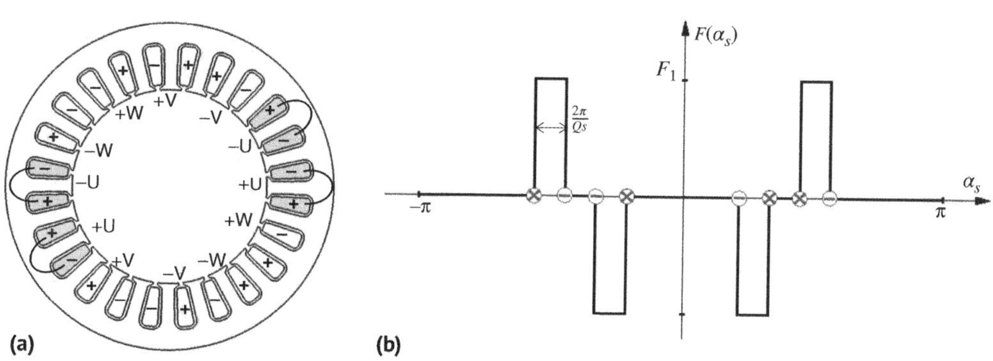

Figure 2.12 (a) Three-phase SL-FSCW with $z = 2$, $n_c = 2$, and opposite polarity coils. (b) MMF distribution of phase-j.

for the coil polarities within each zone, that is, by substituting Equations (2.23) or (2.27) into Equation (2.29). This yields:

$$F_j(t,\alpha_S) = F_{j,z1}(t,\alpha_S) + F_{j,z2}(t,\alpha_S - \pi)$$
$$= \begin{cases} \dfrac{2zn_cN_wi_j(t)}{\pi}\displaystyle\sum_\nu \dfrac{{}^\nu\xi_p\,{}^\nu\xi_p\,{}^\nu\xi_z}{\nu}\sin\left(\nu\left(\alpha_S-(n_c-1)\dfrac{2\pi}{Q_S}\right)\right), & \text{same polarity} \\ \dfrac{2zn_cN_wi_j(t)}{\pi}\displaystyle\sum_\nu \dfrac{{}^\nu\xi_p\,{}^\nu\xi_p\,{}^\nu\xi_z}{\nu}\sin\left(\nu\left(\alpha_S-(n_c-1)\dfrac{2\pi}{Q_S}\right)-(n_c-1)\dfrac{\pi}{2}\right), & \text{opposite polarity} \end{cases} \tag{2.32}$$

In the above expression, the resulting winding factor is expressed as the product of the coil-pitch factor ${}^\nu\xi_p$, distribution factor ${}^\nu\xi_d$, and the zone factor ${}^\nu\xi_z$.

- For the case with same polarity coils (e.g. Figure 2.11), the winding factor parameters for coil-pitch, distribution factor, and zone factor are given by Equations (2.15), (2.24), and (2.31), respectively.
- For the case with opposite polarity coils (e.g. Figure 2.12), the corresponding winding factor parameters are provided in Equations (2.15), (2.28), and (2.31), respectively.

2.2.2 Double-layer FSCWs

In a double-layer FSCW (DL-FSCW), each stator tooth is wound with a coil, while each stator slot contains two coil sides, which may belong to either the same phase or to two different phases. As described in Section 1.3, depending on the coil polarities and zone combinations, this winding type allows the realization of various topologies.

2.2.2.1 DL-FSCW with $z = 1$ and $n_c = 1$

The simplest DL-FSCW configuration is the winding design with one coil per zone $(n_c = 1)$ and one zone per phase $(z = 1)$, as illustrated in Figure 2.13a. In this arrangement, the three-phase base winding consists of only three coils. This winding type is widely used in many applications and is commonly referred to as a 3-slot/2-pole FSCW with $q = 0.5$. The MMF distribution of phase-j (Figure 2.13b) for this winding configuration can be directly determined from the single-coil MMF function derived in (2.14), i.e.,

$$F_j(t,\alpha_S) = F_{\text{Coil}}(t,\alpha_S)$$
$$= \frac{2N_wi_j(t)}{\pi}\sum_\nu \frac{{}^\nu\xi_d}{\nu}\cos\left(\nu\left(\alpha_S - \varphi_{\text{coil}}\right)\right) \tag{2.33}$$

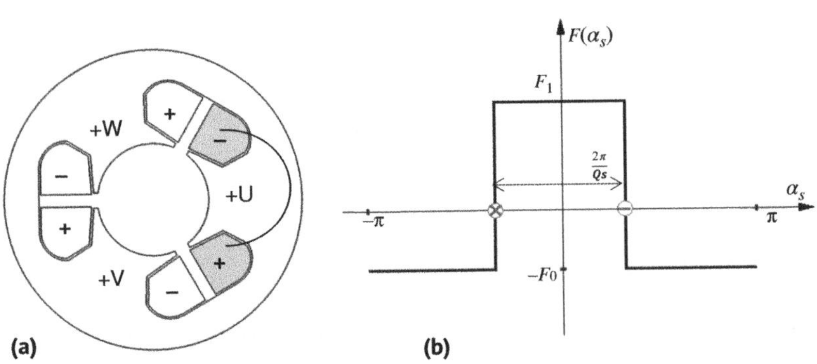

Figure 2.13 (a) Three-phase DL-FSCW with $z = 1$ and $n_c = 1$. (b) MMF distribution of phase-j.

According to (2.15), the coil-pitch factor for this configuration is:

$$ {}^{\nu}\xi_p\Big|_{Q_s=3} = \sin\left(\nu\frac{\pi}{3}\right) \tag{2.34}$$

2.2.2.2 DL-FSCW with $z = 1$ and $n_c > 1$

Beyond the simple example discussed above, a zone may contain more than one coil (i.e. $n_c > 1$), and the coils within the zone may be arranged with either the same or opposite polarity. Figures 2.14a and 2.15a illustrate winding topologies with two and three coils per zone, respectively. In both cases, each phase consists of a single zone, and the coils have opposite polarity. The corresponding phase-j MMF waveforms are presented in Figures 2.14b and 2.15b. An additional example of a single-zone winding with the same polarity coils was previously presented in Figure 1.3c.

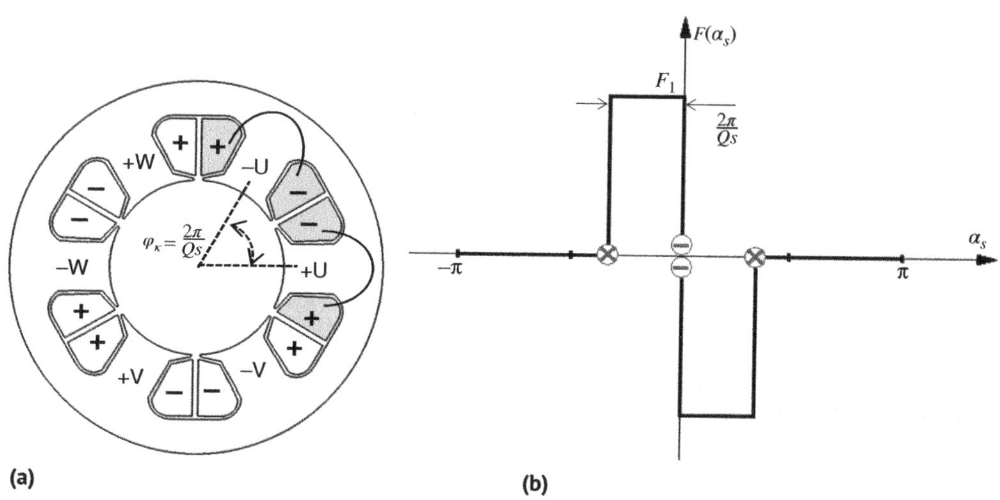

Figure 2.14 (a) Three-phase DL-FSCW with $z = 1, n_c = 2$, and opposite polarity coils. (b) MMF distribution of phase-j.

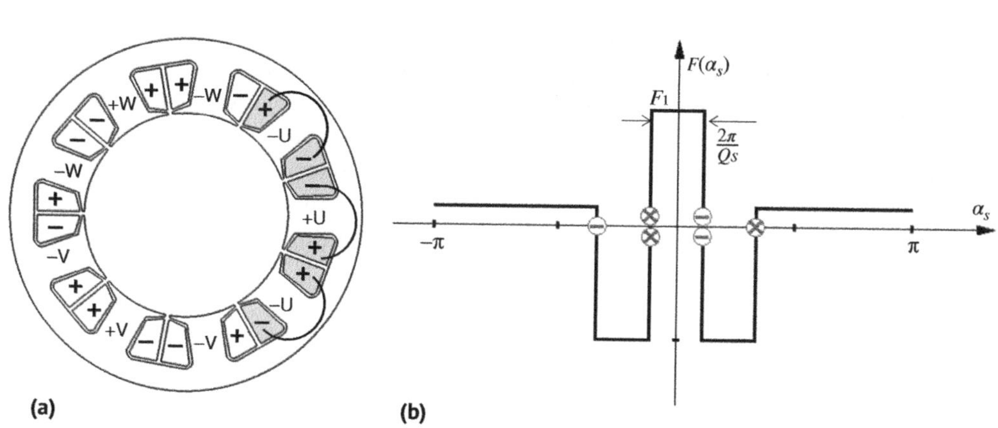

Figure 2.15 (a) Three-phase DL-FSCW with $z = 1, n_c = 3$, and opposite polarity coils. (b) MMF distribution of phase-j.

Case 1: Opposite polarity adjacent coils

According to Figure 2.14a, the spatial displacement between the adjacent coils in a DL-FSCW stator is $2\pi/Q_S$. Taking into account the coil displacement and polarity, the single-phase spatial MMF distribution is obtained by summing the MMF contributions of all coils, as follows:

$$\begin{aligned}
F_j(t,\alpha_S) &= F_{j,1}(t,\alpha_S) + F_{j,2}(t,\alpha_S - \varphi_\kappa) + F_{j,3}(t,\alpha_S - 2\varphi_\kappa) + \dots \\
&= \frac{N_w i_j(t)}{\pi} \sum_\nu \frac{{}^\nu\xi_p}{\nu} \\
&\quad \cdot \left\{ \cos(\nu\alpha_S) - \cos\left(\nu\left(\alpha_S - \frac{2\pi}{Q_S}\right)\right) + \cos\left(\nu\left(\alpha_S - \frac{2\cdot 2\pi}{Q_S}\right)\right) - \dots \right\} \\
&= \frac{2N_w i_j(t)}{\pi} \sum_\nu \frac{{}^\nu\xi_p}{\nu} \cdot \sum_{\kappa=0}^{n_c-1} \cos\left(\nu\left(\alpha_S - \kappa\frac{2\pi}{Q_S}\right) - \kappa\pi\right) \\
&= \frac{2N_w i_j(t)}{\pi} \sum_\nu \frac{{}^\nu\xi_p}{\nu} \cdot \sum_{\kappa=0}^{n_c-1} \cos\left(\nu\alpha_S - \kappa\left(\nu\frac{2\pi}{Q_S} + \pi\right)\right)
\end{aligned} \tag{2.35}$$

Using the trigonometric sum relation:

$$\sum_{\kappa=0}^{n_c-1} \cos\left(\nu\alpha_S - \kappa\left(\nu\frac{2\pi}{Q_S} + \pi\right)\right) = \frac{\sin\left(\nu n_c\frac{\pi}{Q_S} + n_c\frac{\pi}{2}\right)}{\sin\left(\nu\frac{\pi}{Q_S} + \frac{\pi}{2}\right)} e^{j\left(\nu\left(\alpha_S - (n_c-1)\frac{\pi}{Q_S}\right) - (n_c-1)\frac{\pi}{2}\right)} \tag{2.36}$$

the MMF distribution of phase-j can be expressed as:

$$F_j(t,\alpha_S) = \frac{2n_c N_w i_j(t)}{\pi} \sum_\nu \frac{{}^\nu\xi_p\,{}^\nu\xi_d}{\nu} \cdot \cos\left(\nu\left(\alpha_S - (n_c-1)\frac{\pi}{Q_S}\right) - (n_c-1)\frac{\pi}{2}\right) \tag{2.37}$$

where the distribution factor ${}^\nu\xi_d$ for this winding group is defined as:

$${}^\nu\xi_d = \frac{\sin\left(\nu n_c\frac{\pi}{Q_S} + n_c\frac{\pi}{2}\right)}{n_c \cdot \cos\left(\nu\frac{\pi}{Q_S}\right)} \tag{2.38}$$

Case 2: Same polarity adjacent coils

In contrast to the previous winding group, a DL-FSCW configuration in which adjacent coils have the same polarity is inherently inefficient in terms of magnetic field production. This becomes evident when examining the example winding in Figure 1.3c, where such an arrangement results in slots that contain coil sides belonging to the same phase but with opposite polarities. As a result, these slots contribute zero current-linkage and therefore generate no net magnetic field. Furthermore, as the number of coils per zone increases, the number of ineffective slots grows linearly, $Q_{s,k}(\Theta_k = 0) \sim m \cdot (n_c - 1)$. Due to this inherent limitation, this winding group is not examined in detail in this section. The resulting expression for the MMF distribution is:

$$F_j(t,\alpha_S) = \frac{2n_c N_w i_j(t)}{\pi} \sum_\nu \frac{{}^\nu\xi_p\,{}^\nu\xi_d}{\nu} \cdot \cos\left(\nu\left(\alpha_S - (n_c-1)\frac{\pi}{Q_S}\right)\right) \tag{2.39}$$

and the respective distribution factor is:

$$^{\nu}\xi_d = \frac{\sin\left(\nu n_c \frac{\pi}{Q_S}\right)}{n_c \cdot \sin\left(\nu \frac{\pi}{Q_S}\right)} \tag{2.40}$$

2.2.2.3 DL-FSCW with $z = 2$ and $n_c = 1$ and $n_c > 1$

In the configuration with two zones per phase, the winding zones in a base winding are spaced 180 electrical degrees from each other and have opposite polarity. An example of this arrangement for $z = 2$ and $n_c = 1$ is shown in Figure 2.16a, while a second configuration with $z = 2$, $n_c = 2$, and opposite coil polarity is illustrated in Figure 2.17a. The corresponding MMF waveforms (phase-j) are presented in Figures 2.16b and 2.17b, respectively.

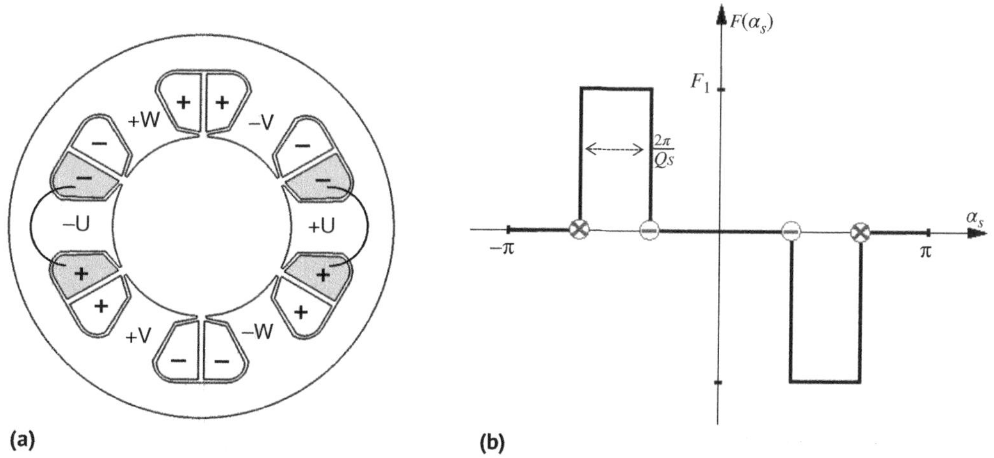

Figure 2.16 (a) Three-phase DL-FSCW with $z = 2$ and $n_c = 1$. (b) MMF distribution of phase-j.

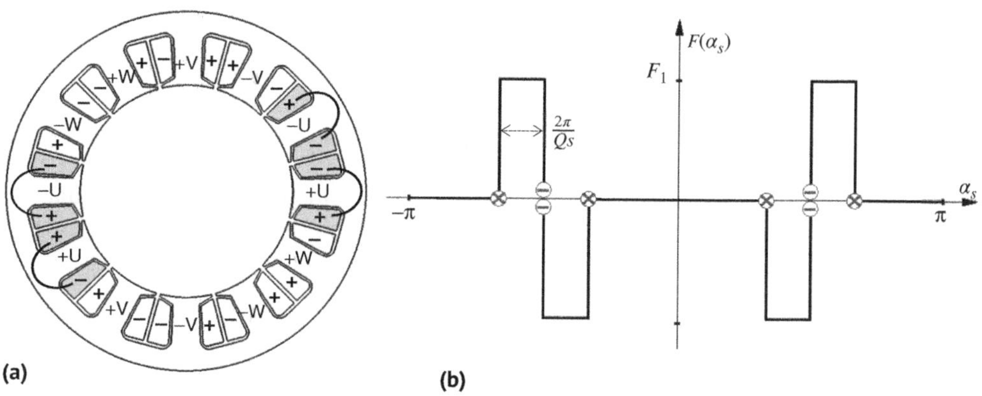

Figure 2.17 (a) Three-phase DL-FSCW with $z = 2$, $n_c = 2$, and opposite polarity coils. (b) MMF distribution of phase-j.

By substituting the respective winding zone MMF functions, i.e. the Equations (2.33), (2.37) and (2.39), into (2.29), the final expressions for the MMF distribution for the two-zone DL-FSCW configurations are given in (2.41). The corresponding coil-pitch, distribution, and zone factors have already been defined by Equations (2.15), (2.31), (2.38), and (2.40), respectively.

$$F_j(t,\alpha_S) = F_{j,z1}(t,\alpha_S) + F_{j,z2}(t,\alpha_S - \pi)$$
$$= \begin{cases} \dfrac{2zN_w i_j(t)i}{\pi} \displaystyle\sum_\nu \dfrac{{}^\nu\xi_p {}^\nu\xi_z}{\nu} \sin\left(\nu\left(\alpha_S - \dfrac{\pi}{2}\right)\right), \; z=2,\; n_c = 1 \\ \dfrac{2zn_c N_w i_j(t)}{\pi} \displaystyle\sum_\nu \dfrac{{}^\nu\xi_p {}^\nu\xi_p {}^\nu\xi_z}{\nu} \sin\left(\nu\left(\alpha_S - (n_c-1)\dfrac{\pi}{Q_S}\right)\right), \\ \qquad z=2,\; n_c>1, \text{ same polarity} \\ \dfrac{2zn_c N_w i_j(t)}{\pi} \displaystyle\sum_\nu \dfrac{{}^\nu\xi_p {}^\nu\xi_p {}^\nu\xi_z}{\nu} \sin\left(\nu\left(\alpha_S - (n_c-1)\dfrac{\pi}{Q_S}\right) - (n_c-1)\dfrac{\pi}{2}\right), \\ \qquad z=2,\; n_c>1, \text{ opposite polarity} \end{cases} \tag{2.41}$$

2.3 Winding Factors

Winding function theory for SL- and DL-FSCW categories, introduced in Sections 2.2.1 and 2.2.2, leads to the definition of winding factors. These are expressed as the product of the coil-pitch, distribution, and the zone factor:

$$^\nu\xi_w = {}^\nu\xi_p {}^\nu\xi_d {}^\nu\xi_z \tag{2.42}$$

Table 2.1 summarizes the analytical expressions for the winding factors of all standard FSCWs. In their generalized form, these factors are expressed as a function of the adjacent coil displacement angle φ_κ. Depending on the winding configuration, this angle is defined by Equation (2.43), while the corresponding number of stator slots in the base winding is given by Equation (1.6).

$$\varphi_\kappa = \begin{cases} 2\alpha_\tau = \dfrac{4\pi}{Q_S}, & \text{for single-layer FSCW} \\ \alpha_\tau = \dfrac{2\pi}{Q_S}, & \text{for double-layer FSCW} \end{cases} \tag{2.43}$$

The winding factors for various configurations are calculated using the analytical formulas in Table 2.1, and summarized in Tables 2.2–2.9. To emphasize the most favorable topologies, the maximum winding factors are highlighted in bold. Based on the obtained results, the following conclusions can be drawn:

- All FSCWs are characterized by a high harmonic content.
- The harmonic content for the winding group with $z = 1$ is higher than that for $z = 2$.
- Coil polarity and coil combination strongly influence the maximum value of winding factor.
- Due to the periodicity of winding factors, the listed values are periodic in harmonic order ν, with a period of Q_S.

Table 2.1 Generalized winding factor formulations for standard FSCWs.

	$z = 1$	$z = 2$
$n_c = 1$	$^{\nu}\xi_w = \sin\left(\nu\frac{\pi}{Q_S}\right)$	$^{\nu}\xi_w = \sin\left(\nu\frac{\pi}{Q_S}\right)\sin\left(\nu\frac{\pi}{2}\right)$
$n_c > 1$, *same polarity*	$^{\nu}\xi_w = \sin\left(\nu\frac{\pi}{Q_S}\right)\frac{\sin\left(\nu n_c\frac{\varphi_\kappa}{2}\right)}{n_c\cdot\sin\left(\nu\frac{\varphi_\kappa}{2}\right)}$	$^{\nu}\xi_w = \sin\left(\nu\frac{\pi}{Q_S}\right)\frac{\sin\left(\nu n_c\frac{\varphi_\kappa}{2}\right)}{n_c\cdot\sin\left(\nu\frac{\varphi_\kappa}{2}\right)}\sin\left(\nu\frac{\pi}{2}\right)$
$n_c > 1$, *opposite polarity*	$^{\nu}\xi_w = \sin\left(\nu\frac{\pi}{Q_S}\right)\frac{\sin\left(\nu n_c\frac{\varphi_\kappa}{2}+\frac{n_c}{2}\pi\right)}{n_c\cdot\cos\left(\nu\frac{\varphi_\kappa}{2}\right)}$	$^{\nu}\xi_w = \sin\left(\nu\frac{\pi}{Q_S}\right)\frac{\sin\left(\nu n_c\frac{\varphi_\kappa}{2}+\frac{n_c}{2}\pi\right)}{n_c\cdot\cos\left(\nu\frac{\varphi_\kappa}{2}\right)}\sin\left(\nu\frac{\pi}{2}\right)$

2.3.1 Maximum Winding Factors

Figures 2.18 and 2.19 show the maximum winding factor values for various winding configurations. When $n_c = 1$ (one coil per zone), the SL and DL configurations exhibit identical behavior for both polarity cases, resulting in the same winding factor. However, as n_c increase (i.e. when $n_c \geq 2$), the following conclusions can be made:

- SL-FSCW: Opposite polarity coil arrangements have lower winding factors, decreasing slightly as n_c increase. Same polarity coil arrangements remain essentially constant with n_c. Configurations with two zones per phase yield a higher winding factor than those with one zone per phase.
- DL-FSCW: Same polarity coil arrangements are less efficient; their winding factors decrease rapidly with n_c (approximately as a function of $1/n_c$). Opposite polarity arrangements offer higher winding factors in both one-zone and two-zone cases, with maxima alternating with n_c. In particular, in the one-zone configuration, the maximum winding factors occur at odd values of n_c, while for two-zone configurations, the maximum winding factors occur at even values of n_c.

Depending on the winding type, the MMF harmonics with the highest winding factor are:

SL-FSCW (same polarity):

$$\nu = \frac{Q_S}{2} \pm 1 \tag{2.44}$$

DL-FSCW (opposite polarity):

$$\nu = \begin{cases} \dfrac{Q_S \pm 1}{2} & (z = 1 \text{ and } n_c = \text{odd}) \\ \dfrac{Q_S}{2} \pm 1 & (z = 2 \text{ and } n_c = \text{even}) \end{cases} \tag{2.45}$$

As mentioned earlier, due to the periodic nature of the winding factors, multiple harmonic orders in the MMF spectrum reach maximum winding factor values, repeating with a periodicity corresponding to the number of stator slots Q_S.

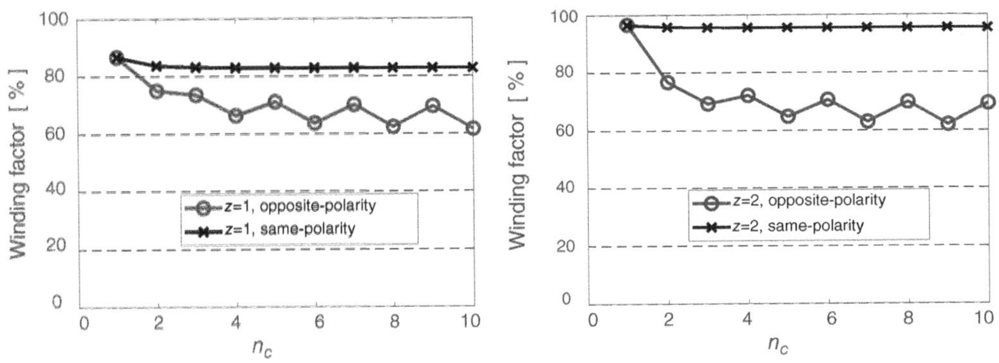

Figure 2.18 Maximum winding factor for SL-FSCWs as a function of z, n_c, and coil polarities.

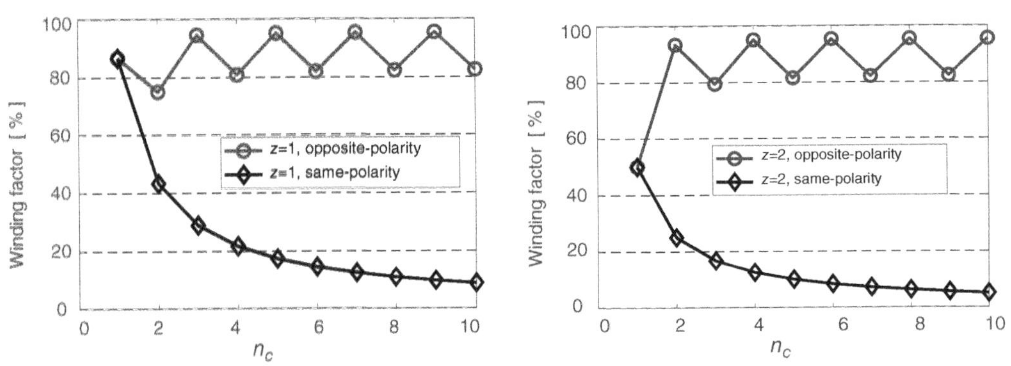

Figure 2.19 Maximum winding factor for DL-FSCWs as a function of z, n_c, and coil polarities.

2.3.2 Selection of Working Harmonic

Because FSCWs have rich MMF spectra and some harmonics have higher winding factors, multiple harmonic orders ν can serve as the operating (working) wave. However, based on the fact that only a suitable slot/pole combination can ensure the best magnetic utilization, the selection of the optimal working harmonic should be defined according to the following key guidelines:

- Select a harmonic component with the highest winding factor to maximize torque density.
- Avoid harmonics whose order is much larger or much smaller than half the number of stator slot. They can negatively affect magnetic-path utilization, leakage fluxes, and thermal behavior.
- The best FSCW slot/pole combinations are usually $Q_S = 2p \pm 1$ or $Q_S = 2p \pm 2$.

Based on these criteria and the values listed in Tables 2.2–2.9, the most suitable working harmonic is the first harmonic order within the first period of ν that gives the maximum winding factor, as defined by:

$$\nu_p = p = \frac{Q_S \pm 1}{2}, \text{or} \frac{Q_S}{2} \pm 1 \tag{2.46}$$

Furthermore, for a given FSCW, the base winding can be repeated $\wp$ times to obtain a higher pole-pair number. Referring to Equation (1.3), this leads to:

$$Q_S = \wp Q_b,\ p = \wp p_b, \text{ and } \nu_p = \wp \nu_b \tag{2.47}$$

Among various winding configurations, the DL-FSCW with $Q_S = 3$, $n_c = 1$, and $z = 1$ is unique in that it operates using the fundamental (first) MMF harmonic as working wave. In contrast, all other FSCW types have their highest winding factors at higher harmonic orders, thus they operate with a higher harmonic order. Notable examples include:

- DL winding with $Q_S = 9$, $n_c = 3$, and $z = 1$: $\nu_p = 4$th or 5th
- DL winding with $Q_S = 12$, $n_c = 2$, and $z = 2$: $\nu_p = 5$th or 7th
- SL winding with $Q_S = 12$, $n_c = 1$, and $z = 2$: $\nu_p = 5$th or 7th

In addition to the commonly used windings listed above, Tables 2.2–2.9 also highlight a wide variety of FSCW topologies, offering numerous options suitable for different machine applications.

2.3.3 Winding Factors for Different SL-FSCWs

Table 2.2 Winding factors (in %) for the SL winding with $z = 1$ and opposite polarity coils.

n_c	Q_s	$\nu=1$	2	3	4	5	6	7	8	9	10	11	12	13	14	15	16	17	
1	6	50.0	**86.6**	—	**86.6**	50.0	—	50.0	**86.6**	—	**86.6**	50.0	—	50.0	**86.6**	—	**86.6**	50.0	—
2	12	12.9	43.3	—	75.0	48.3	—	48.3	75.0	—	43.3	12.9	—	12.9	43.3	—	75.0	48.3	—
3	18	3.1	7.4	—	61.7	73.5	—	20.4	17.5	—	17.5	20.4	—	73.5	61.7	—	7.4	3.1	—
4	24	2.9	6.5	—	21.7	50.9	—	66.4	37.5	—	24.1	22.2	—	22.2	24.1	—	37.5	66.4	—
5	30	1.1	2.3	—	6.1	10.0	—	64.0	71.1	—	17.3	13.7	—	10.7	10.2	—	10.2	10.7	—
6	36	1.3	2.7	—	6.4	9.5	—	24.2	53.4	—	63.7	34.6	—	20.4	17.7	—	15.1	14.6	—

Table 2.3 Winding factors (in %) for the SL winding with $z = 1$ and same polarity coils.

n_c	Q_s	$\nu=1$	2	3	4	5	6	7	8	9	10	11	12	13	14	15	16	17
1	6	50.0	**86.6**	—	**86.6**	50.0	—	50.0	**86.6**	—	**86.6**	50.0	—	50.0	**86.6**	—	**86.6**	50.0
2	12	22.4	25.0	—	43.3	**83.7**	—	**83.7**	43.3	—	25.0	22.4	—	22.4	25.0	—	43.3	**83.7**
3	18	14.7	15.4	—	18.8	22.5	—	42.2	**83.1**	—	**83.1**	42.2	—	22.5	18.8	—	15.4	14.7
4	24	10.9	11.2	—	12.5	13.6	—	17.8	21.7	—	41.8	**82.9**	—	**82.9**	41.8	—	21.7	17.8
5	30	8.7	8.9	—	9.5	10.0	—	11.7	12.9	—	17.3	21.3	—	41.7	**82.9**	—	**82.9**	41.7
6	36	7.2	7.3	—	7.7	8.0	—	8.8	9.4	—	11.2	12.6	—	17.1	21.1	—	41.6	**82.8**

Table 2.4 Winding factors (in %) for the SL winding with $z = 2$ and opposite polarity coils.

n_c	Q_s	$\nu=1$	2	3	4	5	6	7	8	9	10	11	12	13	14	15	16	17
1	12	25.9	—	—	—	**96.6**	—	**96.6**	—	—	—	25.9	—	25.9	—	—	—	**96.6**
2	24	3.4	—	—	—	58.8	—	76.6	—	—	—	25.7	—	25.7	—	—	—	76.6
3	36	2.6	—	—	—	19.0	—	48.4	—	—	—	69.1	—	40.7	—	—	—	29.2
4	48	0.8	—	—	—	5.1	—	9.1	—	—	—	63.1	—	72.0	—	—	—	18.4
5	60	0.9	—	—	—	5.2	—	8.4	—	—	—	23.2	—	52.4	—	—	—	64.7
6	72	0.4	—	—	—	2.0	—	3.1	—	—	—	6.7	—	10.6	—	—	—	64.6

2.3.4 Winding Factors for Different DL-FSCWs

Table 2.5 Winding factors (in %) for the SL winding with $z = 2$ and same polarity coils.

n_c	Q_s	$v=1$	2	3	4	5	6	7	8	9	10	11	12	13	14	15	16	17
1	12	25.9	—	—	—	**96.6**	—	**96.6**	—	—	—	25.9	—	25.9	—	—	—	**96.6**
2	24	12.6	—	—	—	15.8	—	20.5	—	—	—	**95.8**	—	**95.8**	—	—	—	20.5
3	36	8.4	—	—	—	9.2	—	10.2	—	—	—	14.5	—	19.7	—	—	—	**95.6**
4	48	6.3	—	—	—	6.6	—	7.0	—	—	—	8.3	—	9.5	—	—	—	14.1
5	60	5.0	—	—	—	5.2	—	5.4	—	—	—	6.0	—	6.4	—	—	—	7.9
6	72	4.2	—	—	—	4.3		4.4	—	—	—	4.7		4.9	—	—	—	5.7

Table 2.6 Winding factors (in %) for the DL winding with $z = 1$ and opposite polarity coils.

n_c	Q_s	$v=1$	2	3	4	5	6	7	8	9	10	11	12	13	14	15	16	17
1	3	**86.6**	**86.6**	—	**86.6**	**86.6**	—	**86.6**	**86.6**	—	**86.6**	**86.6**	—	**86.6**	**86.6**	—	**86.6**	**86.6**
2	6	25.0	75.0	—	75.0	25.0	—	25.0	75.0	—	75.0	25.0	—	25.0	75.0	—	75.0	25.0
3	9	6.1	14.0	—	**94.5**	**94.5**	—	14.0	6.1	—	6.1	14.0	—	**94.5**	**94.5**	—	14.0	6.1
4	12	5.8	12.5	—	37.5	80.8	—	80.8	37.5	—	12.5	5.8	—	5.8	12.5	—	37.5	80.8
5	15	2.1	4.5	—	11.1	17.3	—	**95.1**	**95.1**	—	17.3	11.1	—	4.5	2.1	—	2.1	4.5
6	18	2.5	5.3	—	12.1	17.2	—	39.7	81.9	—	81.9	39.7	—	17.2	12.1	—	5.3	2.5

Table 2.7 Winding factors (in %) for the DL winding with $z = 1$ and same polarity coils.

n_c	Q_s	$v=1$	2	3	4	5	6	7	8	9	10	11	12	13	14	15	16	17
1	3	**86.6**	**86.6**	—	**86.6**	**86.6**	—	**86.6**	**86.6**	—	**86.6**	**86.6**	—	**86.6**	**86.6**	—	**86.6**	**86.6**
2	6	43.3	43.3	—	43.3	43.3	—	43.3	43.3	—	43.3	43.3	—	43.3	43.3	—	43.3	43.3
3	9	28.9	28.9	—	28.9	28.9	—	28.9	28.9	—	28.9	28.9	—	28.9	28.9	—	28.9	28.9
4	12	21.7	21.7	—	21.7	21.7	—	21.7	21.7	—	21.7	21.7	—	21.7	21.7	—	21.7	21.7
5	15	17.3	17.3	—	17.3	17.3	—	17.3	17.3	—	17.3	17.3	—	17.3	17.3	—	17.3	17.3
6	18	14.4	14.4	—	14.4	14.4	—	14.4	14.4	—	14.4	14.4	—	14.4	14.4	—	14.4	14.4

Table 2.8 Winding factors (in %) for the DL winding with $z = 2$ and opposite polarity coils.

n_c	Q_s	$v=1$	2	3	4	5	6	7	8	9	10	11	12	13	14	15	16	17
1	6	50.0	—	—	—	50.0	—	50.0	—	—	—	50.0	—	50.0	—	—	—	50.0
2	12	6.7	—	—	—	**93.3**	—	**93.3**	—	—	—	6.7	—	6.7	—	—	—	**93.3**
3	18	5.1	—	—	—	34.4	—	79.3	—	—	—	79.3	—	34.4	—	—	—	5.1
4	24	1.6	—	—	—	9.6	—	16.3	—	—	—	**94.9**	—	**94.9**	—	—	—	16.3
5	30	1.8	—	—	—	10.0	—	15.6	—	—	—	38.9	—	81.5	—	—	—	81.5
6	36	0.7	—	—	—	3.9	—	5.8	—	—	—	11.9	—	17.9	—	—	—	**95.3**

Table 2.9 Winding factors (in %) for the DL winding with $z = 2$ and same polarity coils.

n_c	Q_s	$v=1$	2	3	4	5	6	7	8	9	10	11	12	13	14	15	16	17
1	6	50.0	—	—	—	50.0	—	50.0	—	—	—	50.0	—	50.0	—	—	—	50.0
2	12	25.0	—	—	—	25.0	—	25.0	—	—	—	25.0	—	25.0	—	—	—	25.0
3	18	16.7	—	—	—	16.7	—	16.7	—	—	—	16.7	—	16.7	—	—	—	16.7
4	24	12.5	—	—	—	12.5	—	12.5	—	—	—	12.5	—	12.5	—	—	—	12.5
5	30	10.0	—	—	—	10.0	—	10.0	—	—	—	10.0	—	10.0	—	—	—	10.0
6	36	8.3	—	—	—	8.3	—	8.3	—	—	—	8.3	—	8.3	—	—	—	8.3

2.4 Three-phase MMF Distribution

A three-phase symmetric FSCW consists of three identical windings, each spatially shifted by $2\pi/3$ electrical degrees. The phase-j current is expressed as:

$$i_j(t) = \hat{i}\cos\left(\omega t - (j-1)\frac{2\pi}{m} + \delta\right),\ j = 1,2,3, \text{ and } m = 3 \tag{2.48}$$

where $\hat{i}$ is the phase current amplitude, δ the load angle that control the current vector, and ω the angular electrical frequency.

The single-phase MMF, which varies in space and time, can be resolved into forward and backward rotating waves:

$$\begin{aligned} F_j(t,\alpha_S) &= \sum_{\nu} {}^{\nu}\hat{F}_j \cos\left(\omega t - (j-1)\frac{2\pi}{m} + \delta\right)\cos\left(\nu\left(\alpha_S - (j-1)\frac{2\pi}{m}\right)\right) \\ &= \frac{1}{2}\sum_{\nu} {}^{\nu}\hat{F}_j \cos\left(\omega t - \nu\alpha_S + (\nu-1)(j-1)\frac{2\pi}{m} + \delta\right) \\ &\quad + {}^{\nu}\hat{F}_j \cos\left(\omega t + \nu\alpha_S - (\nu+1)(j-1)\frac{2\pi}{m} + \delta\right) \end{aligned} \tag{2.49}$$

The peak value of the νth order MMF harmonic is given by:

$${}^{\nu}\hat{F}_j = \frac{2zn_cN_w\hat{i}}{\pi}\frac{{}^{\nu}\xi_w}{\nu} \tag{2.50}$$

Harmonics of the first term in Equation (2.49) rotate in the same direction as the fundamental wave (forward-rotating). Those of the second term rotate in the opposite direction (backward-rotating).

The total MMF produced by a three-phase winding is obtained by summing the individual contributions from each phase:

$$F(t,\alpha_S) = \sum_{j=1}^{m} F_j(t,\alpha_S) = F_U(t,\alpha_S) + F_V(t,\alpha_S) + F_W(t,\alpha_S) \tag{2.51}$$

Taking into account the effects of harmonics on the sum of the ${}^{\nu}F_j$ functions, the following conditions will predict the entire MMF function.

1. For $\nu = mg_1$, $m = 3$ and, $g_1 = 1, 3, 5, \ldots$

$$(\nu - 1)(j - 1)\frac{2\pi}{3} = (j - 1)\left(2\pi g_1 - \frac{2\pi}{3}\right)$$

$$(\nu + 1)(j - 1)\frac{2\pi}{3} = (j - 1)\left(2\pi g_1 + \frac{2\pi}{3}\right).$$

The $\pm 2\pi/3$ phase shift of the respective harmonics cause complete cancellation, resulting in:

$$\sum_{j=1}^{m} {}^{\nu}F_j \cos\left(\omega t - \nu\alpha_S + (\nu - 1)(j - 1)\frac{2\pi}{m} + \delta\right) = 0$$

$$\sum_{j=1}^{m} {}^{\nu}F_j \cos\left(\omega t + \nu\alpha_S - (\nu + 1)(j - 1)\frac{2\pi}{m} + \delta\right) = 0$$

2. For $\nu = mg_2 + 1$, $m = 3$, and $g_2 = 0, 1, 2, 3, \ldots$
Forward-rotating waves:

$$(\nu - 1)(j - 1)\frac{2\pi}{3} = (j - 1)2\pi g_2$$

$$\Rightarrow \sum_{j=1}^{m} {}^{\nu}F_j \cos\left(\omega t - \nu\alpha_S + (\nu - 1)(j - 1)\frac{2\pi}{m} + \delta\right) = m{}^{\nu}F_j \cos(\omega t - \nu\alpha_S + \delta)$$

Backward-rotating waves:

$$(\nu + 1)(j - 1)\frac{2\pi}{3} = (j - 1)\left(2\pi g_2 + \frac{4\pi}{3}\right)$$

$$\Rightarrow \sum_{j=1}^{m} {}^{\nu}F_j \cos\left(\omega t + \nu\alpha_S - (\nu + 1)(j - 1)\frac{2\pi}{m} + \delta\right) = 0$$

3. For $\nu = mg_3 - 1$, $m = 3$, and $g_3 = 1, 2, 3, \ldots$
Forward-rotating waves:

$$(\nu - 1)(j - 1)\frac{2\pi}{3} = (j - 1)\left(2\pi g_3 - \frac{4\pi}{3}\right)$$

$$\Rightarrow \sum_{j=1}^{m} {}^{\nu}F_j \cos\left(\omega t - \nu\alpha_S + (\nu - 1)(j - 1)\frac{2\pi}{m} + \delta\right) = 0$$

Backward-rotating waves:

$$(\nu + 1)(j - 1)\frac{2\pi}{3} = (j - 1)2\pi g_3$$

$$\Rightarrow \sum_{j=1}^{m} {}^{\nu}F_j \cos\left(\omega t + \nu\alpha_S - (\nu + 1)(j - 1)\frac{2\pi}{m} + \delta\right) = m{}^{\nu}F_j \cos(\omega t + \nu\alpha_S + \delta)$$

Based on cases 1–3, the total MMF function of a three-phase winding is given by:

$$\begin{aligned} F(\alpha_S, t) &= \sum_{j=1}^{m} F_j(\alpha_S, t) \\ &= \frac{m}{2} \sum_{\nu = mg_2 + 1} {}^{\nu}\hat{F}_j \cos(\omega t - \nu\alpha_S + \delta) + \frac{m}{2} \sum_{\nu = mg_3 - 1} {}^{\nu}\hat{F}_j \cos(\omega t + \nu\alpha_S + \delta) \end{aligned} \tag{2.52}$$

By representing the backward rotating waves as harmonics with negative order, we can rewrite the corresponding MMF function as:

$$\sum_{\nu=mg_3-1} {}^{\nu}\widehat{F}_j \cos(\omega t + \nu\alpha_S + \delta) = \sum_{\nu=-(mg_3-1)} {}^{\nu}\widehat{F}_j \cos(\omega t - \nu\alpha_S + \delta) \tag{2.53}$$

Thus the general three-phase MMF distribution can be expressed as:

$$F(\alpha_S, t) = \frac{m}{2}\frac{2zn_cN_w\hat{i}}{\pi}\sum_{\nu}\frac{{}^{\nu}\xi_w}{\nu}\cos(\omega t - \nu\alpha_S + \delta) \tag{2.54}$$

where

$$\nu = mg + 1, \quad g = 0, \pm 1, \pm 2, \pm 3, \dots \tag{2.55}$$

In these equations, the positive order harmonics $\nu = +1, +4, +7, \dots$, are forward-rotating waves, while the negative order harmonics $\nu = -2, -5, -8, \dots$, are backward-rotating waves. Each harmonic has its own magnitude and phase angle.

According to (2.54), the resulting νth-order MMF magnitude is $m/2$ higher than that of the single-phase MMF:

$${}^{\nu}\widehat{F} = \frac{m}{2}{}^{\nu}\widehat{F}_j = \frac{m}{2}\frac{2zn_cN_w\hat{i}}{\pi}\frac{{}^{\nu}\xi_w}{\nu} \tag{2.56}$$

The angular speed of the ν order harmonic is:

$$\omega_\nu = \begin{cases} \dfrac{\omega}{\nu} & \text{for forward-rotating waves} \\ -\dfrac{\omega}{\nu} & \text{for backward-rotating waves} \end{cases} \tag{2.57}$$

Equation (2.57) indicates that the MMF harmonics rotate at different speeds and in different directions. Among these, only the harmonic that produce the main magnetic field and interacts with the rotor poles, i.e. $\nu_p = p$, rotates synchronously with the rotor and produces constant torque. All other spatial MMF harmonics $(\nu \neq p)$ rotate asynchronously with respect to the rotor, increasing stator/rotor core iron losses, magnet eddy-current losses, and acoustic noise.

2.5 MMF Characteristics of Various Three-phase Standard FSCWs

The general expression for the MMF function of three-phase windings is derived in Equation (2.54), whereas the corresponding winding factor formulations are provided in Table 2.1. Based on this analytical approach, several representative FSCW configurations are examined below. The winding factors for the operating harmonic and the total harmonic distortion (THD) are summarized in Table 2.10.

2.5.1 SL-FSCWs with $z = 1$

Example 2.1 6-slot/4-pole FSCW with $z = 1$ and $n_c = 1$

The winding layout for an SL-FSCW with $z = 1$ and $n_c = 1$ consists of 6 stator slots $(Q_s = 6)$ and employs a +U+V+W winding layout (Figure 2.20a). The resulting three-phase spatial MMF

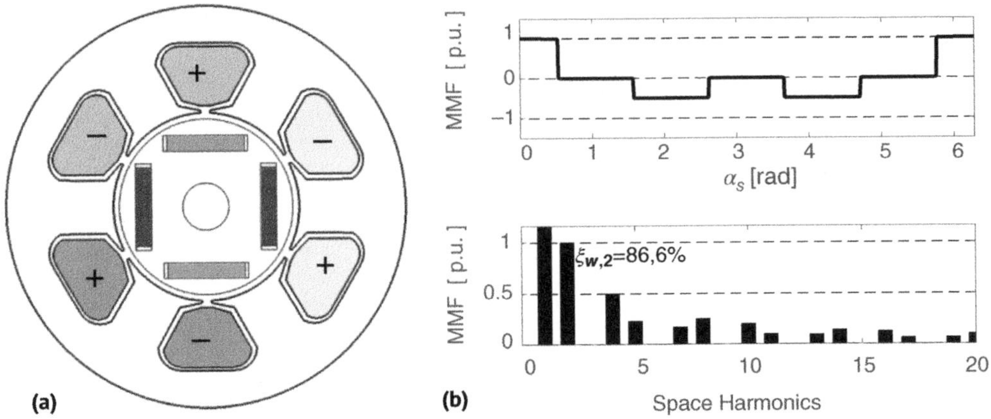

Figure 2.20 SL-FSCW with $z = 1$, $n_c = 1$, and $Q_S = 6$. (a) Winding layout. (b) Three-phase spatial MMF distribution and harmonic content.

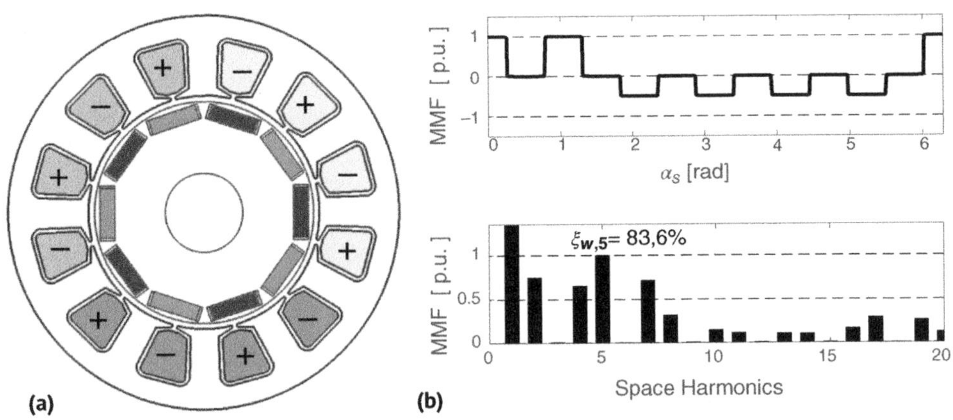

Figure 2.21 SL-FSCW with $z = 1$, $n_c = 2$, and $Q_s = 12$. (a) Winding layout. (b) Three-phase spatial MMF distribution and harmonic content.

distribution and its corresponding harmonic spectrum are presented in Figure 2.20b. This configuration produces both even and odd harmonics in the MMF spectrum, except harmonics of order $3k$th $(k = 1, 2, 3, \ldots)$. The dominant MMF harmonic, i.e., the one with the highest winding factor, is the 2nd harmonic $(\xi_{w,2} = 86.6\%)$, which is therefore selected as the operating harmonic for the 4-pole PM rotor under consideration.

Example 2.2 12-slot/10-pole FSCW with $z = 1$ and $n_c = 2$
For an SL-FSCW with $z = 1$ and $n_c = 2$, the stator comprises 12 stator slots $(Q_s = 12)$ and 6 winding coils with a winding sequence of +U+U+V+V+W+W, as illustrated in Figure 2.21a. The corresponding three-phase MMF characteristics are shown in Figure 2.21b. In this configuration, the 5th harmonic has the highest winding factor $(\xi_{w,5} = 83.6\%)$, which is chosen as the operating harmonic, resulting in a PM motor design with 10 rotor poles $(2p = 10)$.

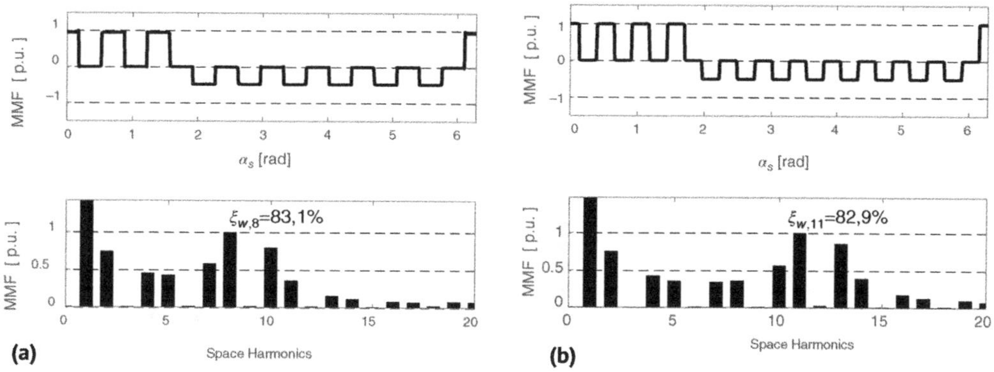

Figure 2.22 MMF characteristic for the SL-FSCWs with $z = 1$. (a) Winding topology with $n_c = 3$, $v_p = 8$, and $Q_S = 18$. (b) Winding topology with $n_c = 4$, $v_p = 11$, and $Q_S = 24$.

Example 2.3 18-slot/16-pole and 24-slot/22-pole FSCW
The MMF characteristics of two additional SL-FSCWs with $z = 1$ and higher number of coils per zone (i.e. with $n_c = 3$ and $n_c = 4$) are shown in Figure 2.22:

- 18-slot/16-pole FSCW with $n_c = 3$
  - Stator slot number: $Q_s = 18$
  - Operating harmonic: 8th harmonic
  - Winding factor: 83.1%
- 24-slot/22-pole FSCW with $n_c = 4$
  - Stator slot number: $Q_s = 24$
  - Operating harmonic: 11th harmonic
  - Winding factor: 82.9%

2.5.2 SL-FSCWs with $z = 2$

Example 2.4 12-slot/14-pole FSCW with $z = 2$ and $n_c = 1$
Figure 2.23a illustrates an SL-FSCW with $z = 2$ and $n_c = 1$. The stator comprises 12 slots $(Q_s = 12)$ and uses the winding arrangement +U−V+W−U+V−W. The corresponding three-phase MMF characteristics are shown in Figure 2.23b. Here, the winding factor for both the 5th and 7th harmonics is 96.6%. In this example, the 7th harmonic is selected as the operating harmonic, resulting in a rotor with seven pole pairs, although the higher winding factor of the 5th harmonic also makes this winding type suitable for 10-pole machines $(2p = 10)$.

Example 2.5 24-slot/22-pole FSCW with $z = 2$ and $n_c = 2$
Another example of a two-zone SL-FSCW with two coils per zone $(n_c = 2)$ is illustrated in Figure 2.24a. The stator has 24 slots $(Q_s = 24)$, its winding arrangement is +U+U−V−V+W+W−U−U+V+V−W−W, and the three-phase spatial MMF distribution and harmonic content are depicted in Figure 2.24b. The 11th harmonic, with a winding factor of 95.7%, is selected as working harmonic for the 22-pole PM machine.

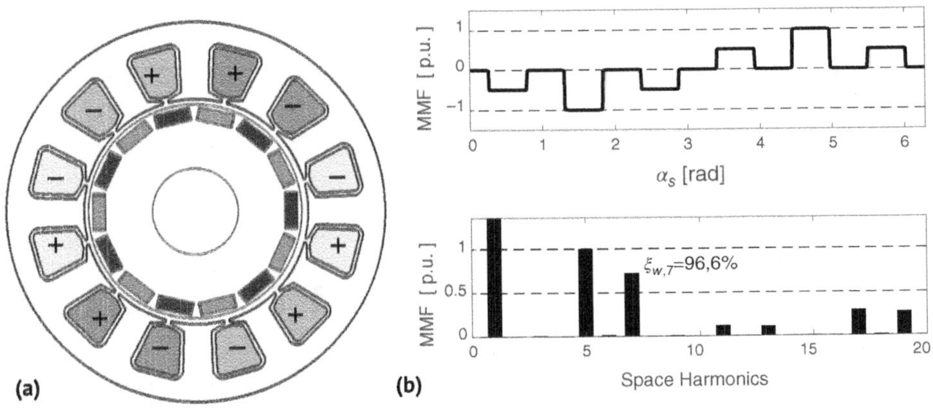

Figure 2.23 SL-FSCW with $z = 2$, $n_c = 1$, and $Q_S = 12$. (a) Winding layout. (b) Three-phase spatial MMF distribution and harmonic content.

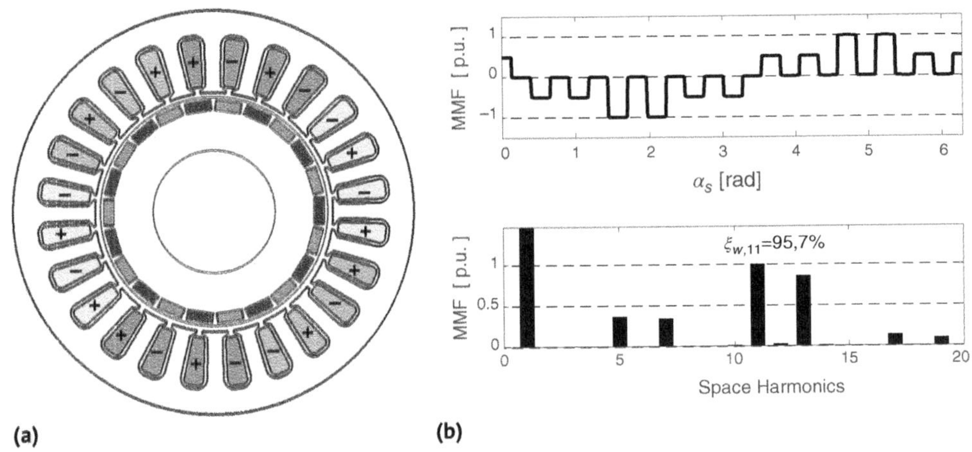

Figure 2.24 SL-FSCW with $z = 2$, $n_c = 2$, and $Q_S = 24$. (a) Winding layout. (b) Three-phase spatial MMF distribution and harmonic content.

Example 2.6 36-slot/34-pole and 48-slot/46-pole FSCW

Two additional SL-FSCW with $z = 2$ and $n_c = 3$, $n_c = 4$ are investigated in Figure 2.25:

- 36-slot/34-pole FSCW with $n_c = 3$
  - Stator slot number: $Q_s = 36$
  - Operating harmonic: 17th harmonic
  - Winding factor: 95.6%
- 48-slot/46-pole FSCW with $n_c = 4$
  - Stator slot number: $Q_s = 48$
  - Operating harmonic: 23rd harmonic
  - Winding factor: 95.6%

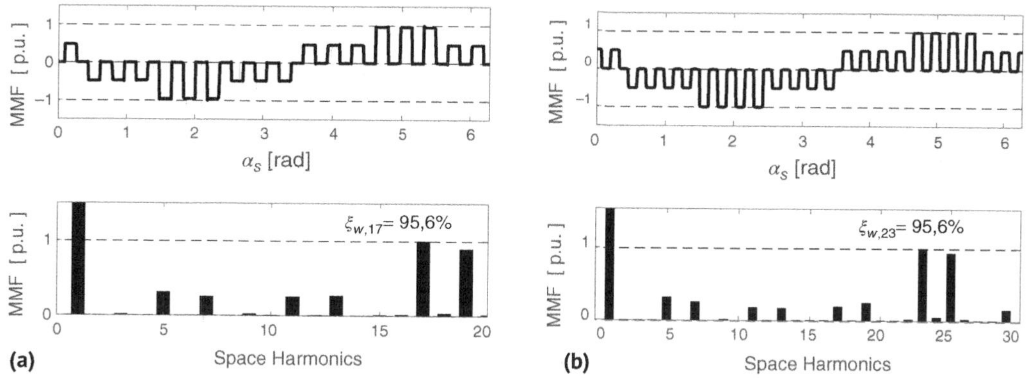

Figure 2.25 MMF characteristic for the SL-FSCWs with $z = 2$. (a) Winding topology with $n_c = 3$, $v_p = 17$, and $Q_S = 36$. (b) Winding topology with $n_c = 4$, $v_p = 23$, and $Q_S = 48$.

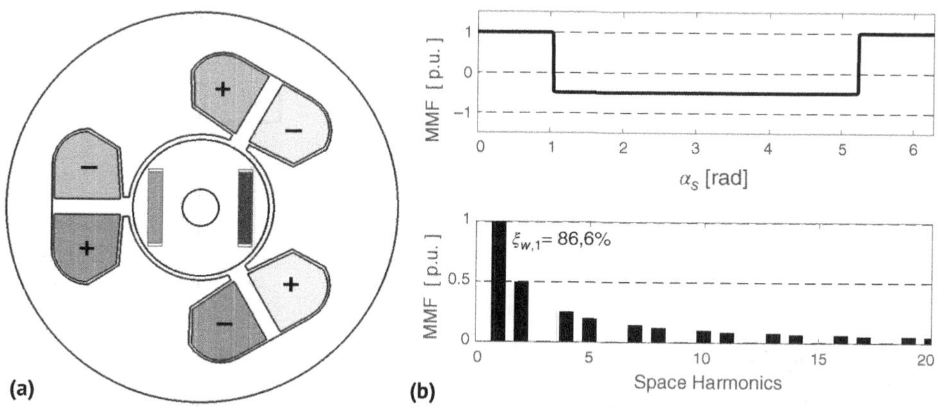

Figure 2.26 DL-FSCW with $z = 1$, $n_c = 1$, and $Q_S = 3$. (a) Winding layout. (b) Three-phase spatial MMF distribution and harmonic content.

2.5.3 DL-FSCWs with $z = 1$

Example 2.7 3-slot/2-pole FSCW with $z = 1$ and $n_c = 1$

The DL-FSCW with $z = 1$ and $n_c = 1$ consists of 3 stator slots ($Q_s = 3$) and 3 winding coils arranged as +U+V+W. Figure 2.26a,b shows the winding layout and corresponding three-phase MMF characteristics. It is noteworthy that this winding topology is the only FSCW where the first MMF harmonic is used as the working harmonic with a winding factor of 86.6%. By repeating its base winding p-times, any desired number of pole pairs can be achieved, as is demonstrated in Figure 2.27 for a 4-pole PMSM example.

Example 2.8 6-slot/4-pole FSCW with $z = 1$ and $n_c = 2$

A 4-pole PM machine with a DL-FSCW and with winding parameters $z = 1$ and $n_c = 2$ is examined in Figure 2.28. The stator has 6 slots ($Q_s = 6$) and uses the winding layout +U−U+V−V+W−W. The winding factor for the operating harmonic (i.e. 2nd harmonic) is 75%.

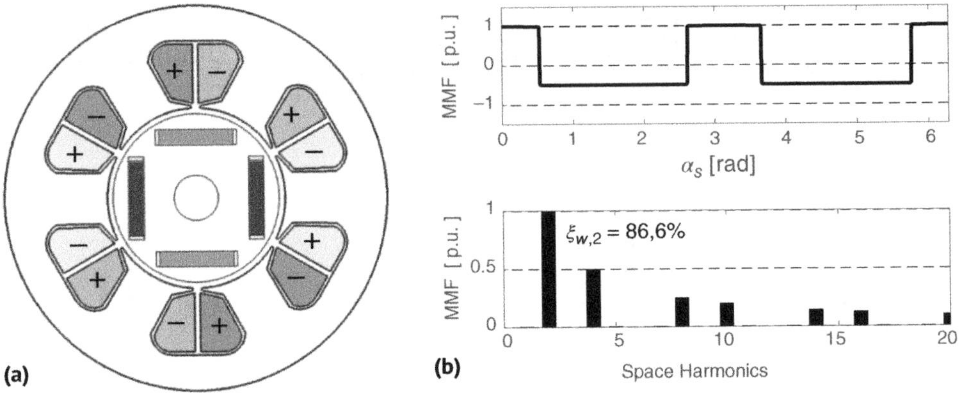

Figure 2.27 DL-FSCW with $z = 1$, $n_c = 1$, and $\wp = 2$. (a) Winding layout for the four pole machine. (b) Three-phase spatial MMF distribution and harmonic content.

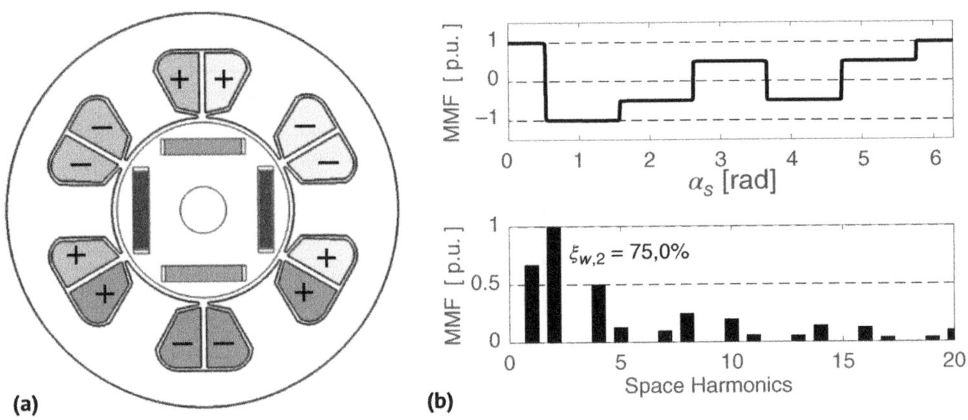

Figure 2.28 DL-FSCW with $z = 1$, $n_c = 2$, and $Q_s = 6$. (a) Winding layout. (b) Three-phase spatial MMF distribution and harmonic content.

Example 2.9 9-slot/8-pole FSCW with $z = 1$ and $n_c = 3$

The 9-slot/8-pole DL-FSCW with $z = 1$ and $n_c = 3$, shown in Figure 2.29a, is one of the most widely used winding type in various electric machine applications. The stator consists of 9 slots ($Q_s = 9$), while the winding pattern is +U−U+U+V−V+V+W−W+W. The corresponding three-phase MMF characteristics are depicted in Figure 2.29b. In this configuration, the winding factor for the 4th and 5th harmonics is 94.5%, with the 4th harmonic being most commonly used as the working harmonic.

Example 2.10 12-slot/10-pole and 15-slot/14-pole FSCW

Two additional DL-FSCW with $z = 1$, and with $n_c = 4$, $n_c = 5$ are analyzed in Figure 2.30:

- 12-slot/10-pole FSCW with $n_c = 4$
  - Stator slot number: $Q_s = 12$
  - Operating harmonic: 5th harmonic
  - Winding factor: 80.8%

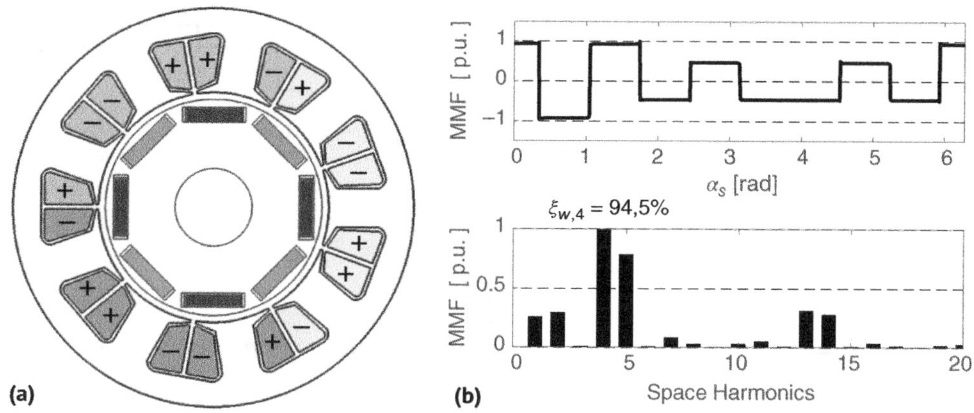

Figure 2.29 DL-FSCW with $z = 1$, $n_c = 3$, and $Q_s = 9$. (a) Winding layout. (b) Three-phase spatial MMF distribution and harmonic content.

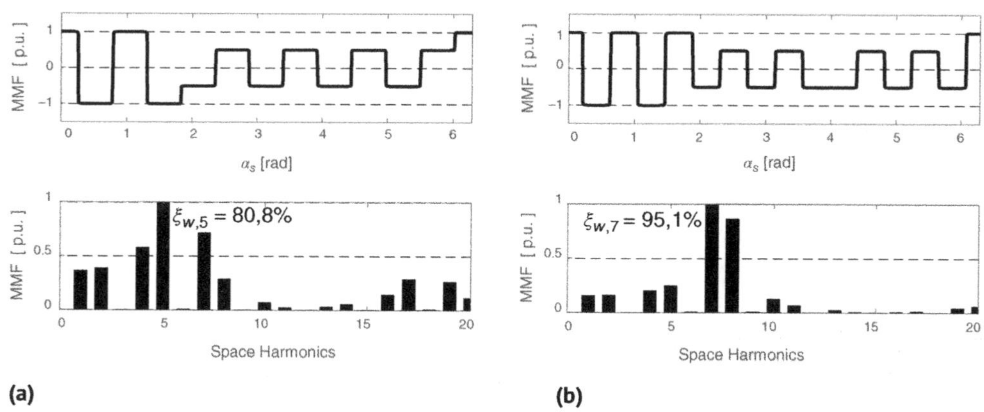

Figure 2.30 MMF characteristic for the DL-FSCWs with $z = 1$. (a) Winding topology with $n_c = 4$, $v_p = 5$, and $Q_S = 12$. (b) Winding topology with $n_c = 5$, $v_p = 7$, and $Q_S = 15$.

- 15-slot/14-pole FSCW with $n_c = 5$
  - Stator slot number: $Q_s = 15$
  - Operating harmonic: 7th harmonic
  - Winding factor: 95.1%

2.5.4 DL-FSCWs with $z = 2$

Example 2.11 6-slot/2-Pole FSCW with $z = 2$ and $n_c = 1$
An alternative to the common 3-slot/2-pole FSCW is the 6-slot/2-pole winding configuration with $z = 2$ and $n_c = 1$, arranged as +U−V+W−U+V−W, as illustrated in Figure 2.31a. The corresponding three-phase spatial MMF distribution and harmonic content are presented in Figure 2.31b. Although the winding factor for the working harmonic is relatively low ($\xi_{w,1} = 50,0\%$), the high quality MMF makes this winding type attractive for certain specific applications.

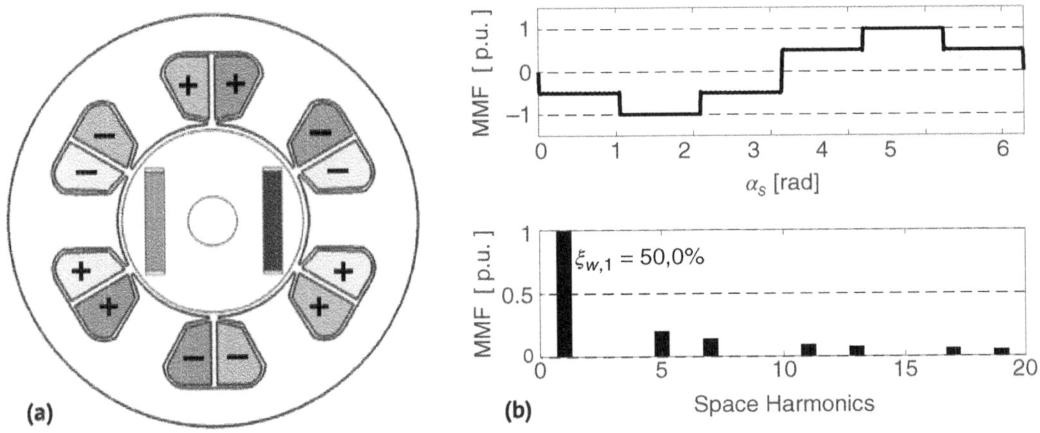

Figure 2.31 DL-FSCW with $z = 2$, $n_c = 1$, and $Q_s = 6$. (a) Winding layout. (b) Three-phase spatial MMF distribution and harmonic content.

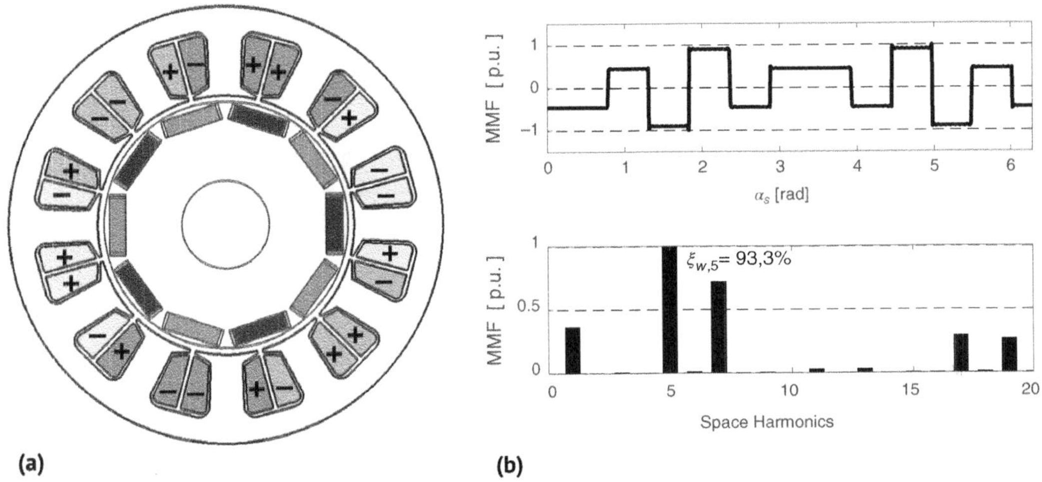

Figure 2.32 DL-FSCW with $z = 2$, $n_c = 2$, and $Q_s = 12$. (a) Winding layout. (b) Three-phase spatial MMF distribution and harmonic content.

Example 2.12 12-slot/10-pole FSCW with $z = 2$ and $n_c = 2$

A widely used configuration is the 12-slot/10-pole winding with $z = 2$ and $n_c = 2$ depicted in Figure 2.32a. The stator has 12 slots ($Qs = 12$), and the winding arrangement is +U−U−V+V+W−W−U+U+V−V−W+W. The corresponding three-phase MMF characteristics are shown in Figure 2.32b. Both the 5th and 7th harmonics have a winding factor of 93.3%, so either of the two harmonics can be selected as the operating wave. In the presented example, the 5th harmonic is chosen, resulting in a 10-pole PM machine.

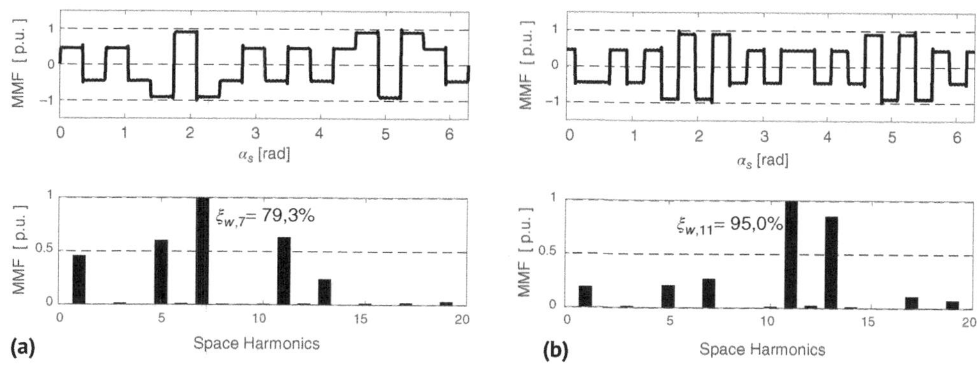

Figure 2.33 MMF characteristic for the DL-FSCWs with $z = 2$. (a) Winding topology with $n_c = 2$, $v_p = 7$, and $Q_S = 18$. (b) Winding topology with $n_c = 4$, $v_p = 11$, and $Q_S = 24$.

Example 2.13 18-slot/14-pole and 24-slot/22-pole FSCW

Figure 2.33 shows the MMF characteristics for two additional DL-FSCWs with $z = 2$:

- 18-slot/14-pole FSCW with $n_c = 3$
  - Stator slot number: $Q_s = 18$
  - Operating harmonic: 7th harmonic
  - Winding factor: 79.3%
- 24-slot/22-pole FSCW with $n_c = 4$
  - Stator slot number: $Q_s = 24$
  - Operating harmonic: 11th harmonic
  - Winding factor: 95.0%

2.5.5 Comparison of Winding Performances

Based on the MMF characteristic of the investigated windings and the summarized results in Table 2.10, the following conclusions can be made:

- One-zone FSCWs contain both even and odd harmonics in the MMF spectrum, with the exception of third-order harmonics.
- Two-zone FSCWs include only odd harmonics.
- For all FSCW configurations, THD increases as n_c increases.
- DL-FSCWs exhibit lower THD than SL-FSCWs.
- In SL windings, configurations with two zones per phase achieve higher winding factors than those with only one zone.
- For DL-FSCWs, both one- and two-zone configurations provide high winding factors.
- SL-FSCWs with opposite polarity coil arrangements and DL-FSCWs with same polarity coils arrangements are inefficient due to their lower winding factors.
- Except for the double-layer winding with:
  - $z = 1$ and $n_c = 1$ (3-slot/2-pole), and
  - $z = 2$ and $n_c = 1$ (6-slot/2-pole),

  all other FSCWs operate with higher-order harmonics and therefore suffer from the MMF subharmonics.

Table 2.10 Winding performances for various common FSCWs.

Layers	z	n_c	Q_b	v_p	$\xi_{w,p}$ (%)	THD (%)
SL	1	1	6	2	86.6	138.7
SL	1	2	12	5	83.6	197.3
SL	1	3	18	8	83.1	215.4
SL	1	4	24	11	82.9	224.1
SL	2	1	12	5	96.6	163.4
SL	2	2	24	11	95.7	187.5
SL	2	3	36	17	95.6	195.1
SL	2	4	48	23	95.6	198.8
DL	1	1	3	1	86.6	67.9
DL	1	2	6	2	75.0	97.4
DL	1	3	9	4	94.5	108.6
DL	1	4	12	5	80.8	127.4
DL	1	5	15	7	95.1	117.1
DL	2	1	6	1	50	31
DL	2	2	12	5	93.3	98.17
DL	2	3	18	7	79.3	116.8
DL	2	4	24	11	95	113.5

3

Fractional Slot Concentrated Windings with Unequal Zones

In the previous chapter, a systematic method for designing standard (common) fractional slot concentrated windings (FSCWs) was presented. These windings typically consist of coils arranged in coil-groups or zones, which are then combined in a phase winding with one or two zones. In the case of a two-zone winding, both zones are equal, i.e. the number of coils per zone is identical, and the zones are spatially offset by 180 electrical degrees. However, beyond these common topologies, there exists a special category of windings, which consist of *unequal zones*. Firstly, this type of winding was often used in fractional slot distributed windings (DW), which exhibit unique properties in terms of magnetomotive force (MMF) harmonics and coil arrangements. The term "*unequal zones*" refers to a configuration where the number of coils per zone differs between the first and the second phase zone. In DWs, the first zone, with n_{c1} coils, is located or distributed on one pole (e.g. pole N), while the second zone, with n_{c2} coils, is located on the adjacent pole (e.g. pole S), Binder [25], Pyrhönen [26].

Similar to DWs, the concept of unequal zones can also be applied to FSCWs, resulting in favorable MMF properties for certain topologies, as reported by Canders and Hülsmann in [49]. In their work, the authors introduced a design methodology, which is applicable to most common FSCWs regardless of the number of phases and layers configuration. As a result, the unequal zones winding arrangements expand the range of feasible slot and pole combinations for FSCWs and, in some cases, also show improved MMF performance. For instance, an 18-slot/14-pole FSCW with an unequal zones exhibits lower harmonic contents compared to a related FSCW with the same number of stator slots. Additionally, this winding group is particularly advantageous for applications requiring stators with a high number of slots, such as machines with large stator diameters, offering an alternative to traditional FSCWs with high winding factors.

3.1 Winding Configurations

Figures 3.1 and 3.2 illustrate the realization of FSCWs with unequal zones using simplified stator winding sketches. For the sake of simplicity, only the coils of a single phase are shown, while the different phases (U, V, and W) are represented with different colors.

In the first example, shown in Figure 3.1, we start from a standard FSCW configuration in which each phase contains a single zone. The transformation process involves splitting this zone into two new zones with unequal number of coils. The second example, depicted in Figure 3.2, demonstrates the transformation of a typical two zone (or one zone-group) FSCW into a winding with two unequal zone-groups. Similar to the first case, the process involves splitting the original zone-group into two new ones, each characterized by a distinct number of coils. Additionally, as illustrated in Figure 3.3, the coil polarity of the resulting zones may either be the same or opposite compared to the initial winding. Therefore, considering the wide variety of common FSCW topologies (see Tables 2.2–2.9), the number of possible FSCW configurations with unequal zones becomes extremely large. Nevertheless, these topologies can be categorized into two main categories, as described below. In the notation "$z_{i,k}$," the first index (i) represents the zone group, while the second index (k) identifies the specific zone number within that group.

1. Phase winding with two unequal zones $z_{1,1}$ and $z_{2,1}$, where
   - the zones $z_{1,1}$ and $z_{2,1}$ are shifted by 180 electrical degrees to each other
   - the number of coils per zone $z_{1,1}$ and $z_{2,1}$ is n_{c1} and n_{c2}, respectively, where $n_{c1} \neq n_{c2}$, and wherein
   - the second zone $z_{2,1}$ of one phase, e.g. phase-U, lies between the first zones $z_{1,1}$ of other phases, e.g. between $z_{1,1}$ of phases V and W, while the first zone $z_{1,1}$ of one phase, e.g. phase-U, lies between the second zones $z_{2,1}$ of other phases, e.g. between $z_{2,1}$ of phases V and W.
2. Phase winding with two unequal zone-groups $z_{1,k}$ and $z_{2,k}$, where
   - one zone-group consists of two zones.
   - the number of coils per zone in a zone-group is same, however the number of coils per zone in the first and the second zone-group is different, i.e. $n_{c1} \neq n_{c2}$
   - the zones belonging to the same zone-group, e.g. $z_{1,1}$ and $z_{1,2}$, are shifted by 180 electrical degrees to each other, and have opposite winding polarity, and wherein
   - the second zone-group $z_{2,k}$ of one phase, e.g. phase-U, lies between the first zone-groups $z_{1,k}$ of other phases, e.g. between $z_{1,k}$ of phases V and W, and the first zone-group $z_{1,k}$ of one phase, e.g. phase-U, lies between the second zone-groups $z_{2,k}$ of other phases, e.g. between $z_{2,k}$ of phases V and W.

The winding topologies for the first and the second winding group are illustrated in Figures 3.1 and 3.2. In these illustrations, all zones belonging to the same phase are represented using the same color. Figure 3.1a shows the initial standard FSCW configuration with one zone per phase and six coils per zone. The modified winding with two unequal zones $z_{1,1}$ and $z_{2,1}$ is illustrated in Figure 3.1b. In this example, the zones $z_{1,1}$ and $z_{2,1}$ are spatially shifted by 180 electrical degrees, the first zone $z_{1,1}$ contains four coils per zone $(n_{c1} = 4)$, and the second zone $z_{2,1}$ has two coils per zone $(n_{c2} = 2)$. Similarly, Figure 3.2 demonstrates the transformation of a standard FSCW with

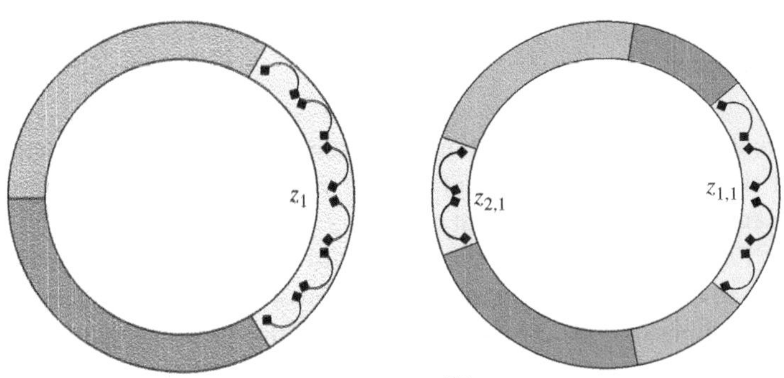

Figure 3.1 (a) Standard FSCW with one zone per phase and with $n_c = 6$ coils per zone. (b) FSCW with two unequal zones per phase and with $n_{c1} = 4$ per zone $z_{1,1}$ and $n_{c2} = 2$ per zone $z_{2,1}$.

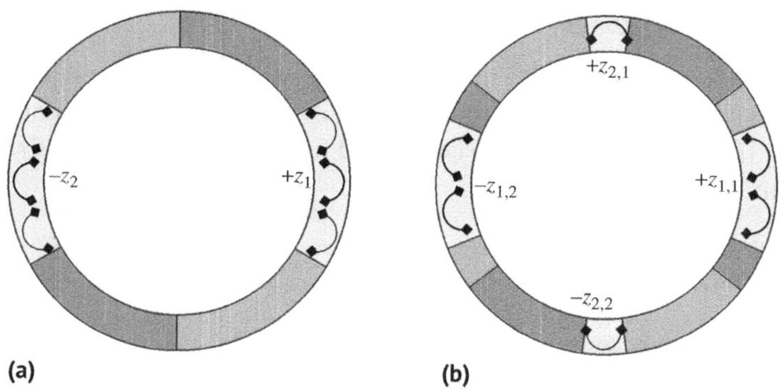

Figure 3.2 (a) Standard FSCW with two zones per phase and with $n_{c1} = 3$ coils per zone. (b) FSCW with unequal zones, two zone-groups per phase, and with $n_{c1} = 2$ per zones of the first zone-group and $n_{c2} = 1$ per zones of the second zone-group.

two zones per phase and three coils per zone into a configuration with unequal zones. Unlike the first example, the resulting winding consists of two zone-groups: the first zone-group $z_{1,k}$ has two coils per zone $(n_{c1} = 2)$, and the second zone-group $z_{2,k}$ has only one coil per zone $(n_{c2} = 1)$.

FSCWs with *unequal zones* can be realized in single-layer or double-layer form, employing different coil polarity arrangements and combinations. However, earlier MMF investigations of standard FSCWs have shown that certain combinations lead to poor winding factors, therefore, such inefficient configurations are not considered in the MMF analysis presented below.

3.2 MMF Analysis

If n_{c1} represents the number of adjacent coils in the first zone $z_{1,k}$ and n_{c2} the number of adjacent coils in the second zone $z_{2,k}$, and if z_g represents the number of zone-groups in the base-winding, then the number of stator slots for an m-phase base winding is given by:

$$Q_b = \begin{cases} 2mz_g(n_{c1} + n_{c2}), & \text{single-layer FSCW} \\ mz_g(n_{c1} + n_{c2}), & \text{double-layer FSCW} \end{cases} \tag{3.1}$$

while the space displacement angle between phase-zones $z_{1,j}$ and $z_{2,j}$ as a function of the number of adjacent coils can be presented as:

$$\varphi_{z_1,z_2} = \begin{cases} 2(2n_{c1} + n_{c2})\dfrac{2\pi}{Q_S}, & \text{single-layer FSCW} \\ (2n_{c1} + n_{c2})\dfrac{2\pi}{Q_S}, & \text{double-layer FSCW} \end{cases} \tag{3.2}$$

When the zone of an initial winding is divided into two zones or zone-groups, the overall winding layout and coil arrangements of the resulting FSCW remain unchanged. The main difference appears only in the winding parameters n_{c1} and n_{c2}, as well as in the polarity of the zone or zone-groups. Consequently, the resulting MMF distribution corresponds either to the sum or difference of the MMF contributions of the individual zones or zone-groups:

$$F_j(t, \alpha_S) = F_{j,zg1}(t, \alpha_S) \pm F_{j,zg2}(t, \alpha_S - \varphi_{z1,z2}) \tag{3.3}$$

where $F_{j,zg1}$ and $F_{j,zg2}$ denote the corresponding MMF functions for the winding zones or zone-groups.

3.2.1 Single-layer FSCWs with Unequal Zones

Based on the results summarized in Tables 2.2–2.5, the single-layer winding topologies that achieve the highest winding factor for the working harmonic are those with coil arrangements of same polarity. Therefore, only these configurations are considered in the subsequent analysis, particularly in combination with:

- two unequal zones per phase and
- two unequal zone-groups per phase.

3.2.1.1 Single-layer FSCW with Two Unequal Zones

Figure 3.3 illustrates the winding scheme for two cases of single-layer FSCW (SL-FSCW) with two unequal zones per phase. Both cases feature the same polarity coil arrangement within each zone, but differ in the polarity between zones. Specifically, Figure 3.3a shows the case where both zones $z_{1,2}$ and $z_{2,1}$ have the same polarity, while Figure 3.3b depicts the case with opposite polarities between the zones.

The winding layout and the corresponding MMF waveform of phase-j for a specific SL-FSCW with two unequal zones is presented in Figure 3.4. In this example, the first zone consists of two coils per zone $(n_{c1} = 2)$ with the same polarity coils, while the second zone has only one coil per zone $(n_{c2} = 1)$. The zones themselves have opposite polarity. In the MMF representation shown in Figure 3.4b, the MMF distribution for the first zone is indicated by a solid line, while the second zone is represented by a dashed line.

According to Equation (2.23), the MMF distribution for a single-layer winding with one-zone and coils of the same polarity can be expressed in complex form as:

$$F_j(a_S, t) = \frac{2n_c N_w i_j(t)}{\pi} \sum_{\nu} \frac{{}^{\nu}\xi_w}{\nu} e^{j\left(\nu\left(a_S-(n_c-1)\frac{2\pi}{Q_S}\right)\right)}, \quad {}^{\nu}\xi_w = {}^{\nu}\xi_p\,{}^{\nu}\xi_d \tag{3.4}$$

where the respective winding factor parameters ${}^{\nu}\xi_p$ and ${}^{\nu}\xi_d$ are defined in Equations (2.15) and (2.24). Since the coil arrangements of the respective zones are identical, but differ only in the parameters n_{c1} and n_{c2}, and are spatially shifted by $\varphi_{z1,z2}$, from Equation (3.4), the corresponding MMF functions can be expressed as:

$$F_{j,z1}(a_S, t) = \frac{2N_w i_j(t)}{\pi} \sum_{\nu} \frac{n_{c1}\,{}^{\nu}\xi_{w1}}{\nu} e^{j\nu\left(a_S-(n_{c1}-1)\frac{2\pi}{Q_S}\right)} \tag{3.5}$$

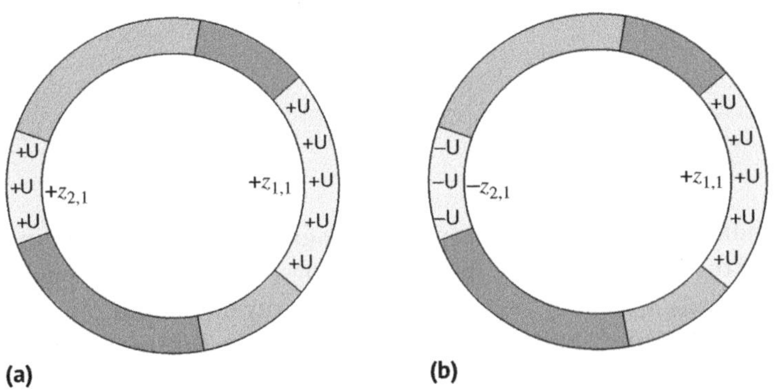

Figure 3.3 FSCW with two unequal zones $z_{1,1}$ and $z_{2,1}$ per phase and with $n_{c1} = 5$ and $n_{c2} = 3$. (a) Case with the same polarity zones. (b) Case with opposite polarity zones.

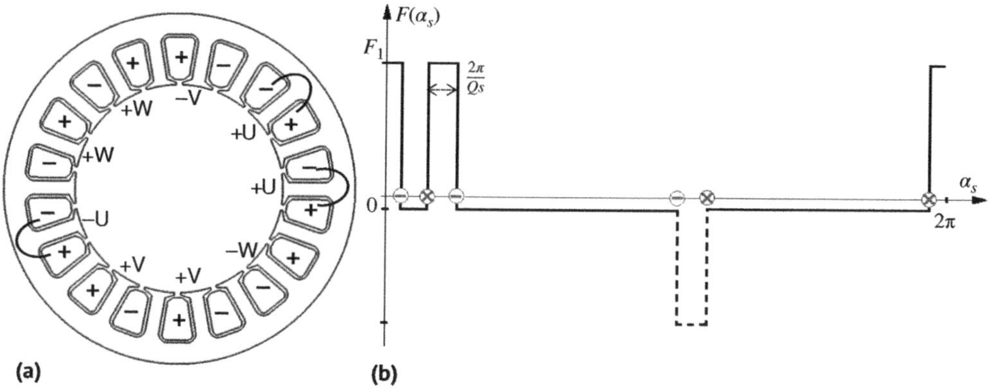

Figure 3.4 (a) SL-FSCW with two unequal zones per phase, with $n_{c1} = 2$ and $n_{c2} = 1$, and with opposite polarity zones. (b) MMF distribution of phase-j.

$$F_{j,z2}(\alpha_S) = \frac{2N_w i_j(t)}{\pi} \sum_\nu \frac{n_{c2}\,{}^\nu\xi_{w2}}{\nu} e^{j\nu\left(\alpha_S - \varphi_{z_1,z_2} - (n_{c2}-1)\frac{2\pi}{Q_S}\right)} \tag{3.6}$$

By substituting $\varphi_{z1,z2}$ into (3.6), we obtain:

$$F_{j,z2}(a_S, t) = \frac{2N_w i_j(t)}{\pi} \sum_\nu \frac{n_{c2}\,{}^\nu\xi_{w2}}{\nu} e^{j\nu\left(\alpha_S - (4n_{c1}+3n_{c2}-1)\frac{2\pi}{Q_S}\right)} \tag{3.7}$$

As defined in Section 2.2.2, the winding factor for a common one-zone SL-FSCW is the product of the coil-pitch and distribution factor. Therefore, the corresponding winding factors for the first and second zones can be expressed as:

$${}^\nu\xi_{w1} = {}^\nu\xi_{p1} \cdot {}^\nu\xi_{d1}, \quad {}^\nu\xi_{w2} = {}^\nu\xi_{p2} \cdot {}^\nu\xi_{d2} \tag{3.8}$$

whereby, according to (2.15) and (2.24), the respective winding factor parameters are:

$${}^\nu\xi_{p1} = {}^\nu\xi_{p2} = \sin\left(\nu\frac{\pi}{Q_S}\right) \tag{3.9}$$

$${}^\nu\xi_{d1} = \frac{\sin\left(\nu n_{c1}\frac{2\pi}{Q_S}\right)}{n_{c1}\cdot\sin\left(\nu\frac{2\pi}{Q_S}\right)}, \quad {}^\nu\xi_{d2} = \frac{\sin\left(\nu n_{c2}\frac{2\pi}{Q_S}\right)}{n_{c2}\cdot\sin\left(\nu\frac{2\pi}{Q_S}\right)} \tag{3.10}$$

Substituting Equations (3.5) and (3. 7) into (3.3), the resulting MMF function is:

$$\begin{aligned} F_j(a_S, t) &= \frac{2N_w i_j(t)}{\pi} \sum_\nu \left\{ \frac{n_{c1}\,{}^\nu\xi_{w1}}{\nu} e^{j\left(-\nu(n_{c1}-1)\frac{2\pi}{Q_S}\right)} \pm \frac{n_{c2}\,{}^\nu\xi_{w2}}{\nu} e^{j\left(-\nu(4n_{c1}+3n_{c2}-1)\frac{2\pi}{Q_S}\right)} \right\} e^{j\nu\alpha_S} \\ &= \frac{2N_w i_j(t)}{\pi} \sum_\nu \left\{ \frac{n_{c1}\,{}^\nu\xi_{w1}}{\nu} \pm \frac{n_{c2}\,{}^\nu\xi_{w2}}{\nu} e^{j\left(-\nu(3n_{c1}+3n_{c2})\frac{2\pi}{Q_S}\right)} \right\} e^{j\nu\left(\alpha_S - (n_{c1}-1)\frac{2\pi}{Q_S}\right)} \end{aligned} \tag{3.11}$$

Using

$$\nu(3n_{c1} + 3n_{c2})\frac{2\pi}{Q_S} = \nu\frac{(3n_{c1} + 3n_{c2})2\pi}{2mz_g(n_{c1} + n_{c2})} = \nu\pi, \quad \text{where } z_g = 1 \text{ and } m = 3$$

Equation (3.11) simplifies to:

$$F_j(\alpha_S, t) = \frac{2N_w i_j(t)}{\pi} \sum_{\nu} \left\{ \frac{n_{c1}\,{}^{\nu}\xi_{w1}}{\nu} \pm \frac{n_{c2}\,{}^{\nu}\xi_{w2}}{\nu} e^{j\nu\pi} \right\} e^{j\nu\left(a_S-(n_{c1}-1)\frac{2\pi}{Q_S}\right)} \tag{3.12}$$

Finally, the MMF function for a SL-FSCW with two unequal zones and same polarity coil arrangement per zone can be written as:

$$F_j(a_S, t) = \frac{2(n_{c1}+n_{c2})N_w i(t)}{\pi} \sum_{\nu} \frac{{}^{\nu}\xi_w}{\nu} \cos\left(\nu a_S - \nu(n_{c1}-1)\frac{2\pi}{Q_S}\right) \tag{3.13}$$

whereas the resulting winding factor is:

$${}^{\nu}\xi_w = \frac{n_{c1}}{n_{c1}+n_{c2}}\,{}^{\nu}\xi_{w1} \pm \frac{n_{c2}}{n_{c1}+n_{c2}}\,{}^{\nu}\xi_{w2}\cos(\nu\pi) \tag{3.14}$$

According to Equations (3.10) and (3.14), the resulting winding factor for this winding group is influenced twice: by the winding parameters n_{c1} and n_{c2}, as well as the zone polarities ($\pm$). First, the parameters n_{c1} and n_{c2} determine the distribution factors of their respective zones and after, the chosen zone polarities determine the final MMF distribution, directly affecting the specific MMF harmonics.

3.2.1.2 SL-FSCW with Two Unequal Zone-groups

The winding schemes for a SL-FSCW with two unequal zone-groups and different zone-group polarities are illustrated in Figure 3.5. As described earlier, in this winding category, the zones within a zone-group have opposite polarity and are 180 electrical degrees shifted in space to each other. Additionally, the second zone-group is offset by 90 electrical degrees from the first zone-group and may have the same or opposite polarity relative to the first group. An exemplary of a single-layer winding based on the winding scheme in Figure 3.5b for the case where $n_{c1} = 2$ and $n_{c2} = 1$ is shown in Figure 3.6a. For this configuration, the required number of stator slots according to Equation (3.1) is $Q_S = 36$. Furthermore, Figure 3.6b presents the corresponding phase-j MMF waveform, which is represented as the sum of the MMFs of the respective zone-groups. The MMF distribution for the first zone-group is depicted with a solid line, while for the second zone-group with a dashed line.

When analyzing the zone-groups separately, the corresponding MMF distribution in complex form can be derived from Equation (2.32) as follows (z is the number of zones in a zone-group):

$$F_{j,zg1}(a_S, t) = \frac{2zN_w i_j(t)}{\pi} \sum_{\nu} \frac{n_{c1}\,{}^{\nu}\xi_{w1}}{\nu} e^{j\left(\nu\left(\alpha_S-(n_{c1}-1)\frac{2\pi}{Q_S}\right)-\frac{\pi}{2}\right)} \tag{3.15}$$

$$F_{j,zg2}(a_S, t) = \frac{2zN_w i_j(t)}{\pi} \sum_{\nu} \frac{n_{c2}\,{}^{\nu}\xi_{w2}}{\nu} e^{j\left(\nu\left(a_S-(4n_{c1}+3n_2-1)\frac{2\pi}{Q_S}\right)-\frac{\pi}{2}\right)} \tag{3.16}$$

Meanwhile, the respective winding factors are:

$${}^{\nu}\xi_{w1} = {}^{\nu}\xi_{p1}\,{}^{\nu}\xi_{d1}\,{}^{\nu}\xi_{z1}, \quad {}^{\nu}\xi_{w2} = {}^{\nu}\xi_{p2}\,{}^{\nu}\xi_{d2}\,{}^{\nu}\xi_{z2} \tag{3.17}$$

The winding factor parameters for the coil-pitch and distribution factor are provided in (3.9) and (3.10), while the zone factor, according to Equation (2.31), is given by:

$${}^{\nu}\xi_{z1} = {}^{\nu}\xi_{z2} = \sin\left(\nu\frac{\pi}{2}\right) \tag{3.18}$$

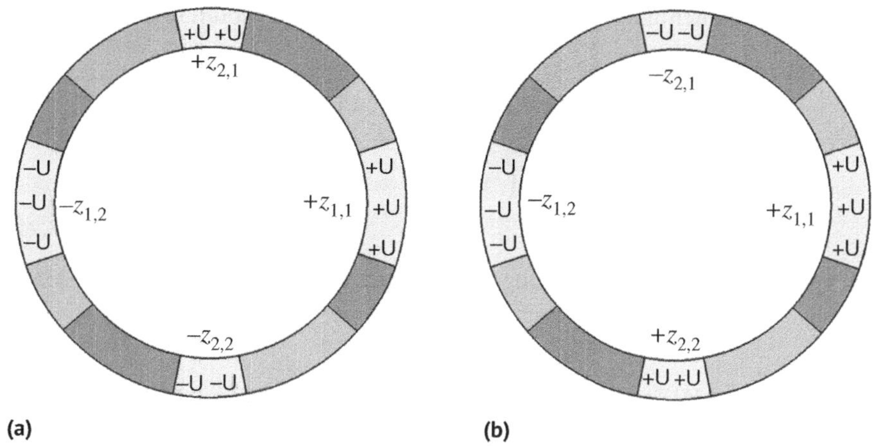

Figure 3.5 FSCW with two unequal zone-groups $z_{1,k}$ and $z_{2,k}$ per phase, and with $n_{c1} = 3$ and $n_{c2} = 2$. (a) Case with the same polarity zone-groups. (b) Case with opposite polarity zone-groups.

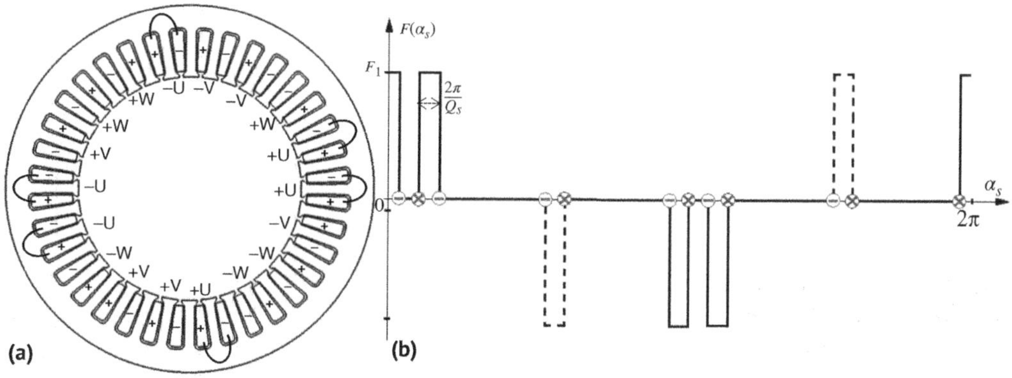

Figure 3.6 (a) SL-FSCW with two unequal zone-groups per phase, with $n_{c1} = 2$ and $n_{c2} = 1$, and with opposite polarity zone-groups. (b) MMF distribution of phase-j.

Substituting Equations (3.15) and (3.16) into (3.3), we obtain:

$$\begin{aligned} F_j(a_S, t) &= \frac{2zN_w i_j(t)}{\pi} \sum_{\nu} \left\{ \frac{n_{c1}{}^{\nu}\xi_{w1}}{\nu} e^{j\left(-\nu(n_{c1}-1)\frac{2\pi}{Q_S}\right)} \pm \frac{n_{c2}{}^{\nu}\xi_{w2}}{\nu} e^{j\left(-\nu(4n_{c1}+3n_{c2}-1)\frac{2\pi}{Q_S}\right)} \right\} e^{j\left(\nu a_S - \frac{\pi}{2}\right)} \\ &= \frac{2zN_w i_j(t)}{\pi} \sum_{\nu} \left\{ \frac{n_{c1}{}^{\nu}\xi_{w1}}{\nu} \pm \frac{n_{c2}{}^{\nu}\xi_{w2}}{\nu} e^{j\left(-\nu(3n_{c1}+3n_{c2})\frac{2\pi}{Q_S}\right)} \right\} e^{j\left(\nu a_S - \nu(n_{c1}-1)\frac{2\pi}{Q_S} - \frac{\pi}{2}\right)} \end{aligned} \tag{3.19}$$

Further, simplifying the term in the exponent function:

$$\nu(3n_{c1} + 3n_{c2})\frac{2\pi}{Q_S} = \nu\frac{(3n_{c1} + 3n_{c2})2\pi}{2mz_g(n_{c1} + n_{c2})} = \nu\frac{\pi}{2}, \text{where } z_g = 2 \text{ and } m = 3$$

Equation (3.19) can be expressed as:

$$F_j(a_S, t) = \frac{2zN_w i_j(t)}{\pi} \sum_{\nu} \left\{ \frac{n_{c1}{}^{\nu}\xi_{w1}}{\nu} \pm \frac{n_{c2}{}^{\nu}\xi_{w2}}{\nu} e^{-j\nu\frac{\pi}{2}} \right\} e^{j\left(\nu a_S - \nu(n_{c1}-1)\frac{2\pi}{Q_S} - \frac{\pi}{2}\right)} \tag{3.20}$$

Since the product of the exponential functions in (3.20) yields different solution depending on the harmonic order:

$$e^{-j\nu\frac{\pi}{2}} \cdot e^{j\left(\nu \alpha_S - \nu(n_{c1}-1)\frac{2\pi}{Q_S} - \frac{\pi}{2}\right)} = \begin{cases} \cos\left(\nu\frac{\pi}{2}\right)\cos\left(\nu\alpha_S - \nu(n_{c1}-1)\frac{2\pi}{Q_S} - \frac{\pi}{2}\right), & \nu = 2,4,6,\ldots \\ \sin\left(\nu\frac{\pi}{2}\right)\sin\left(\nu\alpha_S - \nu(n_{c1}-1)\frac{2\pi}{Q_S} - \frac{\pi}{2}\right), & \nu = 1,3,5,\ldots \end{cases} \tag{3.21}$$

the resulting MMF function and its corresponding winding factors vary depending on the MMF harmonic order.

Case 1: ν = *even number*

For even-order harmonics, the MMF winding function can be simplified to:

$$F_j(\alpha_S,t) = \frac{2z(n_{c1}+n_{c2})N_w\hat{i}}{\pi}\sum_\nu \frac{{}^\nu\xi_w}{\nu}\cos\left(\nu\alpha_S - \nu(n_{c1}-1)\frac{2\pi}{Q_S} - \frac{\pi}{2}\right) \tag{3.22}$$

$\nu = 2,4,6,\ldots$

and the corresponding winding factor is:

$${}^\nu\xi_w = \frac{n_{c1}}{n_{c1}+n_{c2}}{}^\nu\xi_{w1} \pm \frac{n_{c2}}{n_{c1}+n_{c2}}{}^\nu\xi_{w2}\cos\left(\nu\frac{\pi}{2}\right) \tag{3.23}$$

Case 2: ν = *odd number*

For odd-order harmonics, the MMF function in Equation (3.20) represents the sum of two cosine functions that are offset by $\pi/2$ and have different magnitudes.

$$\begin{aligned} F_j(\alpha_S,t) = \frac{2N_w i_j(t)}{\pi}\Bigg\{ &\sum_\nu \frac{n_{c1}{}^\nu\xi_{w1}}{\nu}\cos\left(\nu\alpha_S - \nu(n_{c1}-1)\frac{2\pi}{Q_S} - \frac{\pi}{2}\right) \\ &\pm \sum_\nu \frac{n_{c2}{}^\nu\xi_{w2}}{\nu}\sin\left(\nu\frac{\pi}{2}\right)\cos\left(\nu\alpha_S - \nu(n_{c1}-1)\frac{2\pi}{Q_S} - \frac{\pi}{2} - \frac{\pi}{2}\right)\Bigg\} \end{aligned} \tag{3.24}$$

This relationship can be simplified to:

$$F_j(\alpha_S,t) = \frac{2z(n_{c1}+n_{c2})N_w\hat{i}}{\pi}\sum_\nu \frac{{}^\nu\xi_w}{\nu}\cos\left(\omega t - \nu\alpha_S + \nu(n_{c1}-1)\frac{2\pi}{Q_S} - \frac{\pi}{2} + a_{z_1,z_2}\right) \tag{3.25}$$

$\nu = 1,3,5,\ldots$

where

$$a_{z_1,z_2} = \operatorname{atan}\left(\frac{\pm n_{c2}{}^\nu\xi_{w2}}{n_{c1}{}^\nu\xi_{w1}}\sin\left(\nu\frac{\pi}{2}\right)\right) \tag{3.26}$$

and the resulting winding factor is:

$${}^\nu\xi_w = \sqrt{\left(\frac{n_{c1}}{(n_{c1}+n_{c2})}{}^\nu\xi_{w1}\right)^2 + \left(\frac{n_{c2}}{(n_{c1}+n_{c2})}{}^\nu\xi_{w2}\sin\left(\nu\frac{\pi}{2}\right)\right)^2} \tag{3.27}$$

Equations (3.23) and (3.27) show that, as in the previously examined winding category, the resulting winding factor is influenced twice: by the winding parameters n_{c1} and n_{c2}, and the zone-group polarities ($\pm$). First, the parameters n_{c1} and n_{c2} determine the individual zone-group winding factors, as described in (3.10), while the polarities of the zone-group ($\pm$) subsequently influence the resulting winding factor in (3.23) and (3.27).

3.2.2 Double-Layer FSCWs with Unequal Zones

Unlike SL-FSCWs, the double-layer winding topologies that achieve the highest winding factor for the working harmonic, as indicated in Tables 2.6–2.9, are those that have coil arrangements of opposite polarity. Therefore, in order to reduce the analysis effort, only these specific coil configurations are examined below, specifically those featuring:

- two unequal zones per phase and
- two unequal zone-groups per phase.

3.2.2.1 DL-FSCW with Two Unequal Zones

The winding schemes for FSCWs with two unequal zones per phase and with different zone polarities have already been illustrated in Figure 3.3. A corresponding double-layer FSCW (DL-FSCW) with three coils in the first zone ($n_{c1} = 3$) and one coil in the second zone ($n_{c2} = 1$) is shown in Figure 3.7a. In this example, the adjacent coils in the first zone have opposite polarity, while the first and second zone have the same polarity. The phase-j MMF waveform for this winding is depicted in Figures 3.7b. According to Equation (3.1), this winding configuration requires 12 stator slots, i.e. $Q_S = 12$.

When analyzing the winding zones separately, the MMF functions for zones $z_{1,1}$ and $z_{1,2}$ in complex form can be expressed as:

$$F_{j,z1}(a_S, t) = \frac{2N_w i_j(t)}{\pi} \sum_{\nu} \frac{n_{c1}{}^{\nu}\xi_{w1}}{\nu} e^{j\left(\nu\left(a_S - \frac{n_{c1}-1}{Q_S}\pi\right) - \frac{n_{c1}-1}{2}\pi\right)} \tag{3.28}$$

$$F_{j,z2}(a_S, t) = \frac{2N_w i_j(t)}{\pi} \sum_{\nu} \frac{n_{c2}{}^{\nu}\xi_{w2}}{\nu} e^{j\left(\nu\left(a_S - \frac{4n_{c1}+3n_2-1}{Q_S}\pi\right) - \frac{n_{c2}-1}{2}\pi\right)} \tag{3.29}$$

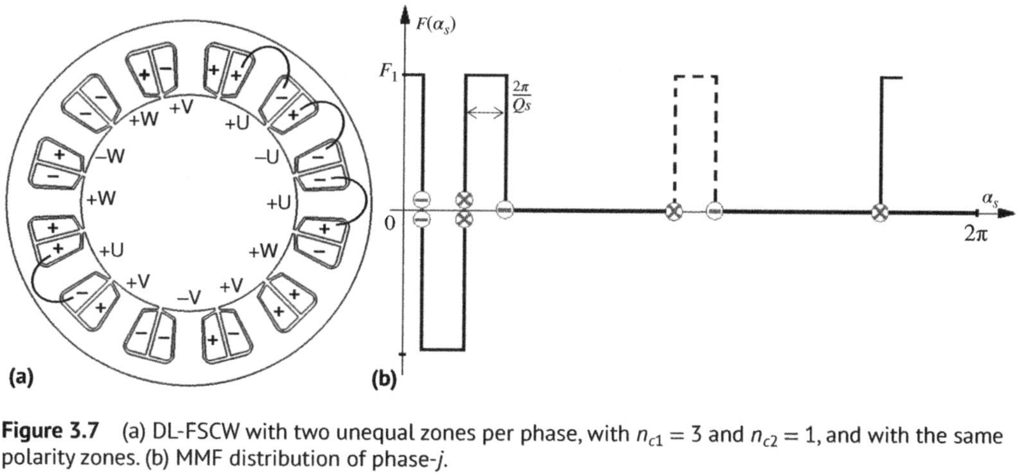

Figure 3.7 (a) DL-FSCW with two unequal zones per phase, with $n_{c1} = 3$ and $n_{c2} = 1$, and with the same polarity zones. (b) MMF distribution of phase-j.

As derived in Section 2.2.2, the winding factor for one-zone DL-FSCWs is the product of the coil-pitch and the distribution factor. Therefore, when the number of coils per zone is different, the respective winding factors for the first and second zone can be presented as follows:

$$^{\nu}\xi_{w1} = {}^{\nu}\xi_{p1}{}^{\nu}\xi_{d1}$$

$$^{\nu}\xi_{w2} = {}^{\nu}\xi_{p2}{}^{\nu}\xi_{d2}$$

The coil-pitch factor is already determined in (2.15), while, according to Equation (2.38), the zone distribution factors for the case when coils have opposite polarities are:

$$^{\nu}\xi_{d1} = \frac{\sin\left(\nu n_{c1}\frac{\pi}{Q_S} + n_{c1}\frac{\pi}{2}\right)}{n_{c1}\cdot\cos\left(\nu\frac{\pi}{Q_S}\right)}, \quad ^{\nu}\xi_{d2} = \frac{\sin\left(\nu n_{c2}\frac{\pi}{Q_S} + n_{c2}\frac{\pi}{2}\right)}{n_{c2}\cdot\cos\left(\nu\frac{\pi}{Q_S}\right)} \tag{3.30}$$

By substituting Equations (3.28) and (3.29) into (3.3), we obtain:

$$\begin{aligned} F_j(\alpha_S,t) &= \frac{2N_w i_j(t)}{\pi}\sum_{\nu}\left\{\frac{n_{c1}{}^{\nu}\xi_{w1}}{\nu}e^{j\left(-\nu\frac{(n_{c1}-1)}{Q_S}\pi - n_{c1}\frac{\pi}{2}\right)} \pm \frac{n_{c2}{}^{\nu}\xi_{w2}}{\nu}e^{j\left(-\nu\frac{4n_{c1}+3n_{c2}-1}{Q_S}\pi - n_{c2}\frac{\pi}{2}\right)}\right\}e^{j\left(\nu\alpha_S+\frac{\pi}{2}\right)} \\ &= \frac{2N_w i_j(t)}{\pi}\sum_{\nu}\left\{\frac{n_{c1}{}^{\nu}\xi_{w1}}{\nu} \pm \frac{n_{c2}{}^{\nu}\xi_{w2}}{\nu}e^{j\left(-\nu\frac{3n_{c1}+3n_{c2}}{Q_S}\pi - (n_{c2}-n_{c1})\frac{\pi}{2}\right)}\right\}e^{j\left(\nu\alpha_S - \nu\frac{(n_{c1}-1)}{Q_S}\pi - n_{c1}\frac{\pi}{2}+\frac{\pi}{2}\right)} \end{aligned} \tag{3.31}$$

For the case where $z_g = 1$ and $m = 3$, the exponent term in the above equation yields to:

$$\begin{aligned} &\nu\frac{(3n_{c1}+3n_{c2})}{Q_S}\pi + (n_{c2}-n_{c1})\frac{\pi}{2} \\ &= \nu\frac{(3n_{c1}+3n_{c2})}{mz_g(n_{c1}+n_{c2})}\pi + (n_{c2}-n_{c1})\frac{\pi}{2} = (2\nu + (n_{c2}-n_{c1}))\frac{\pi}{2} \end{aligned}$$

Consequently, we have:

$$F_j(\alpha_S,t) = \frac{2N_w i_j(t)}{\pi}\sum_{\nu}\left\{\frac{n_{c1}{}^{\nu}\xi_{w1}}{\nu} \pm \frac{n_{c2}{}^{\nu}\xi_{w2}}{\nu}e^{-j(2\nu+(n_{c2}-n_{c1}))\frac{\pi}{2}}\right\}e^{j\left(\nu\alpha_S - \nu\frac{(n_{c1}-1)}{Q_S}\pi - (n_{c1}-1)\frac{\pi}{2}\right)} \tag{3.32}$$

Since the product of the exponential functions in Equation (3.32) gives different solutions depending on the harmonic order, i.e.

$$\begin{aligned} &e^{-j(2\nu+(n_{c2}-n_{c1}))\frac{\pi}{2}}\cdot e^{j\left(\nu\alpha_S-\nu\frac{(n_{c1}-1)}{Q_S}\pi-(n_{c1}-1)\frac{\pi}{2}\right)} \\ &= \begin{cases} \cos\left((2\nu+(n_{c2}-n_{c1}))\frac{\pi}{2}\right)\cos\left(\nu\alpha_S - \nu\frac{(n_{c1}-1)}{Q_S}\pi - (n_{c1}-1)\frac{\pi}{2}\right), \\ \qquad 2\nu + (n_{c2}-n_{c1}) = 2,4,6,\ldots \\ \sin\left((2\nu+(n_{c2}-n_{c1}))\frac{\pi}{2}\right)\sin\left(\nu\alpha_S - \nu\frac{(n_{c1}-1)}{Q_S}\pi - (n_{c1}-1)\frac{\pi}{2}\right), \\ \qquad 2\nu + (n_{c2}-n_{c1})\nu = 1,3,5,\ldots \end{cases} \end{aligned} \tag{3.33}$$

the final MMF function and corresponding winding factor, depending on the MMF harmonic order, can be derived as follows.

Case 1: $(2v + n_{c2} - n_{c1})$ = *even number*

For this condition, the MMF winding function can be expressed as:

$$F_j(\alpha_S, t) = \frac{2N_w i_j(t)}{\pi} \sum_{\nu} \left\{ \frac{n_{c1}{}^{\nu}\xi_{w1}}{\nu} \pm \frac{n_{c2}{}^{\nu}\xi_{w2}}{\nu} \cos\left(2v + (n_{c2} - n_{c1})\right) \frac{\pi}{2} \right\} e^{j\left(\nu\alpha_S - \nu\frac{(n_{c1}-1)}{Q_S}\pi - (n_{c1}-1)\frac{\pi}{2}\right)} \tag{3.34}$$

Equation (3.34) can be simplified to:

$$F_j(\alpha_S, t) = \frac{2(n_{c1} + n_{c2})N_w \hat{i}}{\pi} \sum_{\nu} \frac{{}^{\nu}\xi_w}{\nu} \cos\left(\nu\left(\alpha_S - \frac{(n_{c1}-1)}{Q_S}\pi\right) - (n_{c1}-1)\frac{\pi}{2}\right) \tag{3.35}$$

where the resulting winding factor is:

$${}^{\nu}\xi_w = \frac{n_{c1}}{n_{c1} + n_{c2}} {}^{\nu}\xi_{w1} \pm \frac{n_{c2}}{n_{c1} + n_{c2}} {}^{\nu}\xi_{w2} \cos\left(\left(2v + (n_{c2} - n_{c1})\right)\frac{\pi}{2}\right) \tag{3.36}$$

Case 2: $(2v + n_{c2} - n_{c1})$ = *odd number*

On the other hand, based on Equations (3.32) and (3.33), when the term $(2v + n_{c2} - n_{c1})$ is odd, the corresponding MMF winding function becomes:

$$\begin{aligned} F_j(\alpha_S, t) = \frac{2N_w i_j(t)}{\pi} \Bigg\{ & \sum_{\nu} \frac{n_{c1}{}^{\nu}\xi_{w1}}{\nu} \cos\left(\nu\left(\alpha_S - \frac{(n_{c1}-1)}{Q_S}\pi\right) - (n_{c1}-1)\frac{\pi}{2}\right) \\ & \pm \sum_{\nu} \frac{n_{c2}{}^{\nu}\xi_{w2}}{\nu} \sin\left(\left(2v + (n_{c2} - n_{c1})\right)\frac{\pi}{2}\right) \\ & \cos\left(\nu\left(\alpha_S - \frac{(n_{c1}-1)}{Q_S}\pi\right) - (n_{c1}-1)\frac{\pi}{2} - \frac{\pi}{2}\right) \Bigg\} \end{aligned} \tag{3.37}$$

Simplifying Equation (3.37) leads to:

$$F_j(\alpha_S, t) = \frac{2(n_{c1} + n_{c2})N_w i_j(t)}{\pi} \sum_{\nu} \frac{{}^{\nu}\xi_w}{\nu} \cos\left(\nu\left(\alpha_S - \frac{(n_{c1}-1)}{Q_S}\pi\right) - (n_{c1}-1)\frac{\pi}{2} + \alpha_{Z_1,Z_2}\right) \tag{3.38}$$

where

$$\alpha_{Z_1,Z_2} = \text{atan}\left(\frac{\pm n_{c2}{}^{\nu}\xi_{w2}}{n_{c1}{}^{\nu}\xi_{w1}} \sin\left(\left(2v + (n_{c2} - n_{c1})\right)\frac{\pi}{2}\right)\right) \tag{3.39}$$

and the resulting winding factor is:

$${}^{\nu}\xi_w = \sqrt{\left(\frac{n_{c1}}{(n_{c1} + n_{c2})} {}^{\nu}\xi_{w1}\right)^2 + \left(\frac{n_{c2}}{(n_{c1} + n_{c2})} {}^{\nu}\xi_{w2} \sin\left(\left(2v + (n_{c2} - n_{c1})\right)\frac{\pi}{2}\right)\right)^2} \tag{3.40}$$

3.2.2.2 DL-FSCW with Two Unequal Zone-groups

As explained in Figure 3.2, any common two-zone FSCW can be transformed into a winding with two unequal zone-groups. Furthermore, the zone-groups can be arranged to have either the same or opposite polarities, as illustrated in Figure 3.5. An example of DL-FSCW with two unequal

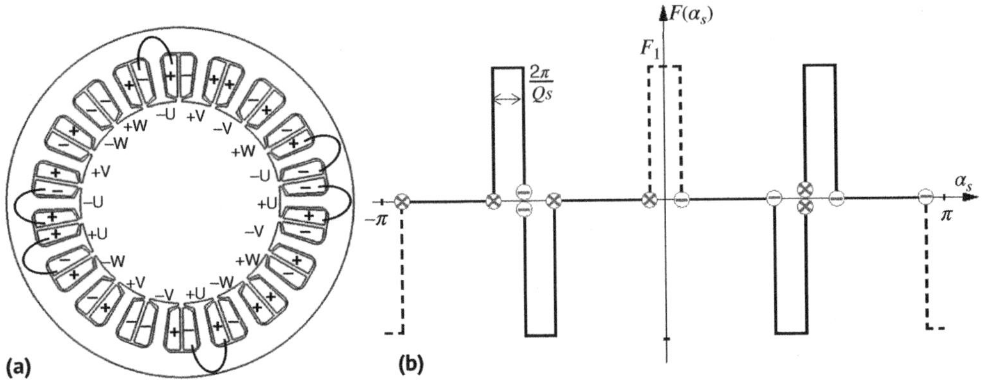

Figure 3.8 (a) DL-FSCW with two unequal zone-groups per phase, with $n_{c1} = 2$ and $n_{c2} = 1$, and with the same polarity zone-groups. (b) MMF distribution of phase-j.

zone-groups is shown in Figure 3.8a. In this configuration, the first zone-group has two coils per zone $(n_{c1} = 2)$, while the second zone-group has one coil per zone $(n_{c2} = 1)$. Moreover, the adjacent coils in zones of the first group have opposite polarity, while the first and the second zone-group share the same polarity. According to Equation (3.1) this winding type requires 18 stator slots $(Q_S = 18)$. The phase-j MMF waveform for this configuration is depicted in Figure 3.8b.

Similar to the previous analysis, the corresponding MMF function for the common two-zone DL-FSCWs was used for the MMF analysis. Therefore, considering the case with opposite polarity adjacent coils in Equation (2.41), the corresponding MMF function in complex form can be written as (z is the number of zones in a zone-group):

$$F_{j,zg1}(\alpha_S, t) = \frac{2zN_w i_j(t)}{\pi} \sum_{\nu} \frac{n_{c1}\,{}^{\nu}\xi_{w1}}{\nu} e^{j\left(\nu\left(\alpha_S - \frac{n_{c1}-1}{Q_S}\pi\right) - \frac{n_{c1}-1}{2}\pi + \frac{\pi}{2}\right)} \tag{3.41}$$

$$F_{j,zg2}(\alpha_S, t) = \frac{2zN_w i_j(t)}{\pi} \sum_{\nu} \frac{n_{c2}\,{}^{\nu}\xi_{w2}}{\nu} e^{j\left(\nu\left(\alpha_S - \frac{4n_{c1}+3n_{2}-1}{Q_S}\pi\right) - \frac{n_{c2}-1}{2}\pi + \frac{\pi}{2}\right)} \tag{3.42}$$

The respective winding factors are defined as:

$$\begin{aligned} {}^{\nu}\xi_{w1} &= {}^{\nu}\xi_{p1}\,{}^{\nu}\xi_{d1}\,{}^{\nu}\xi_{z1} \\ {}^{\nu}\xi_{w2} &= {}^{\nu}\xi_{p2}\,{}^{\nu}\xi_{d2}\,{}^{\nu}\xi_{z2} \end{aligned} \tag{3.43}$$

In (3.43), the coil-pitch factor can be determined from (2.15), the zone distribution factors from (3.30), and the zone factor from (3.18).

From Equations (3.41), (3.42), and (3.3), the resulting MMF function for the DL-FSCW with two unequal zone-groups is:

$$\begin{aligned} F_j(\alpha_S, t) &= \frac{2zN_w i_j(t)}{\pi} \sum_{\nu} \left\{ \frac{n_{c1}\,{}^{\nu}\xi_{w1}}{\nu} e^{j\left(-\nu\frac{(n_{c1}-1)}{Q_S}\pi - n_{c1}\frac{\pi}{2}\right)} \pm \frac{n_{c2}\,{}^{\nu}\xi_{w2}}{\nu} e^{j\left(-\nu\frac{4n_{c1}+3n_{c2}-1}{Q_S}\pi - n_{c2}\frac{\pi}{2}\right)} \right\} e^{j(\nu\alpha_S + \pi)} \\ &= \frac{2zN_w i_j(t)}{\pi} \sum_{\nu} \left\{ \frac{n_{c1}\,{}^{\nu}\xi_{w1}}{\nu} \pm \frac{n_{c2}\,{}^{\nu}\xi_{w2}}{\nu} e^{j\left(-\nu\frac{3n_{c1}+3n_{c2}}{Q_S}\pi - (n_{c2}-n_{c1})\frac{\pi}{2}\right)} \right\} e^{j\left(\nu\alpha_S - \nu\frac{(n_{c1}-1)}{Q_S}\pi - n_{c1}\frac{\pi}{2} + \pi\right)} \end{aligned} \tag{3.44}$$

For the case $z_g = 2$ and $m = 3$, the exponent term in the above expression simplifies to:

$$\nu\frac{(3n_{c1}+3n_{c2})}{Q_S}\pi + (n_{c2}-n_{c1})\frac{\pi}{2} = \nu\frac{(3n_{c1}+3n_{c2})}{mz_g(n_{c1}+n_{c2})}\pi + (n_{c2}-n_{c1})\frac{\pi}{2} = (\nu + (n_{c2}-n_{c1}))\frac{\pi}{2}$$

Consequently, Equation (3.44) yields to:

$$F_j(\alpha_S,t) = \frac{2N_w i_j(t)}{\pi}\sum_\nu\left\{\frac{n_{c1}{}^\nu\xi_{w1}}{\nu} \pm \frac{n_{c2}{}^\nu\xi_{w2}}{\nu}e^{-j(\nu+(n_{c2}-n_{c1}))\frac{\pi}{2}}\right\}e^{j\left(\nu\alpha_S-\nu\frac{(n_{c1}-1)}{Q_S}\pi-\left(\frac{n_{c1}}{2}-1\right)\pi\right)} \quad (3.45)$$

As before, the resulting expression depends on the harmonic order. In particular, the product of the exponential terms in (3.45) leads to different solutions, namely:

$$e^{-j(\nu+(n_{c2}-n_{c1}))\frac{\pi}{2}}\cdot e^{j\left(\nu\alpha_S-\nu\frac{(n_{c1}-1)}{Q_S}\pi-\left(\frac{n_{c1}}{2}-1\right)\pi\right)}$$

$$= \begin{cases} \cos\left((\nu+(n_{c2}-n_{c1}))\frac{\pi}{2}\right)\cos\left(\nu\alpha_S-\nu\frac{(n_{c1}-1)}{Q_S}\pi-\left(\frac{n_{c1}}{2}-1\right)\pi\right), \\ \qquad \nu+(n_{c2}-n_{c1}) = 2,4,6,\ldots \\ \sin\left((\nu+(n_{c2}-n_{c1}))\frac{\pi}{2}\right)\sin\left(\nu\alpha_S-\nu\frac{(n_{c1}-1)}{Q_S}\pi-\left(\frac{n_{c1}}{2}-1\right)\pi\right), \\ \qquad \nu+(n_{c2}-n_{c1})\nu = 1,3,5,\ldots \end{cases} \quad (3.46)$$

Under these conditions, the final MMF function and the associated winding factors can be determined as follows.

Case 1: $\nu+(n_{c2}-n_{c1})$ = *even number*

When the quantity $\nu+(n_{c2}-n_{c1})$ is even, the MMF function can be written as:

$$F_j(\alpha_S,t) = \frac{2zN_w i_j(t)}{\pi}\sum_\nu\left\{\frac{n_{c1}{}^\nu\xi_{w1}}{\nu} \pm \frac{n_{c2}{}^\nu\xi_{w2}}{\nu}\cos(\nu+(n_{c2}-n_{c1}))\frac{\pi}{2}\right\}e^{j\left(\nu\alpha_S-\nu\frac{(n_{c1}-1)}{Q_S}\pi-\left(\frac{n_{c1}}{2}-1\right)\pi\right)} \quad (3.47)$$

In closed-form, Equation (3.47) can be rewritten as:

$$F_j(\alpha_S,t) = \frac{2z(n_{c1}+n_{c2})N_w i_j(t)}{\pi}\sum_\nu\frac{{}^\nu\xi_w}{\nu}\cos\left(\nu\left(\alpha_S-\frac{(n_{c1}-1)}{Q_S}\pi\right)-\left(\frac{n_{c1}}{2}-1\right)\pi\right) \quad (3.48)$$

where the resulting winding factor is given by:

$${}^\nu\xi_w = \frac{n_{c1}}{n_{c1}+n_{c2}}{}^\nu\xi_{w1} \pm \frac{n_{c2}}{n_{c1}+n_{c2}}{}^\nu\xi_{w2}\cos\left((\nu+(n_{c2}-n_{c1}))\frac{\pi}{2}\right) \quad (3.49)$$

Case 2: $\nu+(n_{c2}-n_{c1})$ = *odd number*

When $\nu+(n_{c2}-n_{c1})$ is odd, the corresponding MMF function becomes the sum of two cosine terms with different magnitudes and a phase shift of $\pi/2$:

$$F_j(\alpha_S,t) = \frac{2zN_w i_j(t)}{\pi}\left\{\sum_\nu\frac{n_{c1}{}^\nu\xi_{w1}}{\nu}\cos\left(\nu\left(\alpha_S-\frac{(n_{c1}-1)}{Q_S}\pi\right)-\left(\frac{n_{c1}}{2}-1\right)\pi\right)\right.$$
$$\pm\sum_\nu\frac{n_{c2}{}^\nu\xi_{w2}}{\nu}\sin\left((\nu+(n_{c2}-n_{c1}))\frac{\pi}{2}\right)$$
$$\left.\cos\left(\nu\left(\alpha_S-\frac{(n_{c1}-1)}{Q_S}\pi\right)-\left(\frac{n_{c1}}{2}-1\right)\pi-\frac{\pi}{2}\right)\right\} \quad (3.50)$$

By simplifying Equation (3.50), we obtain:

$$F_j(\alpha_S,t) = \frac{2z(n_{c1}+n_{c2})N_w i_j(t)}{\pi} \sum_\nu \frac{{}^\nu\xi_w}{\nu} \cos\left(\nu\left(\alpha_S - \frac{(n_{c1}-1)}{Q_S}\pi\right) - \left(\frac{n_{c1}}{2}-1\right)\pi + \alpha_{z_1,z_2}\right) \tag{3.51}$$

where

$$\alpha_{z_1,z_2} = \text{atan}\left(\frac{\pm n_{c2}{}^\nu\xi_{w2}}{n_{c1}{}^\nu\xi_{w1}} \sin\left((\nu + (n_{c2}-n_{c1}))\frac{\pi}{2}\right)\right) \tag{3.52}$$

and the resulting winding factor is:

$${}^\nu\xi_w = \sqrt{\left(\frac{n_{c1}}{(n_{c1}+n_{c2})}{}^\nu\xi_{w1}\right)^2 + \left(\frac{n_{c2}}{(n_{c1}+n_{c2})}{}^\nu\xi_{w2}\sin\left((\nu+(n_{c2}-n_{c1}))\frac{\pi}{2}\right)\right)^2} \tag{3.53}$$

3.2.3 FSCWs with Asymmetric Zone Distribution

In all conventional FSCWs and also unequal zone FSCWs studied so far, the zones within each zone-group are typically arranged symmetrically in the machine, i.e. they are spatially shifted by 180 electrical degrees relative to each other. In contrast to these winding configurations, however, the zones within a zone-group can also be distributed asymmetrically around the stator circumference. Moreover, they may be designed with a different number of zones per zone-group, which allows for greater flexibility in the winding design. A practical winding arrangement featuring unequal zones, unequal zone-groups, and asymmetric zone distribution is illustrated in Figure 3.9. The key specifications for this configuration are as follows:

- Winding type: FSCW with unequal zone-groups
- First zone-group: Two zone ($n_{z1} = 2$), n_{c1} coils per zone, and a zone displacement angle of $\varphi_{z,unsm}$
- Second zone-group: Three zone ($n_{z2} = 3$), n_{c2} coils per zone, and a zone displacement angle of $\varphi_{z,unsm}$
- Displacement angle between zone-group 1 and 2 is $\varphi_{z1,z2}$.

This particular configuration is applicable to both single-layer and double-layer winding layouts. Depending on the number of layers, the displacement angle of zones and zone-groups, as well as, the required number of stator slots, are defined in Equations (3.54)–(3.56).

$$\varphi_{z,unsm} = \begin{cases} (n_{c1}+2n_{c2})\dfrac{4\pi}{Q_S}, & \text{single-layer FSCW} \\ (n_{c1}+2n_{c2})\dfrac{2\pi}{Q_S}, & \text{double-layer FSCW} \end{cases} \tag{3.54}$$

$$\varphi_{z_1,z2} = \begin{cases} 3(n_{c1}+n_{c2})\dfrac{4\pi}{Q_S}, & \text{single-layer FSCW} \\ 3(n_{c1}+n_{c2})\dfrac{2\pi}{Q_S}, & \text{double-layer FSCW} \end{cases} \tag{3.55}$$

$$Q_S = \begin{cases} 2m(n_{z1}n_{c1}+n_{z2}n_{c2}), & \text{single-layer FSCW} \\ m(n_{z1}n_{c1}+n_{z2}n_{c2}), & \text{double-layer FSCW} \end{cases} \tag{3.56}$$

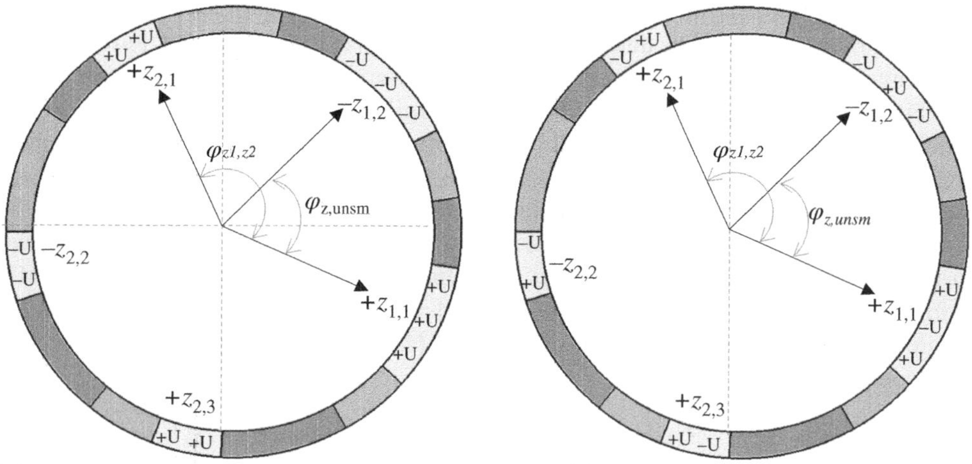

(a) (b)

Figure 3.9 FSCW with two unequal zone-groups $z_{1,k}$ and $z_{2,k}$ per phase, with $n_{c1} = 3$ and $n_{c2} = 2$ coils per zone, and with asymmetrical zone distribution within a zone-group. (a) Zones with the same polarity coil arrangement. (b) Zones with opposite polarity coil arrangement.

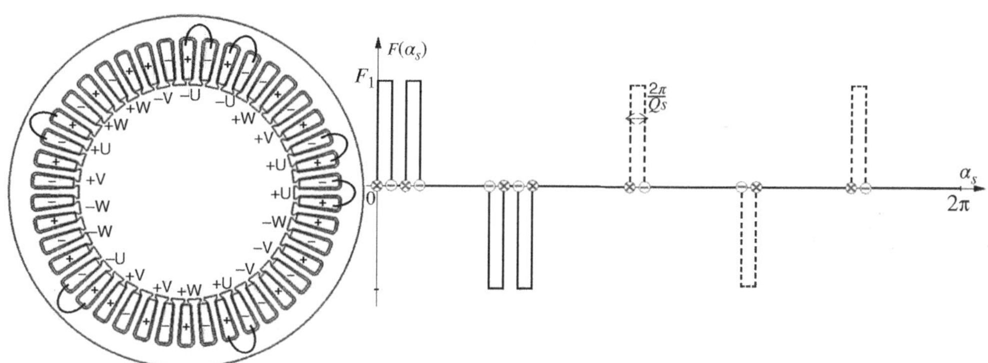

(a) (b)

Figure 3.10 (a) SL-FSCW with two unequal zone-groups per phase, the first zone-group with $n_{z1} = 2$ and $n_{c1} = 2$, the second zone-group with $n_{z2} = 3$ and $n_{c2} = 1$, and with the same polarity zone-groups. (b) MMF distribution of phase-j.

Similar to the previously studied windings, the coil arrangement within a single zone can exhibit either the same or opposite polarity, as illustrated in Figure 3.9a,b. Given that only a limited number of single-layer and double-layer winding configurations have demonstrated effectiveness in achieving a higher winding factor, the following analysis focuses exclusively on these topologies, specifically on:

- SL-FSCW with the same polarity coil arrangement and
- DL-FSCW with opposite polarity coil arrangement.

3.2.3.1 SL-FSCW with Unequal Zone Groups and Asymmetric Zone Distribution

An exemplary single-layer winding with unequal zone groups and asymmetric zone distribution, based on the winding scheme in Figure 3.9a, is depicted in Figure 3.10a. In this example, the winding consists of $n_{c1} = 2$ and $n_{c2} = 1$ coils per zone, and $n_{z1} = 2$, $n_{z2} = 3$ zones per zone-group. Additionally, the adjacent coils within each zone share the same polarity, while the zones within a zone-group alternate in polarity. The zone shift angle within a zone-group and also the zone-group shift within one phase can be determined using Equations (3.54) and (3.55). Furthermore, the required number of stator slots for this winding configuration, according to Equation (3.56), is $Q_S = 42$. Figure 3.10b presents the phase-j MMF waveform, which is obtained as the sum of the corresponding zone-groups MMFs. The solid line represents the first zone-group, while the dashed line represents the second zone-group.

To analyze the resulting MMF function for this winding type, the MMFs of zone-groups with asymmetric zone distribution must first be determined. For the case that the number of zones within a zone group is n_{zi} and the phase shift between zones is $\varphi_{z,unsm}$, the zone-group MMF function for the configuration with the same polarity coil arrangement and zones with alternating polarity can be determined using Equations (2.23) and (2.29), as follows (index i denoted the respective zone-group):

$$\begin{aligned} F_{j,zgi}(\alpha_S, t) &= F_{j,z1}(t, \alpha_S) - F_{j,z2}(t, \alpha_S - \varphi_{z,unsm}) + F_{j,z3}(t, \alpha_S - 2\varphi_{z,unsm}) - \ldots \\ &= \frac{2n_{ci}N_w i_j(t)}{\pi} \sum_{\nu} \frac{{}^{\nu}\xi_p\,{}^{\nu}\xi_d}{\nu} \\ &\quad \cdot \sum_{\kappa=0}^{n_{zi}-1} \cos\left(\nu\left(\alpha_S - (n_{ci}-1)\frac{2\pi}{Q_S}\right) - \kappa(\nu\varphi_{z,unsm} + \pi)\right) \end{aligned} \tag{3.57}$$

Using

$$\begin{aligned} &\sum_{\kappa=0}^{n_{zi}-1} \cos\left(\nu\left(\alpha_S - (n_{ci}-1)\frac{2\pi}{Q_S}\right) - \kappa(\nu \cdot \varphi_{z,unsm} + \pi)\right) \\ &\quad = \frac{\sin\left(\nu n_{zi}\frac{\varphi_{z,unsm}}{2} + \frac{n_{zi}}{2}\pi\right)}{\sin\left(\nu\frac{\varphi_{z,unsm}}{2} + \frac{\pi}{2}\right)} e^{j\left(\nu\left(\alpha_S - (n_{ci}-1)\frac{2\pi}{Q_S} - (n_{zi}-1)\frac{\varphi_{z,unsm}}{2}\right) - (n_{zi}-1)\frac{\pi}{2}\right)} \end{aligned} \tag{3.58}$$

the final expression for the MMF function corresponding to the zone-group with asymmetric zone distributions becomes:

$$\begin{aligned} F_{j,zgi}(\alpha_S, t) &= \frac{2n_{zi}n_{ci}N_w i_j(t)}{\pi} \sum_{\nu} \frac{{}^{\nu}\xi_{p,i}\,{}^{\nu}\xi_{d,i}\,{}^{\nu}\xi_{zgi,unsm}}{\nu} \\ &\quad \cdot \cos\left(\nu\left(\alpha_S - (n_{ci}-1)\frac{2\pi}{Q_S} - (n_{zi}-1)\frac{\varphi_{z,unsm}}{2}\right) - (n_{zi}-1)\frac{\pi}{2}\right) \end{aligned} \tag{3.59}$$

In (3.59), ${}^{\nu}\xi_{zgi,unsm}$ denotes the zone factor for asymmetric zone distribution, which is determined in (3.60). The corresponding coil-pitch and distribution factor parameters have already been defined in Equations (3.9) and (3.10).

$${}^{\nu}\xi_{zgi,unsm} = \frac{\sin\left(\nu \cdot n_{zi}\frac{\varphi_{z,unsm}}{2} + n_{zi}\frac{\pi}{2}\right)}{n_{zi} \cdot \sin\left(\nu\frac{\varphi_{z,unsm}}{2} + \frac{\pi}{2}\right)} \tag{3.60}$$

According to (3.59), the MMF expressions for the individual zone-groups in complex form can be written as ($\varphi_{z1,z2}$ is the shift angle between zone-groups):

$$F_{j,zg1}(\alpha_S) = \frac{2N_w i_j(t)}{\pi} \sum_\nu \frac{n_{z1} n_{c1} {}^\nu\xi_{w1}}{\nu} e^{j\left[\nu\left(\alpha_S - (n_{c1}-1)\frac{2\pi}{Q_S} - (n_{z1}-1)\frac{\varphi_{z,unsm}}{2}\right) - (n_{z1}-1)\frac{\pi}{2}\right]} \tag{3.61}$$

$$F_{j,zg2}(\alpha_S) = \frac{2N_w i_j(t)}{\pi} \sum_\nu \frac{n_{z2} n_{c2} {}^\nu\xi_{w2}}{\nu} e^{j\left[\nu\left(\alpha_S - (n_{c2}-1)\frac{2\pi}{Q_S} - (n_{z2}-1)\frac{\varphi_{z,unsm}}{2} - \varphi_{z1,z2}\right) - (n_{z2}-1)\frac{\pi}{2}\right]} \tag{3.62}$$

Substituting the shift angle $\varphi_{z1,z2}$ for the single-layer winding from (3.55) into (3.62) yields:

$$F_{j,zg2}(\alpha_S) = \frac{2N_w i_j(t)}{\pi} \sum_\nu \frac{n_{z2} n_{c2} {}^\nu\xi_{w2}}{\nu} e^{j\left[\nu\left(\alpha_S - (7n_{c2}+6n_{c1}-1)\frac{2\pi}{Q_S} - (n_{z2}-1)\frac{\varphi_{z,unsm}}{2}\right) - (n_{z2}-1)\frac{\pi}{2}\right]} \tag{3.63}$$

In Equations (3.61) and (3.63), the respective winding factors are:

$${}^\nu\xi_{w1} = {}^\nu\xi_{p1} {}^\nu\xi_{d1} {}^\nu\xi_{zg1,unsm}, \quad {}^\nu\xi_{w2} = {}^\nu\xi_{p2} {}^\nu\xi_{d2} {}^\nu\xi_{zg2,unsm} \tag{3.64}$$

Once the MMF expressions for the individual zone-groups have been determined, the resulting phase-j MMF for the SL-FSCW with two unequal zone-groups and an asymmetric zone distribution follows from (3.61), (3.63), and (3.3), and is given by:

$$\begin{aligned} F_j(\alpha_S, t) = {} & \frac{2N_w i_j(t)}{\pi} \sum_\nu \left\{ \frac{n_{z1} n_{c1} {}^\nu\xi_{w1}}{\nu} \right. \\ & \left. \pm \frac{n_{z2} n_{c2} {}^\nu\xi_{w,z2}}{\nu} e^{j\left(\nu\left(-(7n_{c2}+5n_{c1})\frac{2\pi}{Q_S} - (n_{z2}-n_{z1})\frac{\varphi_{z\_unsm}}{2}\right) - (n_{z2}-n_{z1})\frac{\pi}{2}\right)} \right\} \\ & \cdot e^{j\left[\nu\left(\alpha_S - (n_{c1}-1)\frac{2\pi}{Q_S} - (n_{z1}-1)\frac{\varphi_{z\_unsm}}{2}\right) - (n_{z1}-1)\frac{\pi}{2}\right]} \end{aligned} \tag{3.65}$$

Using the corresponding winding parameters, i.e.,

$Q_S = 2m(n_{z1}n_{c1} + n_{z2}n_{c2})$,
$\varphi_{z1,z2} = (n_{c1} + 2n_{c2})4\pi/Q_S$, and
$n_{z1} = 2$, $n_{z2} = 3$, and $m = 3$

the term in the exponential function can be simplified to:

$$\begin{aligned} & \nu\left((7n_{c2} + 5n_{c1})\frac{2\pi}{Q_S} + (n_{z2} - n_{z1})\frac{\varphi_{z,unsm}}{2}\right) + (n_{z2} - n_{z1})\frac{\pi}{2} \\ & = \left(\nu\frac{2(6n_{c1} + 9n_{c2})}{3(2n_{c1} + 3n_{c2})} + 1\right)\frac{\pi}{2} = (2\nu + 1)\frac{\pi}{2} \end{aligned}$$

Consequently, the MMF function in Equation (3.65) becomes:

$$\begin{aligned} F_j(\alpha_S, t) = {} & \frac{2N_w i_j(t)}{\pi} \sum_\nu \left\{ \frac{n_{z1} n_{c1} {}^\nu\xi_{w1}}{\nu} \pm \frac{n_{z2} n_{c2} {}^\nu\xi_{w,z2}}{\nu} e^{-j\left((2\nu+1)\frac{\pi}{2}\right)} \right\} \\ & \cdot e^{j\left[\nu\left(\alpha_S - (n_{c1}-1)\frac{2\pi}{Q_S} - (n_{z1}-1)\frac{\varphi_{z,unsm}}{2}\right) - (n_{z1}-1)\frac{\pi}{2}\right]} \end{aligned} \tag{3.66}$$

Since the product of the exponential functions in Equation (3.66) depends on the harmonic order, it produces different results, as follows:

$$e^{-j(2\nu+1)\frac{\pi}{2}} \cdot e^{j\left[\nu\left(\alpha_S-(n_{c1}-1)\frac{2\pi}{Q_S}-(n_{z1}-1)\frac{\varphi_{z,unsm}}{2}\right)-(n_{z1}-1)\frac{\pi}{2}\right]}$$
$$= \begin{cases} \cos\left((2\nu+1)\frac{\pi}{2}\right)\cos\left(\nu\left(\alpha_S-(n_{c1}-1)\frac{2\pi}{Q_S}-(n_{z1}-1)\frac{\varphi_{z,unsm}}{2}\right)-(n_{z1}-1)\frac{\pi}{2}\right), \\ \qquad 2\nu+1=2,4,6,\ldots \\ \sin\left((2\nu+1)\frac{\pi}{2}\right)\sin\left(\nu\left(\alpha_S-(n_{c1}-1)\frac{2\pi}{Q_S}-(n_{z1}-1)\frac{\varphi_{z,unsm}}{2}\right)-(n_{z1}-1)\frac{\pi}{2}\right), \\ \qquad 2\nu+1=1,3,5,\ldots \end{cases} \tag{3.67}$$

Accordingly, the following solutions are obtained.

Case 1: $2\nu+1=$ *even number*

If the harmonic order $2\nu+1$ is even, Equation (3.66) can be simplified to:

$$F_j(\alpha_S,t) = \frac{2N_w i_j(t)}{\pi} \sum_\nu \left\{ \frac{n_{z1}n_{c1}{}^{\nu}\xi_{w1}}{\nu} \pm \frac{n_{z2}n_{c2}{}^{\nu}\xi_{w2}}{\nu}\cos\left((2\nu+1)\frac{\pi}{2}\right) \right\} e^{j\left[\nu\left(\alpha_S-(n_{c1}-1)\frac{2\pi}{Q_S}-(n_{z1}-1)\frac{\varphi_{z,unsm}}{2}\right)-(n_{z1}-1)\frac{\pi}{2}\right]} \tag{3.68}$$

From (3.68), the final expression for the phase-j MMF function is expressed as:

$$F_j(\alpha_S,t) = \frac{2(n_{z1}n_{c1}+n_{z2}n_{c2})N_w i_j(t)}{\pi} \cdot \sum_\nu \frac{{}^{\nu}\xi_w}{\nu}\cos\left(\nu\left(\alpha_S-(n_{c1}-1)\frac{2\pi}{Q_S}-(n_{z1}-1)\frac{\varphi_{z,unsm}}{2}\right)-(n_{z1}-1)\frac{\pi}{2}\right) \tag{3.69}$$

where ${}^{\nu}\xi_w$ denotes the resulting winding factor, defined by:

$${}^{\nu}\xi_w = \frac{n_{z1}n_{c1}}{n_{z1}n_{c1}+n_{z2}n_{c2}}{}^{\nu}\xi_{w1} \pm \frac{n_{z2}n_{c2}}{n_{z1}n_{c1}+n_{z2}n_{c2}}{}^{\nu}\xi_{w2}\cos\left((2\nu+1)\frac{\pi}{2}\right) \tag{3.70}$$

Case 2: $2\nu+1=$ *odd number*

However, if the harmonic order $2\nu+1$ is odd, the resulting MMF distribution becomes the sum of two *cosine* functions with different magnitudes and a spatial shift of $\pi/2$, as expressed in Equation (3.71).

$$\begin{aligned} F_j(\alpha_S,t) = \frac{2N_w i_j(t)}{\pi} \Bigg\{ & \sum_\nu \frac{n_{z1}n_{c1}{}^{\nu}\xi_{w1}}{\nu} \cos\left(\nu\left(\alpha_S-(n_{c1}-1)\frac{2\pi}{Q_S}-(n_{z1}-1)\frac{\varphi_{z,unsm}}{2}\right)-(n_{z1}-1)\frac{\pi}{2}\right) \\ & \pm \sum_\nu \frac{n_{z2}n_{c2}{}^{\nu}\xi_{w2}}{\nu}\sin\left((2\nu+1)\frac{\pi}{2}\right) \cos\left(\nu\left(\alpha_S-(n_{c1}-1)\frac{2\pi}{Q_S}-(n_{z1}-1)\frac{\varphi_{z,unsm}}{2}\right)-(n_{z1}-1)\frac{\pi}{2}-\frac{\pi}{2}\right)\Bigg\} \end{aligned} \tag{3.71}$$

The above equation can be rewritten as:

$$
\begin{aligned}
F_j(\alpha_S, t) = {} & \frac{2(n_{z1}n_{c1} + n_{z2}n_{c2})N_w i_j(t)}{\pi} \\
& \cdot \sum_{\nu} \frac{{}^{\nu}\xi_w}{\nu} \cos\left(\nu\left(\alpha_S - (n_{c1} - 1)\frac{2\pi}{Q_S} - (n_{z1} - 1)\frac{\varphi_{z,unsm}}{2}\right)\right. \\
& \left. - (n_{z1} - 1)\frac{\pi}{2} + \alpha_{z_1,z_2}\right)
\end{aligned}
\tag{3.72}
$$

where

$$
\alpha_{z_1,z_2} = a\tan\left(\frac{\pm n_{z2}n_{c2}\,{}^{\nu}\xi_{w2}}{n_{z1}n_{c1}\,{}^{\nu}\xi_{w1}} \sin\left((2\nu + 1)\frac{\pi}{2}\right)\right)
\tag{3.73}
$$

and the resulting winding factor is:

$$
{}^{\nu}\xi_w = \sqrt{\left(\frac{n_{z1}n_{c1}}{(n_{z1}n_{c1} + n_{z2}n_{c2})}\,{}^{\nu}\xi_{w1}\right)^2 + \left(\frac{n_{z1}n_{z2}}{(n_{z1}n_{c1} + n_{z2}n_{c2})}\,{}^{\nu}\xi_{w2} \sin\left((2\nu + 1)\frac{\pi}{2}\right)\right)^2}
\tag{3.74}
$$

3.2.3.2 DL-FSCW with Unequal Zone Groups and Asymmetric Zone Distribution

Another example of a winding with unequal zone-groups, asymmetric zone distribution, and double-layer layout is illustrated in Figure 3.11a. Similar to the previous example, the winding has $n_{c1} = 2$ and $n_{c2} = 1$ coils per zone, $n_{z1} = 2$ and $n_{z2} = 3$ zones per zone-group, and adjacent coils with opposite polarity. Moreover, the zones within a zone-group have alternating polarity, while the zone groups within a phase have opposite polarity. As before, the zone shift angle within a group and the zone-group shift angle within a phase can be determined using Equations (3.54) and (3.55), respectively. Furthermore, the number of stator slots for the studied winding is $Q_S = 21$. The phase-j MMF waveform for this winding is depicted in Figure 3.11b, where the MMFs for the first and second zone groups are represented by solid and dashed lines, respectively.

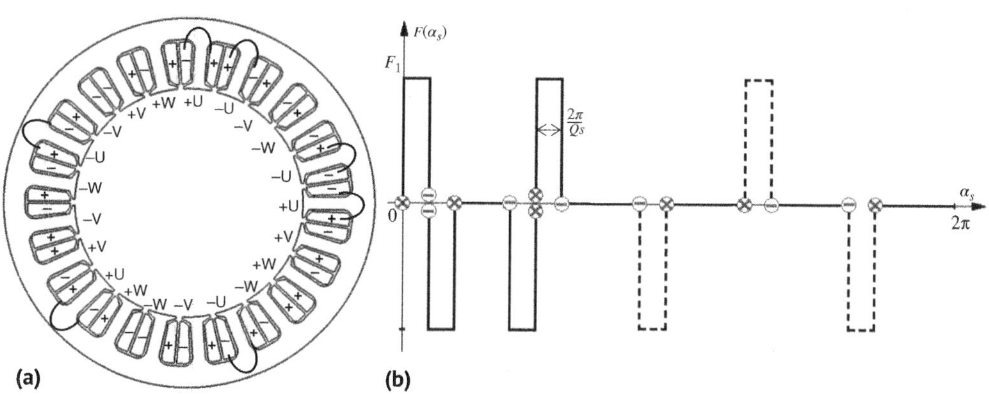

Figure 3.11 (a) DL-FSCW with two unequal zone-groups per phase, the first zone-group with $n_{z1} = 2$ and $n_{c1} = 2$, the second zone-group with $n_{z2} = 3$ and $n_{c2} = 1$, and with opposite polarity zone-groups. (b) MMF distribution of phase-j.

Based on the winding specifications described above, the MMF function for a DL-FSCW with n_{zi} asymmetric zones per zone-group can be derived analogously to (3.57), which leads to the following result (index i denotes the respective zone-group):

$$\begin{aligned} F_{j,zgi}(\alpha_S,t) &= F_{j,z1}(t,\alpha_S) - F_{j,z2}(t,\alpha_S - \varphi_{z,unsm}) + F_{j,z3}(t,\alpha_S - 2\varphi_{z,unsm}) - \ldots \\ &= \frac{2n_{zi}n_{ci}N_w i_j(t)}{\pi} \sum_\nu \frac{{}^\nu\xi_{p,i}\,{}^\nu\xi_{d,i}\,{}^\nu\xi_{zgi,unsm}}{\nu} \\ &\quad \cdot \cos\left(\nu\left(\alpha_S - (n_{ci}-1)\frac{\pi}{Q_S} - (n_{zi}-1)\frac{\varphi_{z\_unsm}}{2}\right) - (n_{ci}+n_{zi}+2)\frac{\pi}{2}\right) \end{aligned} \tag{3.75}$$

Here, ${}^\nu\xi_{zgi,unsm}$ is the zone distribution factor already derived (3.60), but using the corresponding $\varphi_{z,unsm}$ in (3.54) for the double-layer winding. Meanwhile, ${}^\nu\xi_{p,i}$ and ${}^\nu\xi_{d,i}$ represent the coil-pitch and distribution factor parameters derived in Equations (3.9) and (3.30), respectively.

Based on Equation (3.75), the MMF functions for the zone-groups in complex form can be expressed as ($\varphi_{z1,z2}$ is the shift angle between zone-groups):

$$F_{j,zg1}(\alpha_S) = \frac{2N_w i_j(t)}{\pi} \sum_\nu \frac{n_{z1}n_{c1}\,{}^\nu\xi_{w1}}{\nu} e^{j\left[\nu\left(\alpha_S-(n_{c1}-1)\frac{\pi}{Q_S}-(n_{z1}-1)\frac{\varphi_{z,unsm}}{2}\right)-(n_{c1}+n_{z1}+2)\frac{\pi}{2}\right]} \tag{3.76}$$

$$F_{j,zg2}(\alpha_S) = \frac{2N_w i_j(t)}{\pi} \sum_\nu \frac{n_{z2}n_{c2}\,{}^\nu\xi_{w2}}{\nu} e^{j\left[\nu\left(\alpha_S-(7n_{c2}+6n_{c1}-1)\frac{\pi}{Q_S}-(n_{z2}-1)\frac{\varphi_{z,unsm}}{2}\right)-(n_{c2}+n_{z2}+2)\frac{\pi}{2}\right]} \tag{3.77}$$

In the above relations, the corresponding winding factors are given by:

$$\begin{aligned} {}^\nu\xi_{w1} &= {}^\nu\xi_{p1}\,{}^\nu\xi_{d1}\,{}^\nu\xi_{zg1,unsm} \\ {}^\nu\xi_{w2} &= {}^\nu\xi_{p2}\,{}^\nu\xi_{d2}\,{}^\nu\xi_{zg2,unsm} \end{aligned} \tag{3.78}$$

Using the MMF expressions for the individual zone-groups from (3.76) and (3.77), the resulting phase-j MMF function for the DL-FSCW with two unequal zone-groups and asymmetric zone distribution can be written as:

$$\begin{aligned} F_j(\alpha_S,t) &= \frac{2N_w i_j(t)}{\pi} \sum_\nu \left\{ \frac{n_{z1}n_{c1}\,{}^\nu\xi_{w1}}{\nu} \right. \\ &\quad \left. \pm \frac{n_{z2}n_{c2}\,{}^\nu\xi_{w2}}{\nu} e^{j\left(\nu\left(-(7n_{c2}+5n_{c1})\frac{\pi}{Q_S}-(n_{z2}-n_{z1})\frac{\varphi_{z,unsm}}{2}\right)-(n_{c2}-n_{c1}+n_{z2}-n_{z1})\frac{\pi}{2}\right)} \right\} \\ &\quad \cdot e^{j\left[\nu\left(\alpha_S-(n_{c1}-1)\frac{\pi}{Q_S}-(n_{z1}-1)\frac{\varphi_{z,unsm}}{2}\right)-(n_{c1}+n_{z1}+2)\frac{\pi}{2}\right]} \end{aligned} \tag{3.79}$$

Based on the winding parameters for this configuration,

$Q_S = m(n_{z1}n_{c1}+n_{z2}n_{c2})$,
$\varphi_{z1,z2} = (n_{c1}+2n_{c2})2\pi/Q_S$, and
$n_{z1} = 2$, $n_{z2} = 3$, and $m = 3$

the term in the exponential function leads to:

$$\begin{aligned} &\nu\left((7n_{c2}+5n_{c1})\frac{\pi}{Q_S} + (n_{z2}-n_{z1})\frac{\varphi_{z,unsm}}{2}\right) + (n_{c2}-n_{c1}+n_{z2}-n_{z1})\frac{\pi}{2} \\ &= \left(\nu\frac{2(6n_{c1}+9n_{c2})}{3(2n_{c1}+3n_{c2})} + (n_{c2}-n_{c1}+1)\right)\frac{\pi}{2} = (2\nu + (n_{c2}-n_{c1}+1))\frac{\pi}{2} \end{aligned}$$

Consequently, the MMF function (3.79) can be reformulated as:

$$F_j(\alpha_S,t) = \frac{2N_w i_j(t)}{\pi} \sum_{\nu} \left\{ \frac{n_{z1}n_{c1}{}^{\nu}\xi_{w1}}{\nu} \pm \frac{n_{z2}n_{c2}{}^{\nu}\xi_{w2}}{\nu} e^{-j\left((2\nu+(n_{c2}-n_{c1}+1))\frac{\pi}{2}\right)} \right\} \cdot e^{j\left[\nu\left(\alpha_S-(n_{c1}-1)\frac{\pi}{Q_S}-(n_{z1}-1)\frac{\varphi_{z,unsm}}{2}\right)-(n_{c1}+n_{z1}+2)\frac{\pi}{2}\right]} \tag{3.80}$$

Where analogous to Equation (3.67), by considering the harmonic orders of the MMF separately, two distinct solutions for the resulting MMF function can be obtained.

Case 1: $2\nu + (n_{c1} - n_{c2} + 1) =$ *even number*

When the term $2\nu + (n_{c1} - n_{c2} + 1)$ is even, the MMF function in (3.80) simplifies to:

$$F_j(\alpha_S,t) = \frac{2N_w i_j(t)}{\pi} \sum_{\nu} \left\{ \frac{n_{z1}n_{c1}{}^{\nu}\xi_{w1}}{\nu} \pm \frac{n_{z2}n_{c2}{}^{\nu}\xi_{w2}}{\nu} \cos\left(2\nu + (n_{c2} - n_{c1} + 1)\right)\frac{\pi}{2} \right\} \cdot e^{j\left[\nu\left(\alpha_S-(n_{c1}-1)\frac{\pi}{Q_S}-(n_{z1}-1)\frac{\varphi_{z,unsm}}{2}\right)-(n_{c1}+n_{z1}+2)\frac{\pi}{2}\right]} \tag{3.81}$$

This expression can be further rewritten as:

$$F_j(\alpha_S,t) = \frac{2(n_{z1}n_{c1} + n_{z2}n_{c2})N_w i_j(t)}{\pi} \cdot \sum_{\nu} \frac{{}^{\nu}\xi_w}{\nu} \cos\left(\nu\left(\alpha_S - (n_{c1}-1)\frac{\pi}{Q_S} - (n_{z1}-1)\frac{\varphi_{z,unsm}}{2}\right) - (n_{c1}+n_{z1}+2)\frac{\pi}{2}\right) \tag{3.82}$$

where ${}^{\nu}\xi_w$ denotes the resulting winding factor, defined by:

$${}^{\nu}\xi_w = \frac{n_{z1}n_{c1}}{n_{z1}n_{c1} + n_{z2}n_{c2}} {}^{\nu}\xi_{w1} \pm \frac{n_{z2}n_{c2}}{n_{z1}n_{c1} + n_{z2}n_{c2}} {}^{\nu}\xi_{w2} \cos\left((2\nu + (n_{c2} - n_{c1} + 1))\frac{\pi}{2}\right) \tag{3.83}$$

Case 2: $2\nu + (n_{c1} - n_{c2} + 1) =$ *odd number*

When $2\nu + (n_{c1} - n_{c2} + 1)$ is odd, the MMF distribution can be represented as the sum of two *cosine* functions with different magnitudes and a spatial phase shift of $\pi/2$:

$$F_j(\alpha_S,t) = \frac{2N_w i_j(t)}{\pi} \left\{ \sum_{\nu} \frac{n_{z1}n_{c1}{}^{\nu}\xi_{w1}}{\nu} \cos\left(\nu\left(\alpha_S - (n_{c1}-1)\frac{\pi}{Q_S} - (n_{z1}-1)\frac{\varphi_{z,unsm}}{2}\right) - (n_{c1}+n_{z1}+2)\frac{\pi}{2}\right) \pm \sum_{\nu} \frac{n_{z2}n_{c2}{}^{\nu}\xi_{w2}}{\nu} \sin\left((2\nu + (n_{c2} - n_{c1} + 1))\frac{\pi}{2}\right) \cdot \cos\left(\nu\left(\alpha_S - (n_{c1}-1)\frac{\pi}{Q_S} - (n_{z1}-1)\frac{\varphi_{z,unsm}}{2}\right) - (n_{c1}+n_{z1}+2)\frac{\pi}{2} - \frac{\pi}{2}\right) \right\} \tag{3.84}$$

This can be further expressed as:

$$F_j(\alpha_S,t) = \frac{2(n_{z1}n_{c1}+n_{z2}n_{c2})N_w i_j(t)}{\pi}$$
$$\cdot \sum_\nu \frac{{}^\nu\xi_w}{\nu}\cos\left(\nu\left(\alpha_S-(n_{c1}-1)\frac{\pi}{Q_S}-(n_{z1}-1)\frac{\varphi_{z,unsm}}{2}\right)\right.$$
$$\left.-(n_{c1}+n_{z1}+2)\frac{\pi}{2}+\alpha_{z_1,z_2}\right) \tag{3.85}$$

where

$$\alpha_{z_1,z_2} = \operatorname{atan}\left(\frac{\pm n_{z2}n_{c2}{}^\nu\xi_{w2}}{n_{z1}n_{c1}{}^\nu\xi_{w1}}\sin\left((2\nu+(n_{c2}-n_{c1}+1))\frac{\pi}{2}\right)\right) \tag{3.86}$$

and the resulting winding factor for this case becomes:

$${}^\nu\xi_w = \sqrt{\left(\frac{n_{z1}n_{c1}}{(n_{z1}n_{c1}+n_{z1}n_{c2})}{}^\nu\xi_{w1}\right)^2+\left(\frac{n_{z1}n_{z2}}{(n_{z1}n_{c1}+n_{z1}n_{c2})}{}^\nu\xi_{w2}\sin\left((2\nu+(n_{c2}-n_{c1}+1))\frac{\pi}{2}\right)\right)^2} \tag{3.87}$$

3.3 Winding Factors

For the first group of FSCWs with unequal zones, analyzed in Sections 3.2.1 and 3.2.2, the equivalent winding factor is obtained by combining the individual winding factors of the sub-windings, weighted by their respective coefficients $(n_{c,i}/(n_{c1}+n_{c2}))$, and accounting for the polarity of each sub-winding system. The analytical expressions for the equivalent winding factors corresponding to the investigated windings are summarized in Tables 3.1 and 3.2, while the winding factors of the individual sub-winding are calculated using Equations (3.8)–(3.10), as well as (3.17), (3.18), and (3.30).

Applying an analogous modeling approach to the second winding group, characterized with asymmetric zone distribution and unequal zone configuration (studied in section 3.2.3), it is shown that the resulting equivalent winding factors also here are influenced by the respective sub-winding weight coefficients $(n_{ci}\cdot n_{zi}/(n_{z1}\cdot n_{c1}+n_{z2}\cdot n_{c2}))$, as shown in Table 3.3. The respective sub-winding factor parameters are derived using (3.9), (3.10), (3.30), and (3.60).

Table 3.1 Winding factor formulae for SL-FSCW with the same polarity adjacent coils.

Unequal zones		${}^\nu\xi_w = \frac{n_{c1}}{n_{c1}+n_{c2}}{}^\nu\xi_{w1} \pm \frac{n_{c2}}{n_{c1}+n_{c2}}{}^\nu\xi_{w2}\cos(\nu\pi)$
unequal zone-groups	$\nu = even$	${}^\nu\xi_w = \frac{n_{c1}}{n_{c1}+n_{c2}}{}^\nu\xi_{w1} \pm \frac{n_{c2}}{n_{c1}+n_{c2}}{}^\nu\xi_{w2}\cos\left(\nu\frac{\pi}{2}\right)$
	$\nu = odd$	${}^\nu\xi_w = \sqrt{\left(\frac{n_{c1}}{(n_{c1}+n_{c2})}{}^\nu\xi_{w1}\right)^2+\left(\frac{n_{c2}}{(n_{c1}+n_{c2})}{}^\nu\xi_{w2}\sin\left(\nu\frac{\pi}{2}\right)\right)^2}$

Table 3.2 Winding factor formulae for DL-FSCW with opposite polarity adjacent coils.

Unequal zones	$\nu' =$ *even*	$^{\nu}\xi_{w} = \frac{n_{c1}}{n_{c1}+n_{c2}}\,{}^{\nu}\xi_{w1} \pm \frac{n_{c2}}{n_{c1}+n_{c2}}\,{}^{\nu}\xi_{w2}\cos\left(\nu'\frac{\pi}{2}\right)$ $\nu' = 2\nu + (n_{c2} - n_{c1})$
	$\nu' =$ *odd*	$^{\nu}\xi_{w} = \sqrt{\left(\frac{n_{c1}}{(n_{c1}+n_{c2})}\,{}^{\nu}\xi_{w1}\right)^2 + \left(\frac{n_{c2}}{(n_{c1}+n_{c2})}\,{}^{\nu}\xi_{w2}\sin\left(\nu'\frac{\pi}{2}\right)\right)^2}$ $\nu' = 2\nu + (n_{c2} - n_{c1})$
Unequal zone-groups	$\nu'' =$ *even*	$^{\nu}\xi_{w} = \frac{n_{c1}}{n_{c1}+n_{c2}}\,{}^{\nu}\xi_{w1} \pm \frac{n_{c2}}{n_{c1}+n_{c2}}\,{}^{\nu}\xi_{w2}\cos\left(\nu''\frac{\pi}{2}\right)$ $\nu'' = \nu + (n_{c2} - n_{c1})$
	$\nu'' =$ *odd*	$^{\nu}\xi_{w} = \sqrt{\left(\frac{n_{c1}}{(n_{c1}+n_{c2})}\,{}^{\nu}\xi_{w1}\right)^2 + \left(\frac{n_{c2}}{(n_{c1}+n_{c2})}\,{}^{\nu}\xi_{w2}\sin\left(\nu''\frac{\pi}{2}\right)\right)^2}$ $\nu'' = \nu + (n_{c2} - n_{c1})$

Table 3.3 Winding factor formulae for SL- and DL-FSCW with asymmetric zone distribution.

SL-FSCW same polarity coils	$\nu' =$ *even*	$^{\nu}\xi_{w} = \frac{n_{z1}n_{c1}}{n_{z1}n_{c1}+n_{z2}n_{c2}}\,{}^{\nu}\xi_{w1} \pm \frac{n_{z2}n_{c2}}{n_{z1}n_{c1}+n_{z2}n_{c2}}\,{}^{\nu}\xi_{w2}\cos\left(\nu'\frac{\pi}{2}\right)$ $\nu' = 2\nu + 1$
	$\nu' =$ *odd*	$^{\nu}\xi_{w} = \sqrt{\left(\frac{n_{z1}n_{c1}}{(n_{z1}n_{c1}+n_{z2}n_{c2})}\,{}^{\nu}\xi_{w1}\right)^2 + \left(\frac{n_{z2}n_{c2}}{(n_{z1}n_{c1}+n_{z2}n_{c2})}\,{}^{\nu}\xi_{w2}\sin\left(\nu'\frac{\pi}{2}\right)\right)^2}$ $\nu' = 2\nu + 1$
DL-FSCW opposite polarity coils	$\nu'' =$ *even*	$^{\nu}\xi_{w} = \frac{n_{z1}n_{c1}}{n_{z1}n_{c1}+n_{z2}n_{c2}}\,{}^{\nu}\xi_{w1} \pm \frac{n_{z2}n_{c2}}{n_{z1}n_{c1}+n_{z2}n_{c2}}\,{}^{\nu}\xi_{w2}\cos\left(\nu''\frac{\pi}{2}\right)$ $\nu'' = 2\nu + (n_{c2} - n_{c1} + 1)$
	$\nu'' =$ *odd*	$^{\nu}\xi_{w} = \sqrt{\left(\frac{n_{z1}n_{c1}}{(n_{z1}n_{c1}+n_{z1}n_{c2})}\,{}^{\nu}\xi_{w1}\right)^2 + \left(\frac{n_{z2}n_{c2}}{(n_{z1}n_{c1}+n_{z1}n_{c2})}\,{}^{\nu}\xi_{w2}\sin\left(\nu''\frac{\pi}{2}\right)\right)^2}$ $\nu'' = 2\nu + (n_{c2} - n_{c1} + 1)$

To evaluate the performance of various FSCWs in terms of harmonic content and working harmonics, the resulting winding factors were investigated and compared for different combinations of n_{c1} and n_{c2} in Tables 3.4–3.9. The highest winding factors, corresponding to the most favorable topologies, are highlighted in bold. The results indicated that, for single-layer windings with unequal zones, only configurations with opposite zone polarity yield increased winding factors, as shown in Table 3.4. In contrast, for double-layer windings with unequal zones, a wide range of combinations across all investigated categories show higher winding factor values.

Table 3.4 MMF harmonic winding factors (in %) for SL-FSCW with opposite zone polarity and with the same adjacent coil polarity.

n_{c1}	n_{c2}	Q_s	$v = 1$	2	3	4	5	6	7	8	9	10	11	12	13	14	15	16	17	18	19	20	21	22	23	24	25
2	1	18	17	6	—	14	17	—	17	**95**	—	**95**	17	—	17	14	—	6	17	—	17	6	—	14	17	—	17
3	1	24	12	6	—	13	4	—	5	22	—	24	**93**	—	**93**	24	—	22	5	—	4	13	—	6	12	—	12
3	2	30	10	2	—	4	10	—	10	11	—	17	10	—	10	**95**	—	**95**	10	—	10	17	—	11	10	—	10
4	1	30	10	6	—	10	0	—	8	9	—	0	23	—	28	**91**	—	**91**	28	—	23	0	—	9	8	—	0
4	2	36	8	3	—	6	6	—	3	11	—	13	5	—	13	16	—	16	**94**	—	**94**	16	—	16	13	—	5
4	3	42	7	1	—	2	7	—	7	5	—	7	7	—	7	12	—	18	7	—	7	**95**	—	**95**	7	—	7
5	1	36	8	5	—	9	2	—	8	4	—	4	11	—	3	24	—	31	**90**	—	**90**	31	—	24	3	—	11
5	2	42	7	3	—	6	3	—	0	8	—	8	8	—	12	0	—	8	19	—	21	**93**	—	**93**	21	—	19
5	3	48	6	2	—	3	5	—	4	6	—	8	1	—	1	10	—	11	9	—	15	12	—	12	**95**	—	**95**
5	4	54	6	1	—	1	6	—	6	3	—	4	6	—	6	6	—	7	6	—	6	13	—	19	6	—	6
6	1	42	7	5	—	7	2	—	7	1	—	5	6	—	2	12	—	6	24	—	33	**89**	—	**89**	33	—	24
6	2	48	6	3	—	6	2	—	2	6	—	4	8	—	9	5	—	11	4	—	5	21	—	24	**92**	—	**92**

Table 3.5 MMF harmonic winding factors (in %) for DL-FSCW with opposite zone polarity and with opposite adjacent coil polarity.

n_{c1}	n_{c2}	Q_s	$\nu = 1$	2	3	4	5	6	7	8	9	10	11	12	13	14	15	16	17	18	19	20	21	22	23	24	25
2	1	9	14	35	—	73	73	—	35	14	—	14	35	—	73	73	—	35	14	—	14	35	—	73	73	—	35
3	1	12	11	13	—	65	42	—	42	65	—	13	11	—	11	13	—	65	42	—	42	65	—	13	11	—	11
3	2	15	4	7	—	28	46	—	71	71	—	46	28	—	7	4	—	4	7	—	28	46	—	71	71	—	46
4	1	15	5	12	—	16	35	—	80	80	—	35	16	—	12	5	—	5	12	—	16	35	—	80	80	—	35
4	2	18	3	2	—	9	13	—	16	**93**	—	**93**	16	—	13	9	—	2	3	—	3	2	—	9	13	—	16
4	3	21	2	5	—	7	8	—	33	49	—	70	70	—	49	33	—	8	7	—	5	2	—	2	5	—	7
5	1	18	5	7	—	24	6	—	61	56	—	56	61	—	6	24	—	7	5	—	5	7	—	24	6	—	61
5	2	21	2	3	—	13	17	—	25	43	—	75	75	—	43	25	—	17	13	—	3	2	—	2	3	—	13
5	3	24	3	2	—	6	13	—	17	32	—	78	21	—	21	78	—	32	17	—	13	6	—	2	3	—	3
5	4	27	1	2	—	7	9	—	7	8	—	35	50	—	70	70	—	50	35	—	8	7	—	9	7	—	2
6	1	21	3	6	—	9	16	—	12	31	—	84	84	—	31	12	—	16	9	—	6	3	—	3	6	—	9
6	2	24	2	2	—	6	2	—	4	19	—	23	**92**	—	**92**	23	—	19	4	—	2	6	—	2	2	—	2

Table 3.6 MMF harmonic winding factors (in %) for DL-FSCW with same zone polarity and with opposite adjacent coil polarity.

n_{c1}	n_{c2}	Q_s	$v = 1$	2	3	4	5	6	7	8	9	10	11	12	13	14	15	16	17	18	19	20	21	22	23	24	25
2	1	9	14	35	—	73	73	—	35	14	—	14	35	—	73	73	—	35	14	—	14	35	—	73	73	—	35
3	**1**	12	2	13	—	22	**90**	—	**90**	22	—	13	2	—	2	13	—	22	90	—	90	22	—	13	2	—	2
3	2	15	4	7	—	28	46	—	71	71	—	46	28	—	7	4	—	4	7	—	28	46	—	71	71	—	46
4	1	15	5	12	—	16	35	—	80	80	—	35	16	—	12	5	—	5	12	—	16	35	—	80	80	—	35
4	2	18	1	10	—	19	26	—	75	28	—	28	75	—	26	19	—	10	1	—	1	10	—	19	26	—	75
4	3	21	2	5	—	7	8	—	33	49	—	70	70	—	49	33	—	8	7	—	5	2	—	2	5	—	7
5	**1**	18	1	5	—	2	20	—	29	**89**	—	**89**	29	—	20	2	—	5	1	—	1	5	—	2	20	—	29
5	2	21	2	3	—	13	17	—	25	43	—	75	75	—	43	25	—	17	13	—	3	2	—	2	3	—	13
5	**3**	24	0	3	—	6	6	—	13	11	—	12	**94**	—	**94**	12	—	11	13	—	6	6	—	3	0	—	0
5	4	27	1	2	—	7	9	—	7	8	—	35	50	—	70	70	—	50	35	—	8	7	—	9	7	—	2
6	1	21	3	6	—	9	16	—	12	31	—	84	84	—	31	12	—	16	9	—	6	3	—	3	6	—	9
6	2	24	1	5	—	6	16	—	27	19	—	70	43	—	43	70	—	19	27	—	16	6	—	5	1	—	1

Table 3.7 MMF harmonic winding factors (in %) for DL-FSCW with opposite zone-groups polarity and with opposite adjacent coil polarity.

n_{c1}	n_{c2}	Q_s	$v = 1$	2	3	4	5	6	7	8	9	10	11	12	13	14	15	16	17	18	19	20	21	22	23	24	25
2	1	18	8	—	—	—	65	—	28	—	—	—	28	—	65	—	—	—	8	—	8	—	—	—	65	—	28
3	1	24	4	—	—	—	17	—	36	—	—	—	77	—	77	—	—	—	36	—	17	—	—	—	4	—	4
3	2	30	2	—	—	—	10	—	7	—	—	—	9	—	**94**	—	—	—	**94**	—	9	—	—	—	7	—	10
4	1	30	3	—	—	—	20	—	10	—	—	—	63	—	50	—	—	—	50	—	63	—	—	—	10	—	20
4	2	36	1	—	—	—	10	—	13	—	—	—	27	—	45	—	—	—	73	—	73	—	—	—	45	—	27
4	3	42	1	—	—	—	8	—	7	—	—	—	10	—	35	—	—	—	79	—	12	—	—	—	12	—	79
5	1	36	2	—	—	—	8	—	15	—	—	—	14	—	33	—	—	—	82	—	82	—	—	—	33	—	14
5	2	42	1	—	—	—	5	—	0	—	—	—	7	—	16	—	—	—	20	—	**92**	—	—	—	**92**	—	20
5	3	48	1	—	—	—	2	—	4	—	—	—	12	—	13	—	—	—	31	—	48	—	—	—	72	—	72
5	4	54	0	—	—	—	3	—	3	—	—	—	4	—	10	—	—	—	16	—	5	—	—	—	5	—	**95**
6	1	42	2	—	—	—	10	—	7	—	—	—	25	—	3	—	—	—	59	—	60	—	—	—	60	—	59
6	2	48	0	—	—	—	5	—	5	—	—	—	15	—	19	—	—	—	22	—	41	—	—	—	77	—	77

Table 3.8 MMF harmonic winding factors (in %) for DL-FSCW with same zone-groups polarity and with opposite adjacent coil polarity.

n_{c1}	n_{c2}	Q_s	$v = 1$	2	3	4	5	6	7	8	9	10	11	12	13	14	15	16	17	18	19	20	21	22	23	24	25
2	1	18	4	—	—	—	14	—	90	—	—	—	90	—	14	—	—	—	4	0	4	—	—	—	14	—	90
3	1	24	4	—	—	—	17	—	36	—	—	—	77	—	77	—	—	—	36	0	17	—	—	—	4	—	4
3	2	30	2	—	—	—	10	—	28	—	—	—	76	—	17	—	—	—	17	0	76	—	—	—	28	—	10
4	1	30	1	—	—	—	0	—	17	—	—	—	26	—	**89**	—	—	—	**89**	0	26	—	—	—	17	—	0
4	2	36	1	—	—	—	10	—	13	—	—	—	27	—	45	—	—	—	73	0	73	—	—	—	45	—	27
4	3	42	1	—	—	—	3	—	7	—	—	—	14	—	6	—	—	—	7	0	**95**	—	—	—	**95**	—	7
5	1	36	2	—	—	—	8	—	15	—	—	—	14	—	33	—	—	—	82	0	82	—	—	—	33	—	14
5	2	42	1	—	—	—	2	—	14	—	—	—	24	—	23	—	—	—	72	0	37	—	—	—	37	—	72
5	3	48	1	—	—	—	2	—	4	—	—	—	12	—	13	—	—	—	31	0	48	—	—	—	72	—	72
5	4	54	1	—	—	—	3	—	7	—	—	—	13	—	7	—	—	—	8	0	37	—	—	—	81	—	10
6	1	42	1	—	—	—	1	—	7	—	—	—	5	—	21	—	—	—	32	0	**88**	—	—	—	**88**	—	32
6	2	48	0	—	—	—	5	—	5	—	—	—	15	—	19	—	—	—	22	0	41	—	—	—	77	—	77

Table 3.9 MMF harmonic winding factors (in %) for DL-FSCW with asymmetric zones, with opposite zone-groups polarity, and with opposite adjacent coil polarity.

n_{c1}	n_{c2}	Q_s	$\nu = 1$	2	3	4	5	6	7	8	9	10	11	12	13	14	15	16	17	18	19	20	21	22	23	24	25
2	1	21	1	6	—	5	5	—	12	**89**	—	7	7	—	**89**	12	—	5	5	—	6	1	—	1	6	—	5
3	1	27	1	7	—	6	5	—	19	39	—	22	12	—	70	70	—	12	22	—	39	19	—	5	6	—	7
3	2	36	1	4	—	3	3	—	10	27	—	6	7	—	75	11	—	8	13	—	13	8	—	11	75	—	7
4	1	33	0	2	—	1	0	—	2	20	—	7	8	—	27	**86**	—	14	14	—	**86**	27	—	8	7	—	20
4	2	42	0	1	—	3	2	—	9	15	—	4	5	—	32	28	—	13	31	—	60	32	—	32	60	—	31
4	3	51	0	1	—	1	1	—	3	7	—	2	2	—	14	3	—	3	5	—	4	3	—	5	**94**	—	3
5	1	39	1	3	—	3	2	—	8	17	—	7	6	—	18	33	—	21	12	—	76	76	—	12	21	—	33
5	2	48	0	2	—	1	1	—	1	16	—	2	3	—	26	17	—	9	13	—	68	20	—	17	29	—	29
5	3	57	0	2	—	1	1	—	2	3	—	1	3	—	11	8	—	4	9	—	20	7	—	22	50	—	23
5	4	66	0	1	—	1	1	—	3	6	—	2	2	—	12	3	—	3	6	—	4	3	—	7	37	—	4
6	1	45	0	1	—	0	0	—	2	8	—	3	2	—	1	25	—	6	8	—	34	**83**	—	17	17	—	83
6	2	54	0	1	—	2	1	—	4	7	—	1	4	—	17	19	—	7	3	—	28	20	—	12	36	—	62

3.3.1 Maximum Winding Factors

Being one of the most crucial parameters in the winding design process, the maximum winding factor was evaluated for various combinations of n_{c1} and n_{c2}, as well as for different zone polarities. The results are summarized in Tables 3.10–3.18 and are visually presented in Figures 3.12 and 3.13.

SL-FSCWs with unequal zones: Tables 3.10 and 3.11 show the maximum winding factors for various SL-FSCWs with unequal zones, in particular for the case $n_{c1} > n_{c2}$. The light gray cells present the configurations where $n_{c2} > n_{c1}$, which are simply symmetric mirror images of the $n_{c1} > n_{c2}$ cases, while the dark gray cell represent the $n_{c1} = n_{c2}$ configurations corresponding to the standard FSCWs. These results indicate that the maximum achievable winding factor depends strongly on the zone polarities, with configurations featuring opposite zone polarity yielding the highest values.

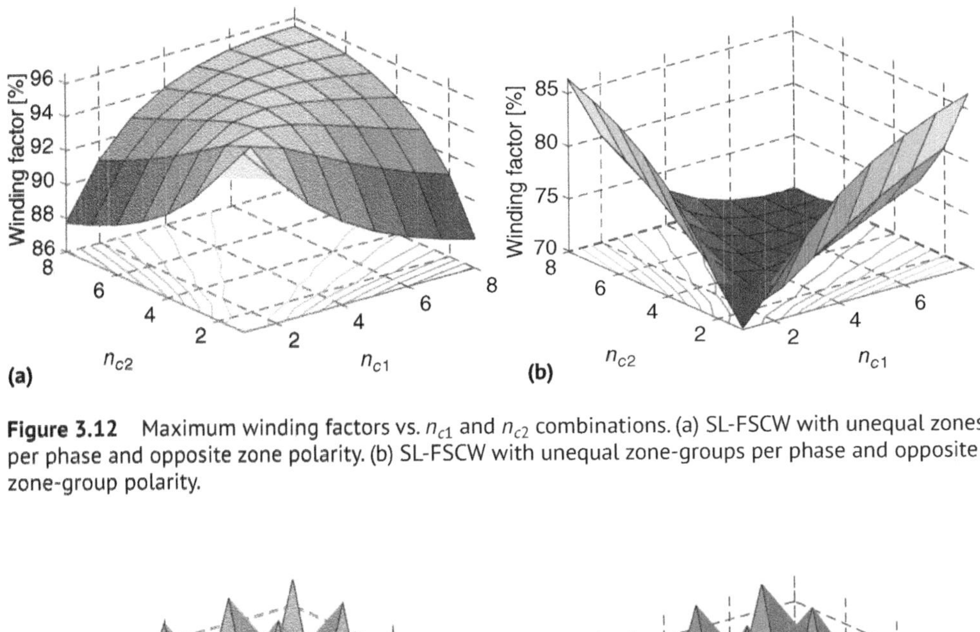

Figure 3.12 Maximum winding factors vs. n_{c1} and n_{c2} combinations. (a) SL-FSCW with unequal zones per phase and opposite zone polarity. (b) SL-FSCW with unequal zone-groups per phase and opposite zone-group polarity.

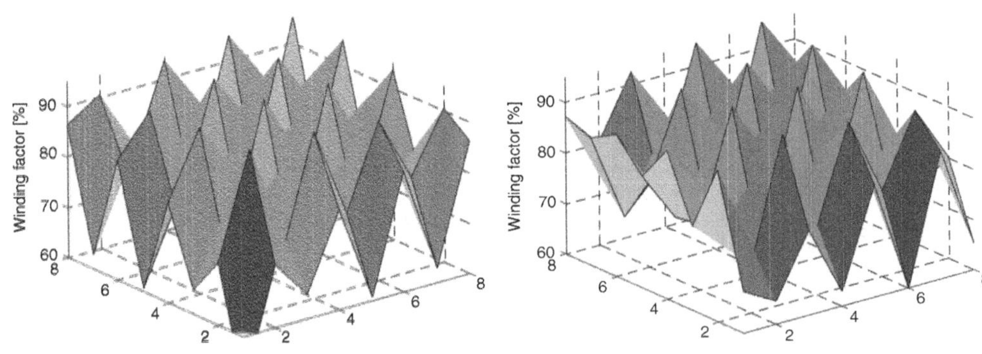

Figure 3.13 Maximum winding factors vs. n_{c1} and n_{c2} combinations. (a) DL-FSCW with unequal zones per phase and opposite zone polarity. (b) DL-FSCW with unequal zone-groups per phase and opposite zone-group polarity.

SL-FSCWs with unequal zone-groups: For the second category of single-layer windings, specifically SL-FSCW with unequal zone-groups, the analysis was conducted only for $n_{c1} > n_{c2}$. Since both zone-group polarities led to similar results, Table 3.12 presents the maximum winding factors only for the case with opposite zone-group polarity. In contrast to SL-FSCWs with unequal zones, this winding category is characterized with low winding factors, making it unsuitable for use in electrical machines.

Table 3.10 Maximum winding factor (in %): SL-FSCW with unequal zones and opposite zone polarity.

	n_{c1}	2	3	4	5	6	7	8
n_{c2}	1	**94.5**	**92.5**	**91.0**	**89.8**	**89.0**	**88.3**	**87.7**
	2	—	**95.1**	**94.2**	**93.2**	**92.3**	**91.5**	**90.9**
	3	—	—	**95.3**	**94.7**	**94.1**	**93.4**	**92.8**
	4	—	—	—	**95.4**	**95.0**	**94.6**	**94.1**
	5	—	—	—	—	**95.4**	**95.2**	**94.8**
	6	—	—	—	—	—	**95.4**	**95.2**
	7	—	—	—	—	—	—	**95.5**

Table 3.11 Maximum winding factor (in %): SL-FSCW with unequal zones and same zone polarity.

	n_{c1}	2	3	4	5	6	7	8
n_{c2}	1	79.3	72.4	67.4	63.7	60.8	63.3	65.5
	2	—	81.5	78.1	74.8	71.8	69.3	67.0
	3	—	—	82.1	80.1	77.9	75.7	73.6
	4	—	—	—	82.3	81.0	79.5	77.8
	5	—	—	—	—	82.4	81.5	80.4
	6	—	—	—	—	—	82.5	81.9
	7	—	—	—	—	—	—	82.6

Table 3.12 Maximum winding factor (in %): SL-FSCW with unequal zone-group and opposite zone-group polarity.

	n_{c1}	2	3	4	5	6	7	8
n_{c2}	1	73.4	77.3	80.2	82.4	84.1	85.4	86.4
	2	—	71.2	73.3	75.4	77.2	78.8	80.2
	3	—	—	70.6	71.8	73.3	74.7	76.0
	4	—	—	—	70.3	71.1	72.2	73.3
	5	—	—	—	—	70.2	70.8	71.5
	6	—	—	—	—	—	70.1	70.5
	7	—	—	—	—	—	—	70.0

DL-FSCWs with unequal zones: Next, we examine the DL-FSCW group with unequal zones. Unlike its single-layer counterpart, this winding category includes various combinations of positive and negative zone polarities. The results for both cases are listed in Tables 3.13 and 3.14. Since identical n_{c1} & n_{c2} relationships correspond to the same topologies, the results appear as mirrored configurations within the n_{c1} & n_{c2} mapping. This winding group provides several topologies with high winding factors for both zone polarity combinations.

DL-FSCWs with unequal zone-groups: On the other hand, for DL-FSCWs with unequal zone-groups, each n_{c1} & n_{c2} combination correspond to a unique winding, while the polarity of zone-groups does not create any additional distinct configuration. Therefore, Table 3.15 presents the winding factors for only one polarity case, specifically windings with the same zone-group polarity. It is evident that several topologies in this category exhibit high winding factors. Notably, this group includes the widely used 18-teeth/14-poles FSCW with winding parameters $n_{c1} = 2$, $n_{c2} = 1$, and a winding factor of 90.2% for the working harmonic.

FSCWs with asymmetric zone distributions: The investigations of winding topologies with asymmetric zone distribution are focused only on a specific zone combination, namely $n_{z1} = 2$, $n_{z2} = 3$. Table 3.16 show the results for SL-FSCW with asymmetric zones and same zone-group polarity (i.e. $z_{g1} + z_{g2}$). For this category, both positive and negative zone-group polarities lead to identical results. It can be observed that all n_{c1} & n_{c2} combinations in this group have low winding factors, making them unsuitable for electrical machine applications. On the other hand, for double-layer windings with asymmetric zones, positive and negative zone-group

Table 3.13 Maximum winding factor (in %): DL-FSCW with unequal zones and opposite zone polarity.

	n_{c1}	2	3	4	5	6	7	8
n_{c2}	1	72.5	65.0	79.9	60.8	83.9	62.9	86.3
	2	—	70.9	**93.1**	75.2	**91.7**	78.7	**90.5**
	3	—	—	70.4	78.0	73.2	74.4	76.0
	4	—	—	—	70.2	**94.6**	72.1	**93.8**
	5	—	—	—	—	70.1	80.6	71.5
	6	—	—	—	—	—	70.1	**95.0**
	7	—	—	—	—	—	—	70.0

Table 3.14 Maximum winding factor (in %): DL-FSCW with unequal zones and same zones polarity.

	n_{c1}	2	3	4	5	6	7	8
n_{c2}	1	72.5	**90.1**	79.9	**88.8**	83.9	**87.7**	86.3
	2	—	70.9	74.5	75.2	70.0	78.7	65.9
	3	—	—	70.4	**94.1**	73.2	**93.1**	76.0
	4	—	—	—	70.2	79.7	72.1	76.9
	5	—	—	—	—	70.1	**94.9**	71.5
	6	—	—	—	—	—	70.1	81.2
	7	—	—	—	—	—	—	70.0

polarities lead to distinct winding configurations. Consequently, the results for both polarity groups are provided in Tables 3.17 and 3.18. In contrast to their single-layer counterparts, certain n_{c1} & n_{c2} combinations in the double-layer category achieve higher winding factors. This leads to the conclusion that, among all winding types with asymmetric zones, only DL-FSCWs with opposite zone-group polarity offer sufficient efficiency for use in electric machines.

Table 3.15 Maximum winding factor (in %): DL-FSCW with unequal zone-group and same zone-group polarity.

	n_{c1}	2	3	4	5	6	7	8
n_{c2}	1	**90.2**	76.8	**89.5**	82.2	**88.2**	85.3	**87.3**
	2	69.5	76.1	73.1	72.3	77.1	67.8	80.1
	3	**93.6**	69.7	**94.5**	71.7	**93.6**	74.6	**92.5**
	4	73.1	79.3	69.8	80.7	71.1	78.4	73.2
	5	**92.4**	71.7	**94.9**	69.8	**95.1**	70.7	**94.6**
	6	77.1	76.3	71.1	81.3	69.9	81.7	70.5
	7	**91.1**	74.6	**94.2**	70.7	**95.2**	69.9	**95.3**

Table 3.16 Maximum winding factor (in %): SL-FSCW with asymmetric zones and same zone-groups polarity.

	n_{c1}	2	3	4	5	6	7	8
n_{c2}	1	61.2	62.3	63.6	64.7	65.5	66.2	66.8
	2	—	61.1	61.0	61.5	62.2	62.8	63.5
	3	—	—	61.4	60.9	61.0	61.3	61.7
	4	—	—	—	61.6	61.0	60.9	60.9
	5	—	—	—	—	61.8	61.2	60.9
	6	—	—	—	—	—	42.2	14.5
	7	—	—	—	—	—	—	14.0

Table 3.17 Maximum winding factor (in %): DL-FSCW with asymmetric zones and opposite zone-groups polarity.

	n_{c1}	2	3	4	5	6	7	8
n_{c2}	1	**89.0**	70.0	**86.2**	76.2	83.4	80.2	81.3
	2	—	75.0	60.5	68.0	61.8	60.3	63.3
	3	—	—	**94.4**	65.1	**92.9**	67.5	**91.0**
	4	—	—	—	80.4	60.9	76.9	60.8
	5	—	—	—	—	**95.1**	64.5	**94.3**
	6	—	—	—	—	—	81.6	61.3
	7	—	—	—	—	—	—	**95.3**

Table 3.18 Maximum winding factor (in %): DL-FSCW with asymmetric zones and same zone-groups polarity.

	n_{c1}	2	3	4	5	6	7	8
n_{c2}	1	61.9	70.0	66.3	76.2	65.1	80.2	63.4
	2	—	68.2	60.5	72.9	61.8	75.4	63.3
	3	—	—	62.8	65.1	67.5	67.5	69.0
	4	—	—	—	66.5	60.9	69.9	60.8
	5	—	—	—	—	61.8	64.5	65.8
	6	—	—	—	—	—	65.8	61.3
	7	—	—	—	—	—	—	61.2

Table 3.19 Working harmonics for SL-FSCW with unequal zones.

n_{c1}	n_{c2}	Opposite polarity	Same polarity
Even	Even and odd	$\nu_p = \frac{Q_S}{2} \pm 1$	Low winding factor
Odd	Even and odd	$\nu_p = \frac{Q_S}{2} \pm 1$	Low winding factor

Table 3.20 Working harmonics for DL-FSCW with unequal zones.

n_{c1}	n_{c2}	Opposite polarity	Same polarity
Even	Even	$\nu_p = \frac{Q_S}{2} \pm 1$	Low winding factor
Odd	Odd	Low winding factor	$\nu_p = \frac{Q_S}{2} \pm 1$
	Others	Low winding factor	Low winding factor

3.3.2 MMF Working Harmonics

Similar to common FSCWs, topologies with unequal zones or zone-groups also exhibit high MMF harmonic content. Each winding type, due to the periodicity of the winding factor, is characterized by multiple harmonics with high winding factors. Therefore, the guidelines outlined in Section 2.3.2 can be applied here as well to select the appropriate operating harmonics.

For single-layer windings, only configurations with unequal zones and opposite zone polarity are associated with high winding factors. As shown in Table 3.10, each combination of n_{c1} and n_{c2} corresponds to a specific winding with a high winding factor. The operating waves for this winding group are defined in Table 3.19.

On the other hand, for double-layer windings, all zone and zone-group combinations include some topologies with higher winding factor. Depending on the n_{c1} and n_{c2} combination, as well as the polarity of the zone or zone-groups, the operating wave can be defined as summarized in Tables 3.20 and 3.21.

Table 3.21 Working harmonics for DL-FSCW with unequal zone-groups.

n_{c1}	n_{c2}	Opposite polarity	Same polarity
Even	Odd	Low winding factor	$\nu_p = \frac{Q_S}{2} \pm 2$
Odd	Even	$\nu_p = \frac{Q_S}{2} \pm 2$	Low winding factor
	Others	Low winding factor	Low winding factor

3.4 MMF Characteristics of Various Three-phase FSCWs with Unequal Zones

Using the general expressions for the MMF function of three-phase windings and the respective winding factors for different winding groups summarized in Tables 3.1–3.3, various FSCWs with unequal zone configurations are examined below. The results, including the MMF winding distribution, winding factor for the operating wave, and winding THD, are presented in Figures 3.14–3.25 and listed in Tables 3.22–3.24.

3.4.1 SL-FSCWs with Unequal Zones

Example 3.1 18-slot/16-pole FSCW with unequal zones and $n_{c1} = 2$, $n_{c2} = 1$
The first example FSCW with unequal zones consists of two coils $(n_{c1} = 2)$ in the first zone and one coil $(n_{c2} = 1)$ in the second zone. These zones are arranged opposite to each other and have same coil polarity. In this configuration, the stator has 18 slots $(Q_S = 18)$, and the winding layout is $+\text{U} + \text{U} - \text{V} + \text{W} + \text{W} - \text{U} + \text{V} + \text{V} - \text{W}$, as illustrated in Figure 3.14a. The corresponding MMF characteristics are shown in Figure 3.14b. For this configuration, the 8th harmonic, which has the highest winding factor $(\xi_{w,8} = 94.5\%)$, was selected as the operating harmonic, resulting in a rotor with 16 PM poles $(2p = 16)$.

Example 3.2 24-slot/22-pole FSCW with unequal zones and $n_{c1} = 3$, $n_{c2} = 1$
In another configuration, the winding parameters for the first and second zone are set to $n_{c1} = 3$ and $n_{c2} = 1$, respectively, while the zone arrangement remains the same as in the first example. The corresponding winding layout and MMF characteristics are shown in Figure 3.15. The results indicate that the dominant MMF harmonic, with the highest winding factor, is the 11th harmonic $(\xi_{w,11} = 92.5\%)$. Therefore, this harmonic was selected as the working harmonic, resulting in a rotor design with 22 poles $(2p = 22)$.

Example 3.3 30-slot/28-pole and 36-slot/34-pole FSCW with unequal zones
The MMF characteristics of two other SL-FSCW configurations, one with zone parameters $n_{c1} = 3$, $n_{c2} = 2$, and the other with $n_{c1} = 4$, $n_{c2} = 2$, are examined in Figure 3.16. In the first topology, the stator contains 30 slots $(Q_S = 30)$, and the operating wave is the 14th harmonic with a winding factor of 95.2%. In the second example, the stator has 36 slots $(Q_S = 36)$, and the operating harmonic is the 17th harmonic with a winding factor of 94.2%.

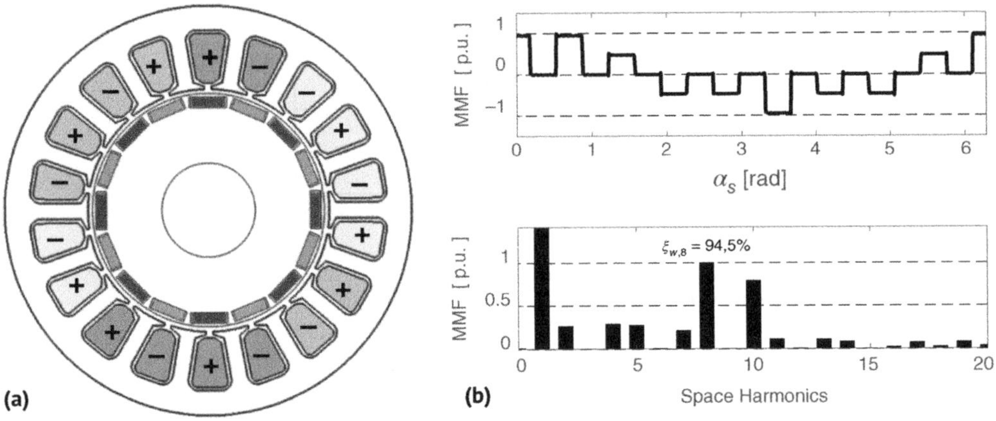

Figure 3.14 SL-FSCW with opposite polarity zones $(z_1 - z_2)$ and with $n_{c1} = 2, n_{c2} = 1$, and $Q_S = 18$. (a) Winding layout. (b) Three-phase spatial MMF distribution and harmonic content.

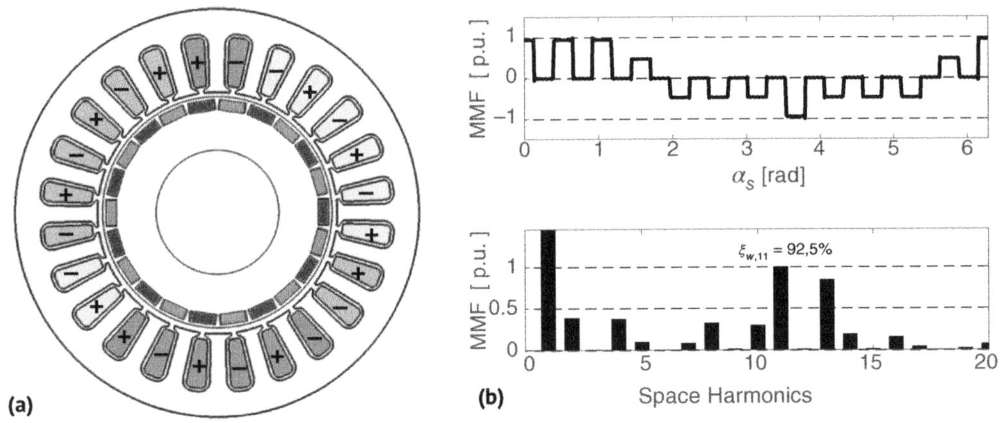

Figure 3.15 SL-FSCW with opposite polarity zones $(z_1 - z_2)$ and with $n_{c1} = 3, n_{c2} = 1$, and $Q_S = 24$. (a) Winding layout. (b) Three-phase spatial MMF distribution and harmonic content.

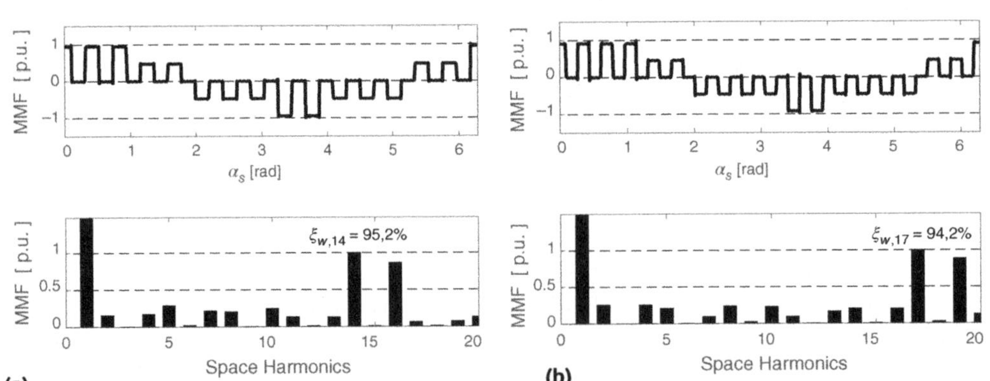

Figure 3.16 MMF characteristic for the SL-FSCWs with opposite polarity zones $(z_1 - z_2)$. (a) Winding topology with $n_{c1} = 3, n_{c2} = 2, \nu_p = 14$, and $Q_S = 30$. (b) Winding topology with $n_{c1} = 4, n_{c2} = 2, \nu_p = 17$, and $Q_S = 36$.

3.4.2 DL-FSCWs with Unequal Zones

For the double-layer winding configuration, both opposite zone $(z_1 - z_2)$ and same zone $(z_1 + z_2)$ polarities are considered. The following examples show the winding layouts and their MMF characteristics for several topologies.

Example 3.4 18-slot/16-pole FSCW with unequal zones $(z_1 - z_2)$ and $n_{c1} = 4$, $n_{c2} = 2$
The winding layout for the first FSCW example with opposite zone configuration, defined by winding parameters $n_{c1} = 4$ and $n_{c2} = 2$ is illustrated in Figure 3.17a, whereas its MMF characteristics are shown in Figure 3.17b. In this topology, the dominant MMF harmonic with the highest winding factor is the 8th harmonic $(\xi_{w,8} = 93.1\%)$. Accordingly, this harmonic was selected as the operating wave, leading to a rotor design with 16 PM poles $(2p = 16)$.

Example 3.5 24-slot/22-pole FSCW with unequal zones $(z_1 - z_2)$ and $n_{c1} = 6$, $n_{c2} = 2$
Another example with opposite zone polarity and winding parameters $n_{c1} = 6$ and $n_{c2} = 2$ is presented in Figure 3.18a. The MMF characteristics, shown in Figure 3.18b, indicate that the first

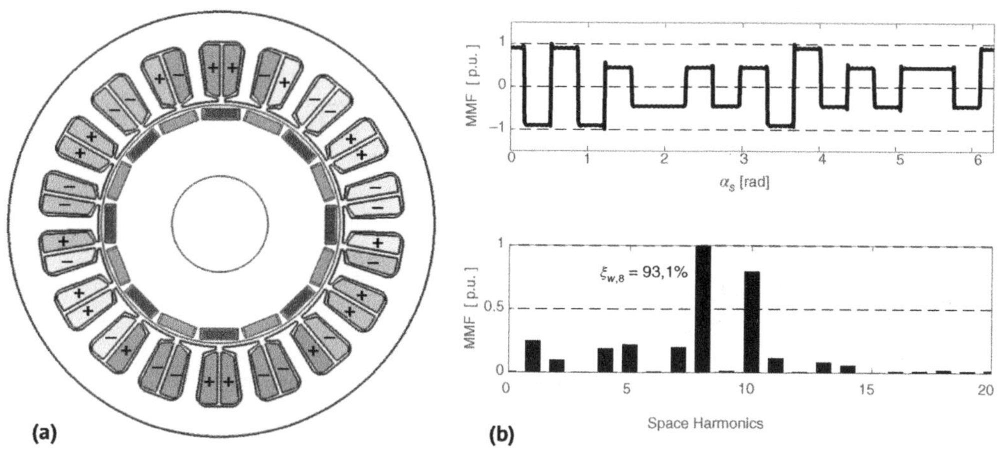

Figure 3.17 DL-FSCW with opposite polarity zones $(z_1 - z_2)$ and with $n_{c1} = 4$, $n_{c2} = 2$, and $Q_S = 18$. (a) Winding layout. (b) Three-phase spatial MMF distribution and harmonic content.

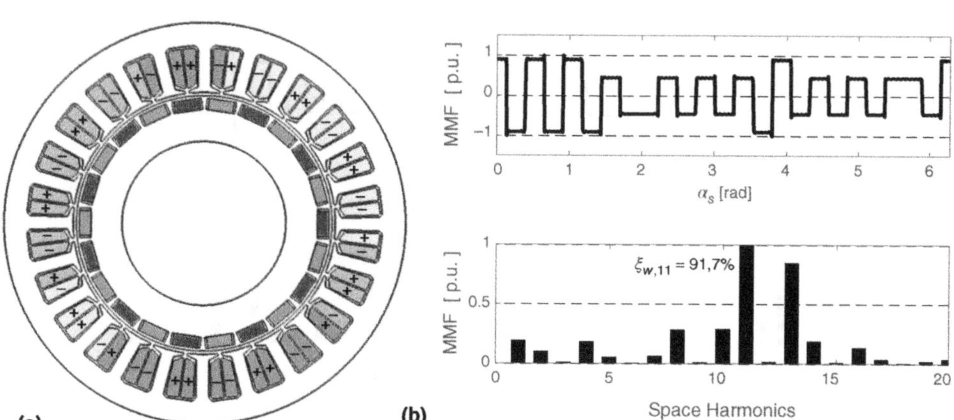

Figure 3.18 DL-FSCW with opposite polarity zones $(z_1 - z_2)$ and with $n_{c1} = 6$, $n_{c2} = 2$, and $Q_S = 24$. (a) Winding layout. (b) Three-phase spatial MMF distribution and harmonic content.

MMF harmonic component with the highest winding factor is the 11th harmonic ($\xi_{w,11} = 91.7\%$). Consequently, this harmonic was selected as the working harmonic, resulting in a rotor design with 22 PM poles ($2p = 22$).

Example 3.6 12-slot/10-pole FSCW with unequal zones ($z_1 + z_2$) and $n_{c1} = 3$, $n_{c2} = 1$
Next, the configuration with same zone polarity ($z_1 + z_2$) is considered. The winding arrangement for $n_{c1} = 3$ and $n_{c2} = 1$ is presented in Figure 3.19a, while its MMF characteristics are shown in Figure 3.19b. For this topology, the dominant MMF harmonic with the highest winding factor is the 5th harmonic ($\xi_{w,5} = 90.1\%$), which was selected as the working harmonic in the 10-pole ($2p = 10$) PM machine.

Example 3.7 18-slot/16-pole and 24-slot/22-pole FSCW with unequal zones
Figure 3.20 analyzes the MMF characteristics of two additional DL-FSCW configurations, both featuring same zone polarities. These are defined by the parameter sets $n_{c1} = 5$, $n_{c2} = 1$ and $n_{c1} = 5$, $n_{c2} = 3$. In the first configuration, the stator has 18 slots ($Q_S = 18$), and the operating wave is the 8th harmonic with a winding factor of 88.8%. In the second topology, the stator has 24 slots ($Q_S = 24$), and the selected working harmonic is the 11th harmonic with a winding factor of 94.1%.

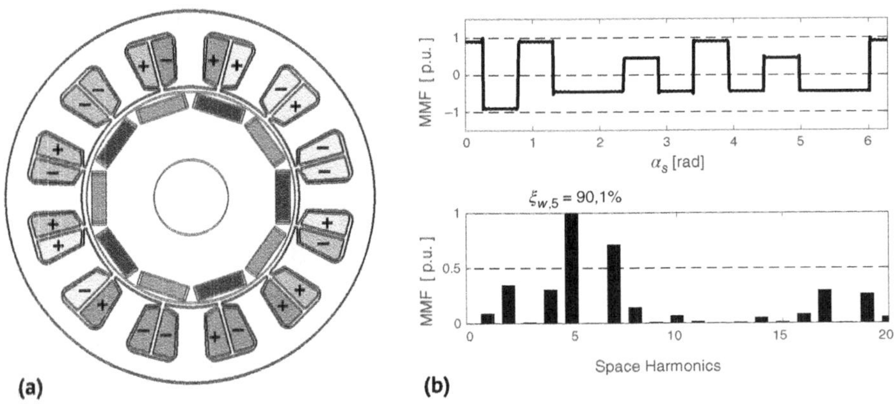

Figure 3.19 DL-FSCW with the same polarity zones ($z_1 + z_2$) and with $n_{c1} = 3$, $n_{c2} = 1$, and $Q_S = 12$. (a) Winding layout. (b) Three-phase spatial MMF distribution and harmonic content.

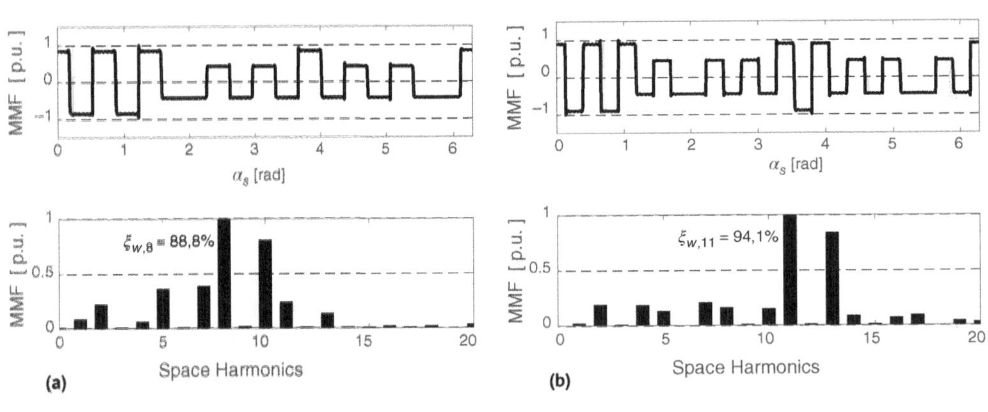

Figure 3.20 MMF characteristic for the DL-FSCWs with the same polarity zones ($z_1 + z_2$). (a) Winding topology with $n_{c1} = 5$, $n_{c2} = 1$, $\nu_p = 8$, and $Q_S = 18$. (b) Winding topology with $n_{c1} = 5$, $n_{c2} = 3$, $\nu_p = 11$, and $Q_S = 24$.

3.4.3 DL-FSCWs with Unequal Zone-groups

Example 3.8 18-slot/14-pole FSCW with unequal zone-groups $(z_{g1} + z_{g2})$ and $n_{c1} = 2$, $n_{c2} = 1$
Figure 3.21a illustrates a winding layout featuring the same zone-group polarities and with winding parameters $n_{c1} = 2$ and $n_{c2} = 1$. The corresponding MMF characteristics for this winding type are presented in Figure 3.21b. The MMF harmonic with the highest winding factor is the 7th harmonic $(\xi_{w,7} = 90.2\%)$, which has been selected as the operating harmonic in the example motor with 14 PM poles $(2p = 14)$.

Example 3.9 30-slot/26-pole and 42-slot/22-pole FSCW with unequal zone-groups $(z_{g1} + z_{g2})$
The MMF characteristics of two additional example DL-FSCW configurations are shown in Figure 3.22. Both examined windings have identical zone-group polarities and are defined by winding parameters $n_{c1} = 4$ and $n_{c2} = 1$ for the first winding, and $n_{c1} = 4$ and $n_{c2} = 3$ for the second one. The first configuration has 30 stator slots $(Q_S = 30)$, and the operating wave is the 13th harmonic with a winding factor of 89.5%. In the second topology, the stator has 42 slots $(Q_S = 42)$, and the operating wave is the 19th harmonic with a winding factor of 94.5%.

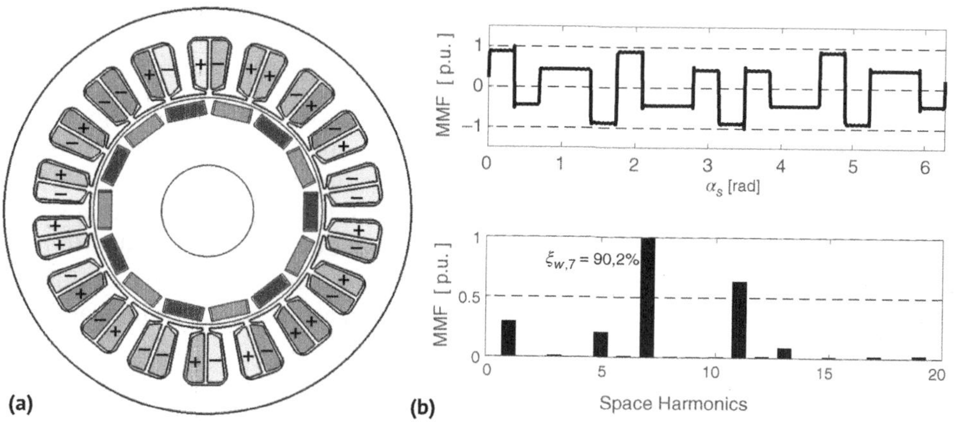

Figure 3.21 DL-FSCWs with the same polarity zone-groups $(z_{g1} + z_{g2})$ and with $n_{c1} = 2$, $n_{c2} = 1$, and $Q_S = 18$. (a) Winding layout. (b) Three-phase spatial MMF distribution and harmonic content.

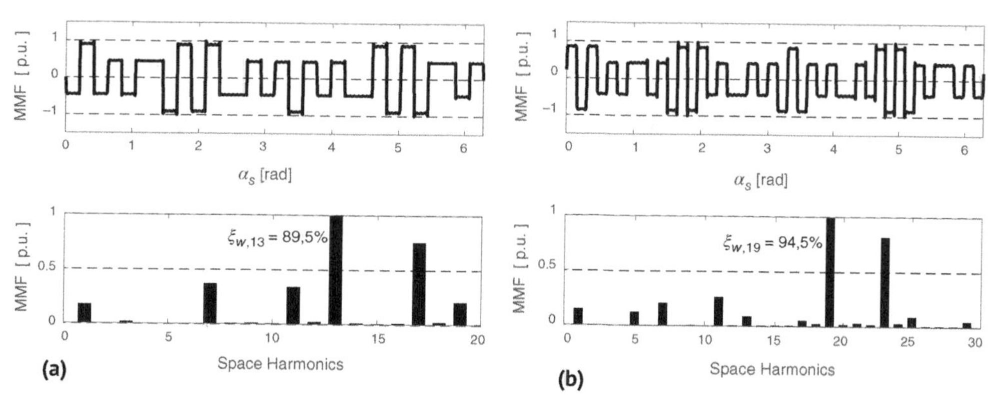

Figure 3.22 MMF characteristic for the DL-FSCWs with the same polarity zone-groups $(z_{g1} + z_{g2})$. (a) Winding topology with $n_{c1} = 4$, $n_{c2} = 1$, $v_p = 13$, and $Q_S = 30$. (b) Winding topology with $n_{c1} = 4$, $n_{c2} = 3$, $v_p = 19$, and $Q_S = 42$.

Example 3.10 30-slot/26-pole and 42-slot/22-pole FSCW with unequal zone-groups $(z_{g1} - z_{g2})$
Similar slot and pole combinations can also be achieved using configurations with opposite zone-group polarities; however, with different parameter combinations for n_{c1} and n_{c2}.

Figure 3.23a shows the MMF characteristics for the first example winding defined by $n_{c1} = 3$ and $n_{c2} = 2$, and featuring opposite zone-group polarities. This winding type requires 30 stator slots $(Q_S = 30)$, and the operating harmonic is the 13th harmonic with a winding factor of 93.6%. On the other side, Figure 3.23b examines the second example winding with parameters $n_{c1} = 5$ and $n_{c2} = 2$. This configuration requires 42 stator slots $(Q_S = 42)$, and operates with the 19th harmonic, which has a winding factor of 92.4%.

3.4.4 DL-FSCWs with Asymmetric Zone-groups

Example 3.11 21-slot/16-pole FSCW with asymmetric zone-groups
An example of a DL-FSCW asymmetric winding with opposite polarity zone-groups is illustrated in Figure 3.24a. As described in Section 3.2.3, this configuration is defined by the winding parameters $n_{z1} = 2$ & $n_{c1} = 2$, and $n_{z2} = 3$ & $n_{c2} = 1$. The corresponding MMF characteristics are presented in Figure 3.24b. In this topology, the 8th harmonic has the highest winding factor $(\xi_{w,8} = 89.0\%)$, and was therefore selected as the working harmonic for the example PM machine, which features 16 rotor poles $(2p = 16)$.

Example 3.12 33-slot/28-pole and 51-slots/46-pole FSCW with asymmetric zone-groups
Figure 3.25 examines two further FSCWs with asymmetric zone-group distributions. The first topology is the 33-slot/28-pole FSCW with winding parameters $n_{z1} = 2$ & $n_{c1} = 4$, and $n_{z2} = 3$ & $n_{c2} = 1$. In this design, the 14th MMF harmonic is selected as the operating wave, corresponding to a winding factor of $\xi_{w,14} = 86.2\%$. The second example is the 51-slot/46-pole FSCW with winding parameters $n_{z1} = 2$ & $n_{c1} = 4$, and $n_{z2} = 3$ & $n_{c2} = 3$. This configuration yields the 23rd MMF harmonic as the working harmonic, achieving a winding factor of $\xi_{w,23} = 94.5\%$.

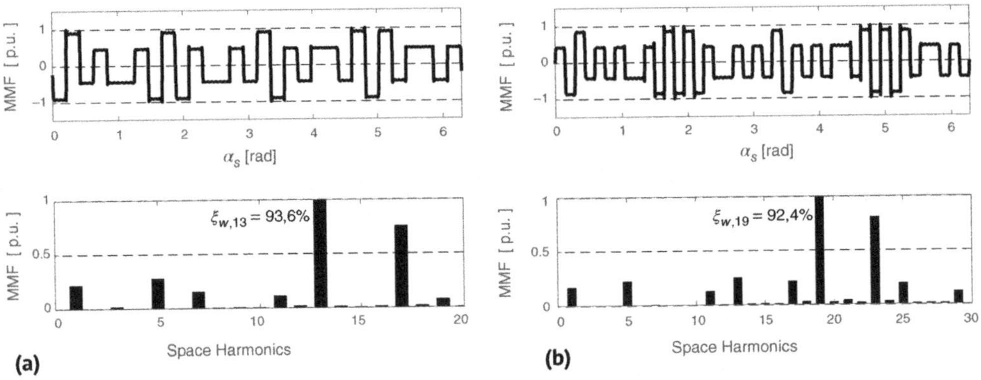

Figure 3.23 MMF characteristic for the DL-FSCWs with opposite polarity zone-groups $(z_{g1} - z_{g2})$. (a) Winding topology with $n_{c1} = 3, n_{c2} = 2, v_p = 13$, and $Q_S = 30$. (b) Winding topology with $n_{c1} = 5, n_{c2} = 2, v_p = 19$, and $Q_S = 42$.

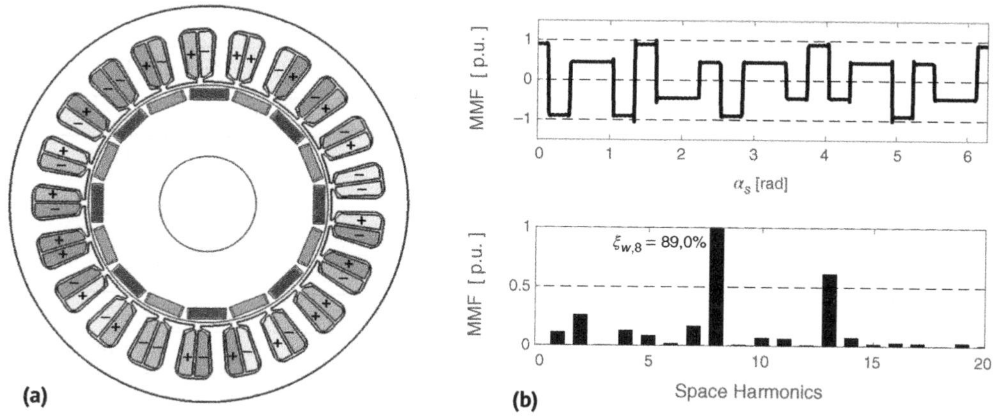

Figure 3.24 MMF characteristic for the DL-FSCWs with asymmetric zone-groups, opposite polarity zone-groups arrangement $(z_{g1} - z_{g2})$, and with $n_{z1} = 2$ & $n_{c1} = 2$, $n_{z2} = 3$ & $n_{c2} = 1$, and $Q_S = 21$, (a) Winding layout. (b) Three-phase spatial MMF distribution and harmonic content.

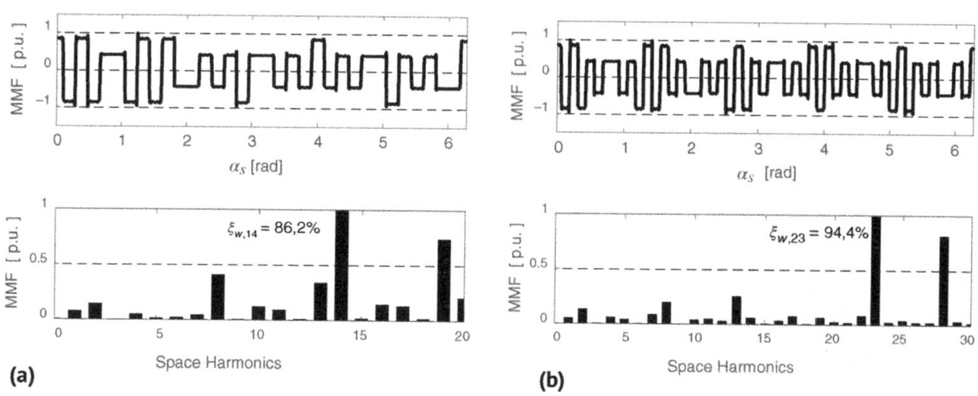

Figure 3.25 MMF characteristic for the DL-FSCWs with asymmetric zone-groups and opposite polarity zone-groups arrangement $(z_{g1} - z_{g2})$. (a) Winding topology with $n_{z1} = 2$ & $n_{c1} = 4$, $n_{z2} = 3$ & $n_{c2} = 1$, and $v_p = 14$, $Q_S = 33$. (b) Winding topology with $n_{z1} = 2$ & $n_{c1} = 4$, $n_{z2} = 3$ & $n_{c2} = 3$, and $v_p = 23$, $Q_S = 51$.

3.4.5 Concluding Remarks

Based on the MMF characteristic of the investigated windings, the following conclusions can be drawn:

- FSCWs with unequal zones, as well as those with asymmetric zones, produce both even- and odd-order spatial harmonics in the MMF spectrum, with the exception of 3rd-order harmonics.
- In contrast, FSCW composed of unequal zone-groups generate exclusively odd-order harmonics.
- For all winding configurations considered, the winding factor for the operating harmonic decreases with increase the n_{c1}/n_{c2} ratio.
- Additionally, an increase in the n_{c1}/n_{c2} ratio results in a degradation of the winding THD.

Table 3.22 Winding factors and THDs for FSCWs with unequal zones.

Layers	z	n_{c1}	n_{c2}	Q_b	v_p	$\xi_{w,p}$ (%)	THD (%)
SL	$z_1 - z_2$	2	1	18	8	94.5	183.2
SL	$z_1 - z_2$	3	1	24	11	92.5	195.8
SL	$z_1 - z_2$	3	2	30	14	95.2	193.2
SL	$z_1 - z_2$	4	2	36	17	94.2	198.6
DL	$z_1 - z_2$	4	2	18	8	93.1	111.4
DL	$z_1 - z_2$	6	2	24	11	91.7	120.5
DL	$z_1 + z_2$	3	1	12	5	90.1	105.1
DL	$z_1 + z_2$	5	1	18	8	88.8	120.9
DL	$z_1 + z_2$	5	3	24	11	94.1	115.2

Table 3.23 Winding factors and THDs for FSCWs with unequal zone-groups.

Layers	z	n_{c1}	n_{c2}	Q_b	v_p	$\xi_{w,p}$ (%)	THD (%)
DL	$z_{g1} - z_{g2}$	3	2	30	13	93.6	105.1
DL	$z_{g1} - z_{g2}$	5	2	42	19	92.4	116.0
DL	$z_{g1} - z_{g2}$	5	4	54	25	94.9	115.0
DL	$z_{g1} + z_{g2}$	2	1	18	7	90.2	91.1
DL	$z_{g1} + z_{g2}$	4	1	30	13	89.5	114.0
DL	$z_{g1} + z_{g2}$	4	3	42	19	94.5	111.4

Table 3.24 Winding factors and THDs for FSCWs with asymmetric zone-groups for $n_{z1} = 2$ and $n_{z2} = 3$.

Layers	z	n_{c1}	n_{c2}	Q_b	v_p	$\xi_{w,p}$	THD
SL	$z_{g1} - z_{g2}$						
DL	$z_{g1} - z_{g2}$	2	1	21	8	89.0	89.57
DL	$z_{g1} - z_{g2}$	4	1	33	14	86.2	117.2
DL	$z_{g1} - z_{g2}$	4	3	51	23	94.4	110.9

4

Fractional Slot Concentrated Windings with Specific Coil Configuration

Despite their simple structure, concentrated windings offer a wide range of possible winding groups and topologies. The winding process typically begins by creating a coil with N_w turns wound around a single tooth. A coil group (or zone) is then constructed by combining n_c adjacent coils. The final winding topology is determined by the specific arrangement of coils within each zone and zones within each phase. Using this structured approach, all fractional slot concentrated windings (FSCWs) presented and studied in the previous chapters can be systematically designed and implemented. Above all, however, a key feature of these windings is their inherently simple and fully symmetrical coil structure. Therefore, initial approaches to the winding optimization process can already be applied at the coil level itself. For instance, replacing the conventional symmetrical coils or coil-groups with specific coil structures introduces an additional degree of freedom for winding design and optimization. In recent years, several methods based on this principle have been developed, primarily to reduce the magnetomotive force (MMF) subharmonics in concentrated windings, as reported in the works of Kometani et al. [50], Cistelecan et al. [42], Dajaku and Gerling [51, 52], and Dotz and Gerling [53, 54]. Based on their implementation principles, these approaches can be classified into two main categories:

1. Coils with unequal turns per coil-side and
2. Zones with unequal turns per coil.

The first technique involves FSCWs in which the number of turns on one side of the coil is different to the other. This particular coil structure is referred to below as an "asymmetric coil." On the other hand, the second technique is based on the zone configuration, where coils with different numbers of turns are arranged alternately within a zone in order to modify the MMF distribution.

Beyond these methods, another crucial factor that influences the MMF properties of the winding is the coil-pitch parameter. Hence, by implementing coils with different coil-pitches within a coil group, an alternative winding configuration can be achieved, which is known as

3. FSCWs with unequal coil-pitch.

4.1 FSCWs with Asymmetric Coils

In order to reduce the subharmonics of double-layer-FSCWs (DL-FSCWs), a simple approach is to employ a coil configuration with unequal turns per coil side, as introduced by Dajaku and Gerling [51, 52]. This asymmetric coil design differs from the standard coil primarily in the conductor distribution between its two sides. In a standard coil, both coil sides contain the same number of conductors, and the coil terminals are located only on one side of the machine. In contrast, the asymmetric coil has an unequal number of conductors in its two sides, and the coil terminals are located on opposite sides of the machine, as illustrated in Figure 4.1. In other words, in a fully wound standard coil, the coil consists of N_w turns, meaning both sides of the coil contain N_w conductors, resulting in an integer number of turns per coil. However, in an asymmetric coil, the last turn is not fully wound, leading to an unequal number of turns per coil-side and a fractional number of turns per coil. Therefore, for this particular configuration:

- the first side of the coil has $N_{w11} = N_w$ turns
- the second side of the coil has $N_{w12} = N_w - 1$ turns
- the total number of turns per coil is: $N_w - 1/2$.

4.1.1 Coil MMF Function

The MMF function for a coil arrangement with unequal turns per coil side can be approximated as the superposition of the MMF distribution of a standard coil and that of a single conductor shifted by π/Q_S, as illustrated in Figure 4.2a. Let N_w denote the number of turns in a standard coil and

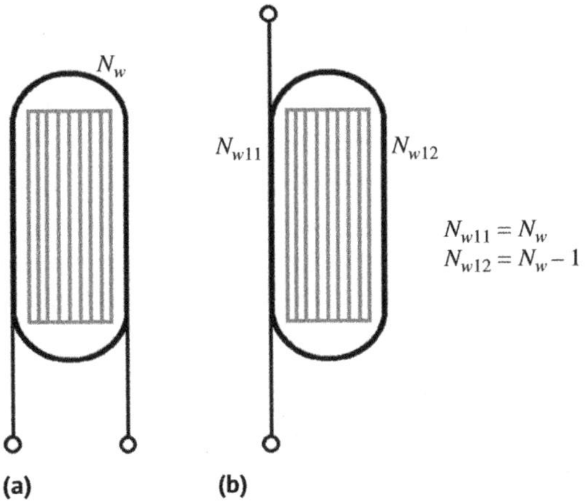

Figure 4.1 Different coil configurations. (a) Standard coil. (b) Coil configuration with unequal turns per coil-side (asymmetric coil).

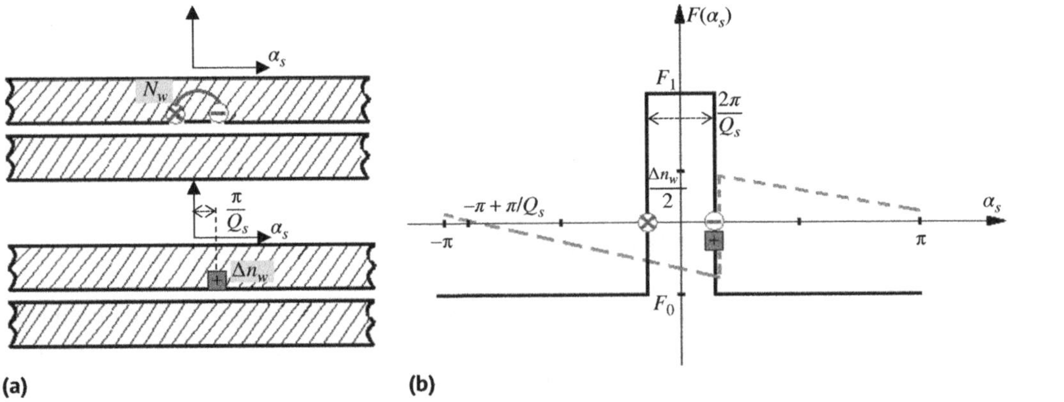

Figure 4.2 (a) A simple approach to a coil arrangement with unequal turns per coil side. (b) MMF distribution for a standard coil (solid line) and for a single conductor (dashed line).

Δn_w the number of additional (or removed) conductors associated with the assumed "go-in" coil configuration. Assuming both winding systems are supplied with the same current, the resulting current-linkage per coil-side for the asymmetric coil is

- 1st slot: $N_w i$
- 2nd slot: $(-N_w + \Delta n_w)i$.

Accordingly, the current linkage in the second slot is reduced by $\Delta n_w i$ relative to the 1st slot. For the coil structure in Figure 4.1b, the parameter Δn_w is always equal to one, i.e.

$$\Delta n_w = N_{w11} - N_{w12} = 1 \tag{4.1}$$

Consequently, the equivalent number of turns per coil can be defined as:

$$N_{w,\mathrm{eq}} = \frac{2N_w - \Delta n_w}{2} = N_w - 1/2, \quad \Delta n_w = 1 \tag{4.2}$$

The corresponding MMF distributions for the conventional coil and for a single conductor placed in a slot are shown in Figure 4.2b, represented by the solid and dashed curves, respectively. As discussed in Chapter 2, a common coil generates an MMF distribution characterized by a square waveform, which can be analytically expressed using Equations (2.14) and (2.15), yielding:

$$F_{\mathrm{coil},j}(t,\alpha_S) = \frac{2N_w i_j(t)}{\pi}\sum_\nu \frac{{}^\nu\xi_{p,1}}{\nu}\cos(\nu\alpha_S), \quad {}^\nu\xi_{p1} = \sin\left(\nu\frac{\pi}{Q_S}\right) \tag{4.3}$$

On the other hand, assuming constant magnetic permeability, the MMF distribution produced by a single conductor can be approximated by a sawtooth waveform, as indicated by the dashed curve in Figure 4.2b. This distribution may be represented by the following Fourier series:

$$F_{\text{go-in},j}(t,\alpha_S) = \frac{\Delta n_w i_j(t)}{\pi}\sum_\nu \frac{1}{\nu}\sin\left(\nu\left(\alpha_S - \frac{\pi}{Q_S}\right)\right) \tag{4.4}$$

The total MMF distribution of the asymmetric coil is obtained by applying the principle of superposition, yielding:

$$F_j(t,\alpha_S) = F_{\mathrm{coil},j}(t,\alpha_S) + F_{\text{go-in},j}(t,\alpha_S) \tag{4.5}$$

By substituting Equations (4.3) and (4.4) into (4.5), the final expression for the MMF function is obtained as:

$$\begin{aligned} F_j(t,\alpha_S) &= \frac{i_j(t)}{\pi}\left\{\sum_\nu \frac{2N_w {}^\nu\xi_{p,1}}{\nu}\cos(\nu\alpha_S) + \sum_\nu \frac{\Delta n_w}{\nu}\sin\left(\nu\left(\alpha_S - \frac{\pi}{Q_S}\right) + \pi\right)\right\} \\ &= \frac{2N_{w,\mathrm{eq}} i_j(t)}{\pi}\left\{\sum_\nu \frac{{}^\nu\xi_{p,1}}{\nu}\frac{N_w}{N_{w,\mathrm{eq}}}\sin\left(\nu\alpha_S + \frac{\pi}{2}\right) + \frac{1}{\nu}\frac{\Delta n_w}{2N_{w,\mathrm{eq}}}\sin\left(\nu\left(\alpha_S - \frac{\pi}{Q_S}\right) + \pi\right)\right\} \\ &= \frac{2N_{w,\mathrm{eq}} i_j(t)}{\pi}\sum_\nu \frac{{}^\nu\xi_{p,\mathrm{eq}}}{\nu}\sin(\nu\alpha_S + \varphi_{1,2}) \end{aligned} \tag{4.6}$$

where

$${}^\nu\xi_{p,\mathrm{eq}} = \frac{1}{N_{w,\mathrm{eq}}}\sqrt{\left(N_w {}^\nu\xi_{p,1}\right)^2 + \left(\frac{\Delta n_w}{2}\right)^2 + N_w \Delta n_w {}^\nu\xi_{p,1}\cos\left(\nu\frac{\pi}{Q_S} + \frac{\pi}{2}\right)} \tag{4.7}$$

$$\varphi_{1,2} = a\tan\left(\frac{2N_w - \Delta n_w \sin\left(v\frac{\pi}{Q_S}\right)}{-\Delta n_w \cos\left(v\frac{\pi}{Q_S}\right)}\right) \quad (4.8)$$

In Equation (4.7), the term ${}^{v}\xi_{p,\mathrm{eq}}$ represents the equivalent coil-pitch factor, which accounts for the impact of the asymmetric coil arrangement on the harmonic winding factors. The dependence of this factor on the number of turns N_w (with the coil-side parameters defined as $N_{w11} = N_w$ and $N_{w12} = N_w - 1$), as well as on the number of stator slots Q_S, is plotted in Figure 4.3. A more detailed examination of MMF modification introduced by asymmetric coils is provided in Figure 4.4. In this analysis:

- the reference coil is taken as a standard coil with $N_w = 2$
- while the asymmetric coil is defined by $N_{w11} = 2$ and $N_{w12} = 1$ $(\Delta n_w = 1)$.

This configuration represents an extreme case, namely a 50% turn imbalance between the coil sides, and as expected, produces a substantial deviation in the resulting MMF distribution

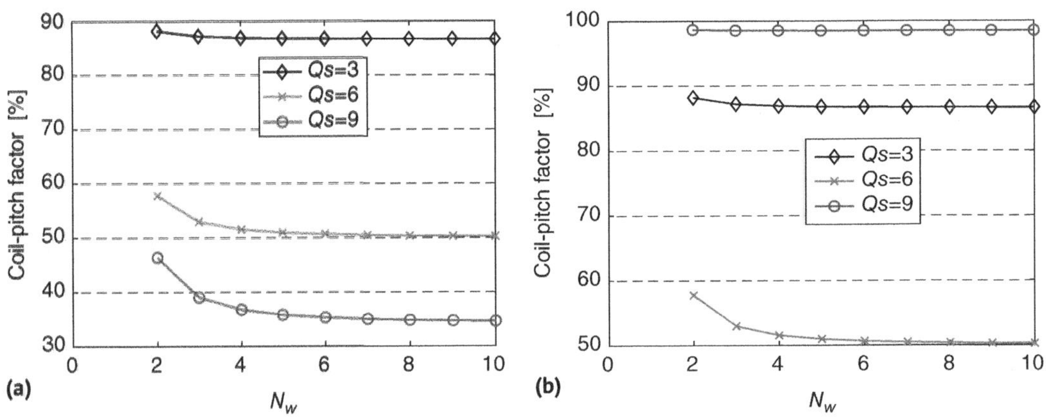

Figure 4.3 Effect of the number of turns N_w on the winding factor (coil-pitch factor) harmonics for various numbers of stator slots Q_S. (a) Winding factor of $v = 1$. (b) Winding factor of $v = 5$.

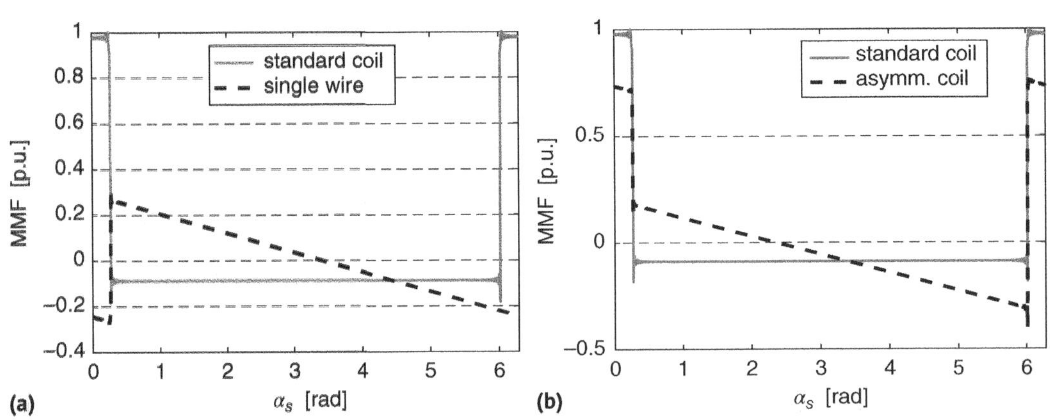

Figure 4.4 (a) Comparison of MMF distribution for a common coil and a single wire. (b) Comparison of MMF distribution for a common coil and an asymmetric coil.

relative to the symmetric case. However, as the number of turns increases, the ratio $(N_w - 1)/N_w$ approaches unity, and the MMF waveform of the asymmetric coil approaches that of the standard coil.

4.1.2 FSCWs with $z = 1$, $n_c > 1$, and Asymmetric Coils

The asymmetric coil concept may be applied to any double-layer concentrated winding topology, including FSCWs with one zone per phase and multiple coils per zone. In order to achieve an unequal number of turns per slot or, equivalently, a nonuniform current-linkage from slot to slot, the adjacent coils within a given zone must have a mirrored (i.e. reversed) turn distribution. Additionally, the number of coils per zone must be even (e.g. $n_c = 2, 4, 6, \ldots$). This arrangement is illustrated in Figure 4.5, which provides representative examples for $n_c = 2$, $n_c = 4$, and $n_c = 6$.

The MMF function for this winding category can be derived using the same superposition procedure outlined earlier, wherein the MMF contributions of the standard coils and the auxiliary

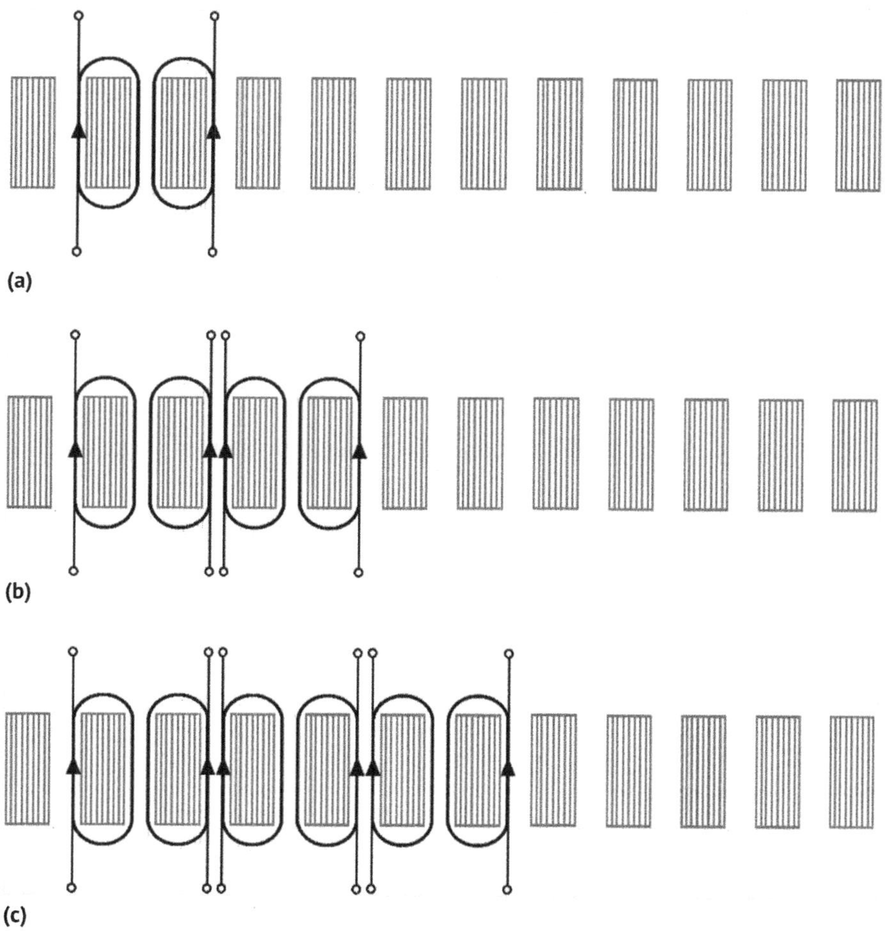

Figure 4.5 Winding topologies with one zone per phase and coils with unequal turns per coil side. (a) FSCW with $z = 1$ and $n_c = 2$. (b) FSCW with $z = 1$ and $n_c = 4$. (c) FSCW with $z = 1$ and $n_c = 6$.

single conductors are summed. Hence, for the winding system employing standard coils, the MMF is given directly by Equations (2.37) and (2.38):

$$\begin{aligned} F_{\text{conv},j}(t,\alpha_S) &= \frac{2N_w i_j(t)}{\pi}\sum_{\nu}\sum_{\kappa=0}^{n_c-1}\frac{{}^{\nu}\xi_{p,1}}{\nu}\cos\left(\nu\left(\alpha_S-\kappa\frac{2\pi}{Q_S}\right)-\kappa\pi\right) \\ &= \frac{2n_c N_w i_j(t)}{\pi}\cdot\sum_{\nu}\frac{{}^{\nu}\xi_{p,1}\,{}^{\nu}\xi_{d1}}{\nu}\cos\left(\nu\left(\alpha_S-(n_c-1)\frac{\pi}{Q_S}\right)-(n_c-1)\frac{\pi}{2}\right) \end{aligned} \tag{4.9}$$

$$\xi_{d,1} = \frac{\sin\left(\nu n_c\frac{\pi}{Q_S}+n_c\frac{\pi}{2}\right)}{n_c\cos\left(\nu\frac{\pi}{Q_S}\right)} \tag{4.10}$$

In contrast, for a winding system in which the "go-in" conductors are placed in every 2nd slot (i.e. slot = 2, 4, 6, ...), the associated MMF follows from the algebraic summation of the current-linkages corresponding to those slots. This leads to the expression:

$$F_{\text{go-in},j}(t,\alpha_S) = \frac{i_j(t)}{\pi}\left\{\sum_{\nu}\sum_{\kappa=0}^{n_c/2-1}\frac{2\Delta n_w}{\nu}\sin\left(\nu\left(\alpha_S-\kappa\frac{4\pi}{Q_S}-\frac{\pi}{Q_S}\right)\right)\right\} \tag{4.11}$$

where the factor "2" accounts for the presence of two single conductors per stator slot in each of the occupied (every second) slot.

Analogous to Equation (2.22), the summation of the sine terms over all slots containing single conductors may be written as:

$$\begin{aligned} \sum_{\kappa=0}^{n_c/2-1}\sin\left(\nu\left(\alpha_S-\kappa\frac{4\pi}{Q_S}-\frac{\pi}{Q_S}\right)\right) &= e^{j\nu\left(\alpha_S-\frac{\pi}{Q_S}\right)}\sum_{\kappa=0}^{n_c/2-1}e^{-j\nu\left(\kappa\frac{4\pi}{Q_S}\right)} \\ &= \frac{\sin\left(\nu n_c\frac{\pi}{Q_S}\right)}{\sin\left(\nu\frac{2\pi}{Q_S}\right)}e^{j\nu\left(\alpha_S-(n_c-1)\frac{\pi}{Q_S}-\pi\right)} \\ &= {}^{\nu}\xi_{d_2}\sin\left(\nu\left(\alpha_S-(n_c-1)\frac{\pi}{Q_S}-\pi\right)\right) \end{aligned} \tag{4.12}$$

From Equations (4.11) and (4.12), the MMF function for a winding consisting of single conductors becomes:

$$F_{\text{go-in},j}(t,\alpha_S) = \frac{2n_c\Delta n_w i_j(t)}{\pi}\cdot\sum_{\nu}\frac{{}^{\nu}\xi_{d,2}}{\nu}\sin\left(\nu\left(\alpha_S-(n_c-1)\frac{\pi}{Q_S}\right)\right) \tag{4.13}$$

where ${}^{\nu}\xi_{d,2}$ denotes the respective distribution factor given by:

$${}^{\nu}\xi_{d,2} = \frac{\sin\left(\nu n_c\frac{\pi}{Q_S}\right)}{n_c\sin\left(\nu\frac{2\pi}{Q_S}\right)} \tag{4.14}$$

Finally, the total MMF is obtained as:

$$\begin{aligned} F_j(t,\alpha_S) &= F_{\text{coil},j}(t,\alpha_S)+F_{\text{go-in},j}(t,\alpha_S) \\ &= \frac{2n_c N_{w,\text{eq}} i_j(t)}{\pi}\cdot\sum_{\nu}\frac{1}{\nu}\left\{\frac{N_w\,{}^{\nu}\xi_{p,1}\,{}^{\nu}\xi_{d,1}}{N_{w,\text{eq}}}\cos\left(\nu\left(\alpha_S-(n_c-1)\frac{\pi}{Q_S}\right)-n_c\frac{\pi}{2}+\frac{\pi}{2}\right)\right. \\ &\quad \left.+\frac{\Delta n_w\,{}^{\nu}\xi_{d,2}}{N_{w,\text{eq}}}\cos\left(\nu\left(\alpha_S-(n_c-1)\frac{\pi}{Q_S}\right)+\frac{\pi}{2}\right)\right\} \end{aligned} \tag{4.15}$$

which is simplified to:

$$F_j(t,\alpha_S) = \frac{2n_c N_{w,\text{eq}} i_j(t)}{\pi} \sum_\nu \frac{{}^\nu\xi_w}{\nu} \cos\left(\nu\left(\alpha_S - (n_c - 1)\frac{\pi}{Q_S}\right) + \frac{\pi}{2}\right) \tag{4.16}$$

$${}^\nu\xi_w = \frac{1}{N_{w,\text{eq}}}\left(N_w {}^\nu\xi_{w,1} \cos\left(n_c\frac{\pi}{2}\right) + \Delta n_w {}^\nu\xi_{w,2}\right), \quad n_c = 2,4,6,\ldots \tag{4.17}$$

where

$${}^\nu\xi_{w,1} = {}^\nu\xi_{p,1}\,{}^\nu\xi_{d,1}, \quad {}^\nu\xi_{w,2} = {}^\nu\xi_{d,2}$$

represents the combined coil-pitch and distribution factors of respective winding sets.

4.1.3 FSCWs with $z = 2, n_c > 1$, and Asymmetric Coils

The winding topology with only one zone per phase and with asymmetrical coils is primarily of theoretical interest. In this configuration, the position of coil terminals is unfavorable, which makes practical implementation difficult. In particular, a series connection of coils within the same zone is not feasible when the input terminals of adjacent coils are located on opposite sides of the machine. However, when the number of zones per phase is two, this winding arrangement becomes practically feasible, allowing straightforward coil connections.

Figure 4.6 illustrates the coil connections (within one zone and between zones) for a conventional FSCW and an FSCW with an asymmetric coil arrangement, both featuring two zones and two coils per zone. As shown, a series connection of coils is also possible in an asymmetric coil configuration when the number of zones is two. In this case, the coil connections are located on both sides of the machine, unlike those in the conventional winding, where they are only placed on one side of the machine.

The MMF distribution corresponding to the configuration in Figure 4.6b, with $z = 2$ and $n_c = 2$, is presented in Figure 4.7. Here, N_{w11} and N_{w12} denote the number of turns on the first and second

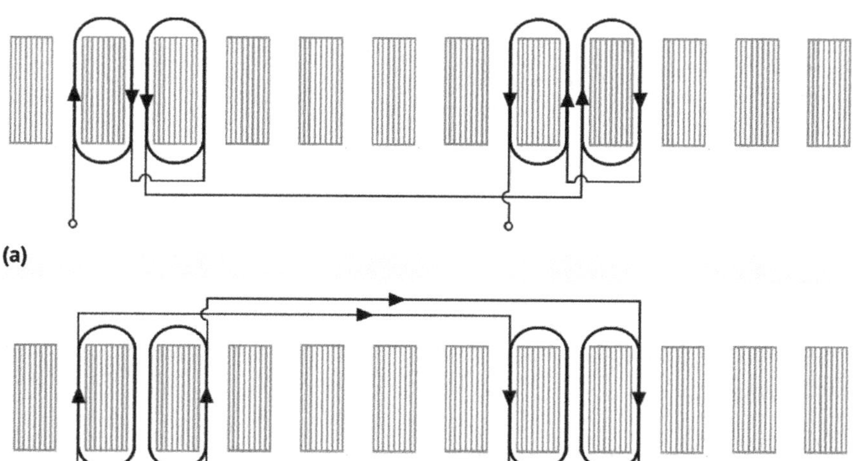

(a)

(b)

Figure 4.6 Winding topologies with $z = 2$ and $n_c = 2$. (a) Conventional FSCW. (b) FSCW with unequal turns per coil side.

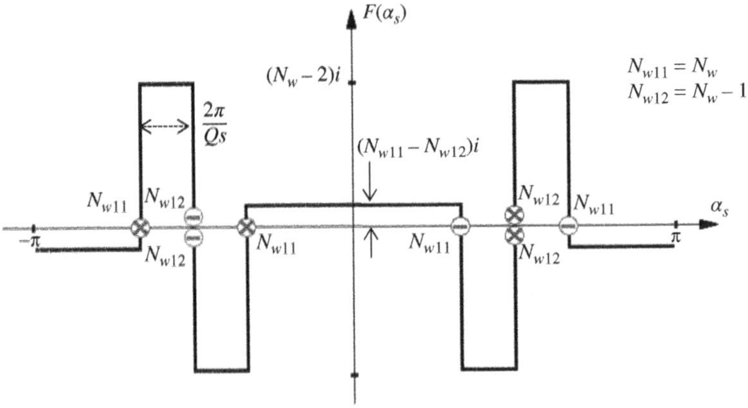

Figure 4.7 MMF distribution of phase-j for the FSCW with $z = 2$, $n_c = 2$, and coil arrangement with unequal turns per coil side.

coil sides, respectively. Analytically, this winding type can be described using the MMF functions derived in the preceding sections, with the winding-zone-related factors from Equation (2.31) incorporated. The resulting MMF contributions from the winding with standard coils and the winding with single conductors are given by:

$$F_{\text{conv},j}(t,\alpha_S) = \frac{2zn_cN_wi_j(t)}{\pi} \cdot \sum_{\nu} \frac{{}^{\nu}\xi_{p,1}{}^{\nu}\xi_{d,1}{}^{\nu}\xi_{z,1}}{\nu} \sin\left(\nu\left(\alpha_S - (n_c-1)\frac{\pi}{Q_S}\right) - (n_c-1)\frac{\pi}{2}\right) \tag{4.18}$$

$$F_{\text{go-in},j}(t,\alpha_S) = \frac{2zn_c\Delta n_wi_j(t)}{\pi} \cdot \sum_{\nu} \frac{{}^{\nu}\xi_{d,2}{}^{\nu}\xi_{z,2}}{\nu} \cos\left(\nu\left(\alpha_S - (n_c-1)\frac{\pi}{Q_S}\right)\right) \tag{4.19}$$

Consequently, the total MMF function can be obtained as:

$$\begin{aligned} F_j(t,\alpha_S) = \frac{2zn_cN_{w,\text{eq}}i_j(t)}{\pi} \cdot \Bigg\{ &\sum_{\nu} \frac{1}{\nu} \frac{N_w{}^{\nu}\xi_{w,1}}{N_{w,\text{eq}}} \cos\left(\nu\left(\alpha_S - (n_c-1)\frac{\pi}{Q_S}\right) - n_c\frac{\pi}{2}\right) \\ &+ \sum_{\nu} \frac{1}{\nu} \frac{\Delta n_w{}^{\nu}\xi_{w,2}}{N_{w,\text{eq}}} \cos\left(\nu\left(\alpha_S - (n_c-1)\frac{\pi}{Q_S}\right)\right) \Bigg\} \end{aligned} \tag{4.20}$$

where

$$ {}^{\nu}\xi_{w,1} = {}^{\nu}\xi_{p,1}{}^{\nu}\xi_{d,1}{}^{\nu}\xi_{z,1}$$
$$ {}^{\nu}\xi_{w,2} = {}^{\nu}\xi_{d,2}{}^{\nu}\xi_{z,2}$$
$$ {}^{\nu}\xi_{z,1} = {}^{\nu}\xi_{z,2} = \sin\left(\nu\frac{\pi}{2}\right)$$

whereas the equivalent number of turns $N_{w,\text{eq}}$ is already defined in Equation (4.2).

Solving Equation (4.20) yields:

$$F_j(t,\alpha_S) = \frac{2zn_cN_{w,\text{eq}}i_j(t)}{\pi} \sum_{\nu} \frac{{}^{\nu}\xi_w}{\nu} \cos\left(\nu\left(\alpha_S - (n_c-1)\frac{\pi}{Q_S}\right)\right) \tag{4.21}$$

from which the corresponding winding factor for this winding type is:

$${}^{\nu}\xi_w = \frac{1}{N_{w,\text{eq}}}\left(N_w{}^{\nu}\xi_{w,1}\cos\left(\left(\frac{n_c}{2}-1\right)\pi\right) + \Delta n_w{}^{\nu}\xi_{w,2}\right), \quad n_c = 2,4,6,\ldots \tag{4.22}$$

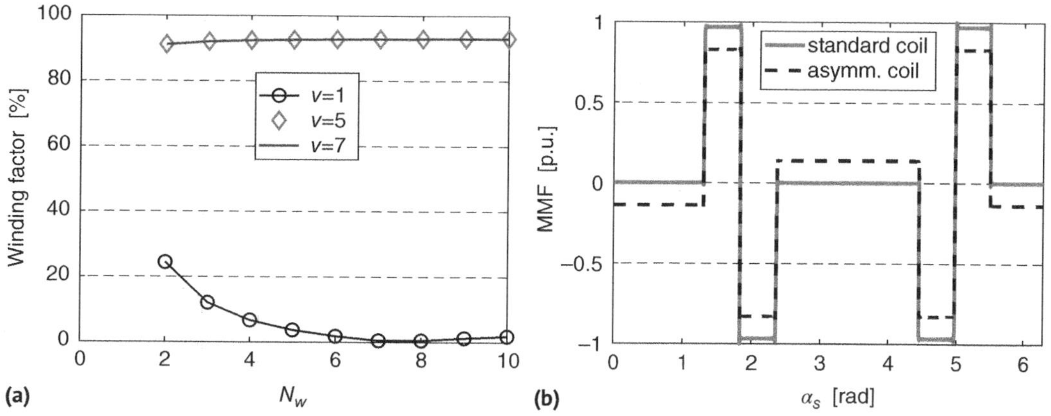

Figure 4.8 (a) The effect of N_w on the winding factor harmonics for the FSCW with $z = 2$ and $n_c = 2$. (b) MMF comparison for a common coil with $N_w = 7$ and an asymmetric coil with $N_{w11} = 7$ and $N_{w12} = 6$.

Equation (4.22) explicitly shows the influence of the asymmetric coil structure through the parameter Δn_w, on the harmonic winding factors. When $\Delta n_w = 0$, the winding is transformed to a conventional configuration with standard coils, and the corresponding winding factor simplifies to:

$$^{v}\xi_w\big|_{\Delta n_w=0} = {^{v}\xi_{w,1}} \tag{4.23}$$

As shown in Section 2.5.4 (see Figure 2.32), a typical FSCW with $z = 2$ and $n_c = 2$ exhibits a pronounced 1st subharmonic when either the 5th or the 7th harmonic is selected as the operating wave. This subharmonic component can adversely affect the electromagnetic performance of the machine. By replacing the standard coils with the asymmetric coils, the MMF harmonic spectrum can be selectively modified. This effect is investigated in Figure 4.8. Figure 4.8a presents the winding factors of the 1st, 5th, and 7th harmonic components as a function of the number of turns N_w, where the second coil side contains $(N_w - 1)$ turns. The results indicate that for suitable values of N_w, such as $N_w = 7$ or $N_w = 8$, the 1st subharmonic component can be effectively suppressed while the desired operating harmonic remains essentially unaffected.

Further investigations, shown in Figure 4.8b, compare the resulting MMF distribution of a conventional winding and an asymmetric-coil winding for the optimal case of $N_w = 7$. In the asymmetric coil, a slight shift of the average MMF level from the zero axis is observed, reflecting the compensating effect introduced by the asymmetric turn distribution, which leads to a reduction of the subharmonic component. This analysis proves that the asymmetric coil configuration is an effective means of adjusting the MMF distribution, thereby enhancing the electromagnetic properties of the winding.

4.2 FSCWs with Unequal Turns per Coil

As demonstrated in the previous section, the asymmetric coil concept provides a simple and effective method for mitigating MMF subharmonics in FSCWs. However, its applicability is limited by practical considerations, in particular, this approach can only be implemented when two zones per phase are available and when each zone contains an even number of coils. As an alternative method, MMF subharmonics in concentrated windings can be influenced (reduced) by employing zones with different number of turns per coil. This concept was first introduced by Heller and Kauders [5] to compensate the current-linkages in the slots containing coils of different phases in distributed windings. With the resurgence of interest in concentrated windings, the same principle was later extended to multilayers FSCWs to enhance their MMF characteristics, as reported by Kometani et al. [50], Ito et al. [41], and Cistelecan et al. [42]. Few years later, a generalized analytical approach for FSCWs with different numbers of turns per coil was subsequently developed by Dotz and Gerling [53, 54].

4.2.1 FSCWs with $z = 1, n_c > 1$, and Unequal Turns per Coil

A typical one-zone FSCW with two coils per zone and unequal turns per coil is illustrated in Figure 4.9a, where N_{w1} and N_{w2} denote the number of turns in the 1st and the 2nd coil, respectively. When the number of coils per zone exceeds two, the winding pattern simply repeats as illustrated in Figure 4.9b,c.

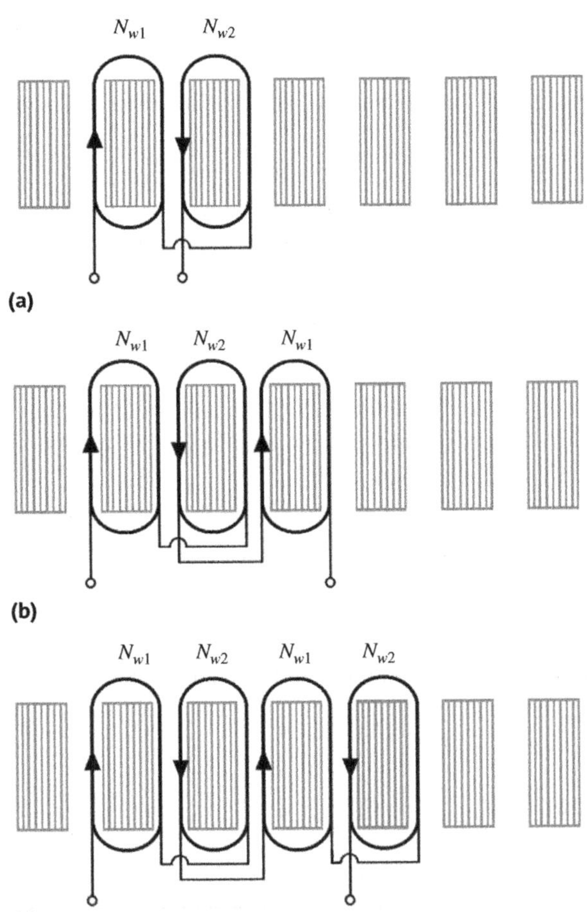

Figure 4.9 Winding topologies with one zone per phase and unequal turns per coil. (a) FSCW with $z = 1$ and $n_c = 2$. (b) FSCW with $z = 1$ and $n_c = 3$. (c) FSCW with $z = 1$ and $n_c = 4$.

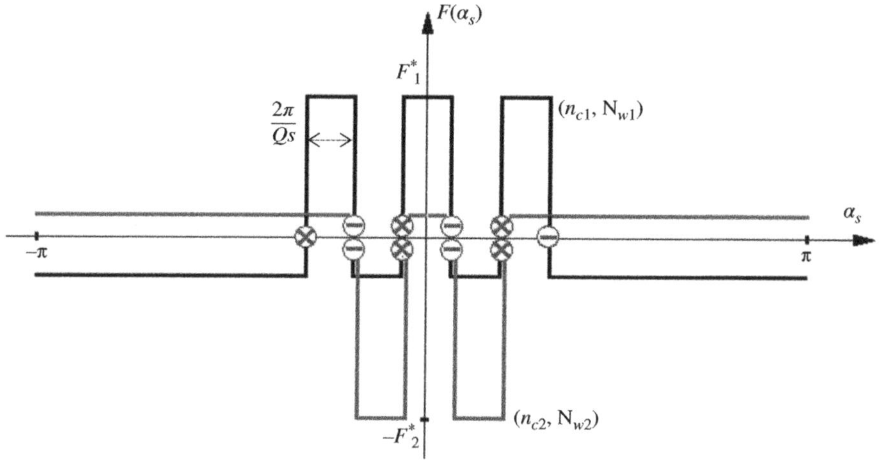

Figure 4.10 MMF functions of FSCW subzones when $z = 1$ and $n_c = 5$.

Let n_{c1} and n_{c2} denote the number of coils with N_{w1} and N_{w2} turns, respectively, the total number of coils per zone is therefore:

$$n_c = n_{c1} + n_{c2} \tag{4.24}$$

and the total number of turns per zone becomes:

$$N_{w,z} = n_{c1}N_{w1} + n_{c2}N_{w2} \tag{4.25}$$

Figure 4.10 shows the corresponding MMF distribution for an example FSCW with five coils per zone, consisting of three coils with N_{w1} turns and two coils with N_{w2} turns. From this plot, the total MMF can be determined by summing the contributions of the two "subzones," each of which behaves as a single-layer winding with n_{c1} and n_{c2} coils per zone. Using the analytical expressions developed in Chapter 2, the MMF contributions of the two subzones characterized by n_{c1}, N_{w1}, and n_{c2}, N_{w2} winding parameters can be written as:

$$\begin{aligned} F_{j,N_{w1}}(t,\alpha_S) &= \frac{2N_{w1}i_j(t)}{\pi}\sum_{\nu}\sum_{\kappa=0}^{n_{c1}-1}\frac{{}^{\nu}\xi_{p,1}}{\nu}\cos\left(\nu\left(\alpha_S-\kappa\frac{4\pi}{Q_S}\right)\right) \\ &= \frac{2n_{c1}N_{w1}i_j(t)}{\pi}\cdot\sum_{\nu}\frac{{}^{\nu}\xi_{p,1}\,{}^{\nu}\xi_{d,1}}{\nu}\cos\left(\nu\left(\alpha_S-(n_{c1}-1)\frac{2\pi}{Q_S}\right)\right) \end{aligned} \tag{4.26}$$

$$\begin{aligned} F_{j,N_{w2}}(t,\alpha_S) &= \frac{2N_{w2}i_j(t)}{\pi}\sum_{\nu}\sum_{\kappa=0}^{n_{c2}-1}\frac{{}^{\nu}\xi_{p,2}}{\nu}\cos\left(\nu\left(\alpha_S-\kappa\frac{4\pi}{Q_S}\right)-\frac{2\pi}{Q_S}\right) \\ &= \frac{2n_{c2}N_{w2}i_j(t)}{\pi}\cdot\sum_{\nu}\frac{{}^{\nu}\xi_{p,2}\,{}^{\nu}\xi_{d,2}}{\nu}\cos\left(\nu\left(\alpha_S-n_{c2}\frac{2\pi}{Q_S}\right)-\pi\right) \end{aligned} \tag{4.27}$$

The total MMF follows directly from superposition of (4.26) and (4.27):

$$\begin{aligned} F_j(t,\alpha_S) = \frac{2N_{w,z}i_j(t)}{\pi}\cdot &\left\{\sum_{\nu}\frac{n_{c1}N_{w1}{}^{\nu}\xi_{w,1}}{\nu\cdot N_{w,z}}\cos\left(\nu\left(\alpha_S-(n_{c1}-1)\frac{2\pi}{Q_S}\right)\right)\right. \\ &\left.+\sum_{\nu}\frac{n_{c2}N_{w2}{}^{\nu}\xi_{w,2}}{\nu\cdot N_{w,z}}\cos\left(\nu\left(\alpha_S-n_{c2}\frac{2\pi}{Q_S}\right)-\pi\right)\right\} \end{aligned} \tag{4.28}$$

Here, ${}^{\nu}\xi_{w,1}$ and ${}^{\nu}\xi_{w,2}$ denote the respective subzone winding factors:

$$ {}^{\nu}\xi_{w,1} = {}^{\nu}\xi_{p,1}\,{}^{\nu}\xi_{d,1},\ {}^{\nu}\xi_{w,2} = {}^{\nu}\xi_{p,2}\,{}^{\nu}\xi_{d,2} $$

The coil-pitch factors of both subwindings are equal:

$$ {}^{\nu}\xi_{p,1} = {}^{\nu}\xi_{p,2} = \sin\left(\nu\frac{\pi}{Q_S}\right) \tag{4.29} $$

whereas their distribution factors may be either identical or different, depending on the number of coils per subzone:

$$ {}^{\nu}\xi_{d,1} = \frac{\sin\left(\nu n_{c1}\frac{2\pi}{Q_S}\right)}{n_{c1}\sin\left(\nu\frac{2\pi}{Q_S}\right)} \tag{4.30} $$

$$ {}^{\nu}\xi_{d,2} = \frac{\sin\left(\nu n_{c2}\frac{2\pi}{Q_S}\right)}{n_{c2}\sin\left(\nu\frac{2\pi}{Q_S}\right)} \tag{4.31} $$

After simplifying Equation (4.28), the final MMF function for a one-zone FSCW with unequal turns per coil becomes:

$$ F_j(t,\alpha_S) = \frac{2N_{w,z}i_j(t)}{\pi}\sum_{\nu}\frac{{}^{\nu}\xi_w}{\nu}\cos\left(\nu\alpha_S - \varphi_{1,2}\right) \tag{4.32} $$

while the resulting winding factor ${}^{\nu}\xi_w$ and phase angle $\varphi_{1,2}$ are:

$$ {}^{\nu}\xi_w = \frac{1}{N_{w,z}}\sqrt{\left(n_{c1}N_{w1}{}^{\nu}\xi_{w,1}\right)^2 + \left(n_{c2}N_{w2}{}^{\nu}\xi_{w,2}\right)^2 - 2n_{c1}n_{c2}N_{w1}N_{w2}{}^{\nu}\xi_{w,1}{}^{\nu}\xi_{w,2}\cos\left(\nu(n_{c2}-n_{c1}+1)\frac{2\pi}{Q_S}\right)} \tag{4.33} $$

$$ \varphi_{1,2} = \mathrm{a}\tan\left(\frac{n_{c1}N_{w1}{}^{\nu}\xi_{w,1}\sin\left(\nu(n_{c1}-1)\frac{2\pi}{Q_S}\right) - n_{c2}N_{w2}{}^{\nu}\xi_{w,2}\sin\left(\nu n_{c1}\frac{2\pi}{Q_S}\right)}{n_{c1}N_{w1}{}^{\nu}\xi_{w,1}\cos\left(\nu(n_{c1}-1)\frac{2\pi}{Q_S}\right) - n_{c2}N_{w2}{}^{\nu}\xi_{w,2}\cos\left(\nu n_{c1}\frac{2\pi}{Q_S}\right)}\right) \tag{4.34} $$

As implied by Equation (4.24), the combination of n_{c1} and n_{c2} determines the total number of coils per zone, and therefore defines the winding type for a double-layer one-zone FSCWs. The analysis in Chapter 2 has shown that the parameter n_c has a direct impact on the MMF function and its harmonic content. However, by varying the number of turns per coil and selecting an optimal combination of N_{w1} and N_{w2}, certain MMF harmonics can be influenced, i.e. optimized, as indicated in Equation (4.33).

To validate the effectiveness of this concept, the unequal turns per coil technique is applied to a common FSCW with $z = 1$ and $n_c = 3$, whose MMF distribution and harmonic content were previously analyzed in Figure 2.29. The corresponding winding layout employing unequal coils is presented in Figure 4.9b. In this configuration, the winding parameters are $n_{c1} = 2$ and $n_{c2} = 1$, while the number of turns N_{w1} and N_{w2} are considered as design variables. Initially, the winding factors for the 1st and the 2nd subharmonics, as well as the 4th harmonic (working harmonic), are evaluated as functions of the turn ratio N_{w1}/N_{w2}. The results, presented in Figure 4.11a, indicate a strong dependence of the subharmonic components on the turns ratio. In particular, the 1st harmonic is completely eliminated when N_{w1} is approximately 60% of N_{w2} while the working harmonic remains largely unaffected. The resulting phase MMF distribution corresponding to this

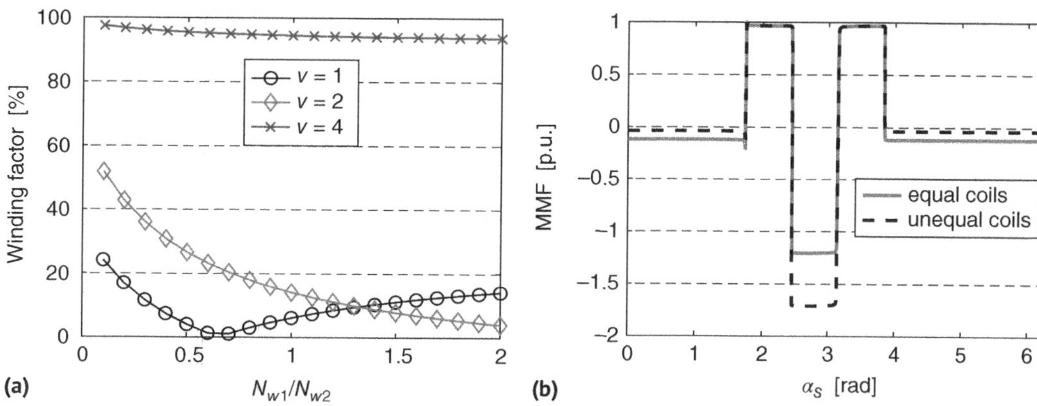

Figure 4.11 (a) Effect of N_{w1}/N_{w2} on the winding factor harmonics for the FSCW with $z = 1$ and $n_c = 3$. (b) MMF distribution of one-phase for the standard and new windings with $N_{w1} = 0.6 \times N_{w2}$ turn ratio.

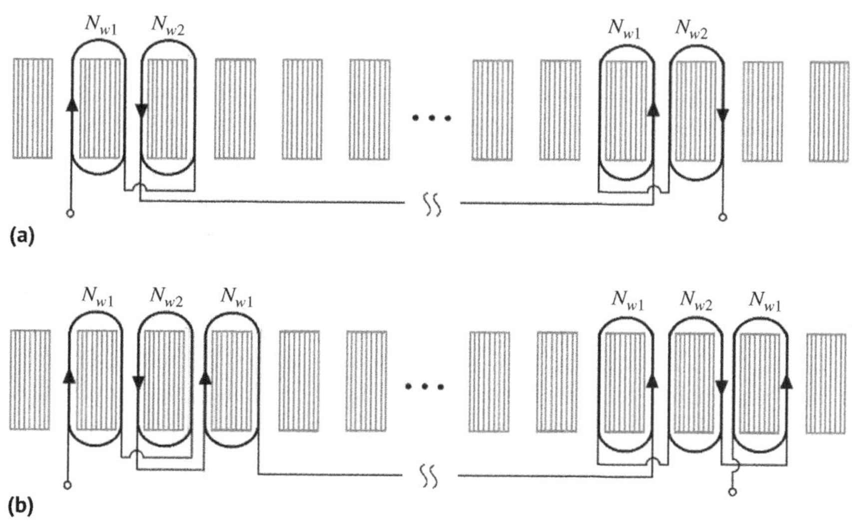

Figure 4.12 Winding topologies with unequal turns per coil. (a) FSCW with $z = 2$ and $n_c = 2$. (b) FSCW with $z = 2$ and $n_c = 3$.

optimal turn ratio is shown in Figure 4.11b. From this example, it can be concluded that employing different numbers of turns per coil provides a flexible and powerful way for subharmonic suppression and MMF waveform modification in FSCWs.

4.2.2 FSCWs with $z = 2$, $n_c > 1$, and Unequal Turns per Coil

As shown previously, the use of unequal turns per coil concept provides an additional degree of freedom for modifying the MMF spectrum in one-zone FSCWs. Of course, this approach can also be easily extended to two-zone windings, as illustrated in Figure 4.12 for two distinct topologies. The analytical procedure follows that of one-zone windings, however in this case, the zone factor introduced in Equation (2.31) must also be included into the calculation of the winding factors for the respective winding sets. Accordingly, the MMF function for a two-zone configuration obtained by applying Equation (4.32) to each set and summing their contributions becomes:

$$F_j(t, \alpha_S) = \frac{2zN_{w,z}i_j(t)}{\pi} \sum_{\nu} \frac{{}^{\nu}\xi_w}{\nu} \sin(\nu\alpha_S - \varphi_{1,2}) \tag{4.35}$$

The corresponding winding factor ${}^{\nu}\xi_w$ and phase angle $\varphi_{1,2}$ follow in the same manner as Equations (4.33) and (4.34), with corresponding winding factors given by:

$${}^{\nu}\xi_{w,1} = {}^{\nu}\xi_{p,1}\,{}^{\nu}\xi_{d,1}\,{}^{\nu}\xi_{z,1},\ {}^{\nu}\xi_{w,2} = {}^{\nu}\xi_{p,2}\,{}^{\nu}\xi_{d,2}\,{}^{\nu}\xi_{z,2}$$

The coil-pitch factor and distribution factors remain as defined in Equations (4.29) to (4.31), while the winding sets zone factors, according to (2.31), are:

$${}^{\nu}\xi_{z,1} = {}^{\nu}\xi_{z,2} = \sin\left(\nu\frac{\pi}{2}\right)$$

The influence of unequal turns per coil approach in the two-zone FSCWs is examined by considering two representative cases. The first winding is an FSCW with $z = 2$ and $n_c = 2$, where the winding factor behavior for some specific harmonics as a function of N_{w1}/N_{w2} ratio are shown in Figure 4.13a. In this configuration, the 1st harmonic acts as a subharmonic, whereas the 5th or the 7th harmonic typically serves as the working harmonic. The results indicate that modifying the number of turns per respective coils has no positive effect on the subharmonic content. In fact, the subharmonic reaches its minimum when $N_{w1}/N_{w2} = 1$, corresponding to the standard winding condition. Therefore, for this topology, or in general for two-zone FSCWs with an even number of coils per zone ($n_c =$ even), no MMF improvement can be obtained through the use of this technique.

A different behavior is observed when the number of coils per zone is odd, as shown in Figure 4.13b. In this case, both the subharmonic components and the operating harmonic are strongly affected by the coil-turn ratio. For example, compared to the standard configuration (i.e. the case with $N_{w1}/N_{w2} = 1$), setting $N_{w1}/N_{w2} = 0.5$ eliminates the 1st harmonic entirely while at the same time enhancing the 7th harmonic (working harmonic). However, this improvement comes at the expense of an increase in the 2nd subharmonic (5th harmonic), which may adversely affect machine performance.

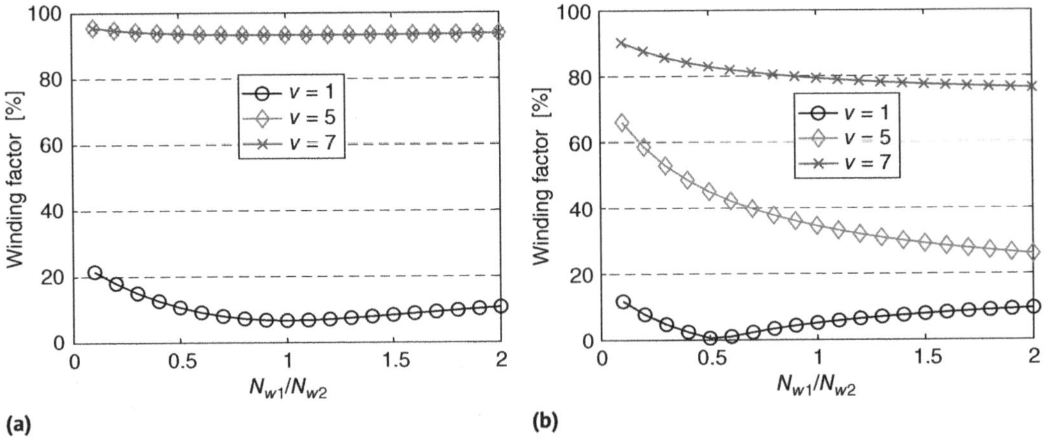

Figure 4.13 Effect of N_{w1}/N_{w2} on the winding factor harmonics for: (a) FSCW with $z = 2$ and $n_c = 2$. (b) FSCW with $z = 2$ and $n_c = 3$.

From these investigations, it may be concluded that the use of the technique with unequal turns per coil offers limited benefits for two-zone DL windings. Nevertheless, as noted by Ito et al. [41] and Dajaku and Gerling [51, 52], this technique can provide substantial improvement in multi-layer FSCW or in two-tooth FSWs, where additional degrees of freedom exist.

4.3 FSCWs with Unequal Coil-pitch

In general, standard concentrated windings as well as all previously studied specific winding configurations are characterized by coils with equal coil-pitch. Therefore, coil-pitch variation may represent an additional design parameter that can be used to further refine the MMF spectrum of FSCWs. Figure 4.14 illustrates two winding arrangements incorporating alternating coil pitches, which is implemented for both equal turns and unequal turns per coil configurations (i.e. $N_{w1} = N_{w2}$ or $N_{w1} \neq N_{w2}$).

Let $\alpha_{\tau1}$ and $\alpha_{\tau2}$ denote the coil-pitch angles of alternating coils. The pitch of two adjacent coils can be expressed as:

$$2\alpha_\tau = \alpha_{\tau1} + \alpha_{\tau2} \tag{4.36}$$

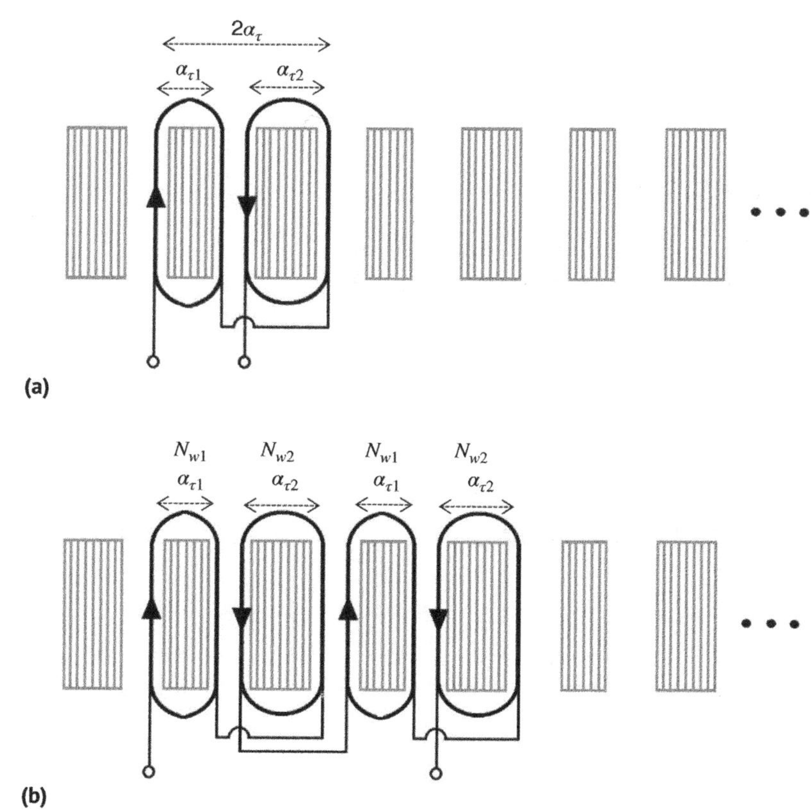

Figure 4.14 Concentrated winding with unequal coil-pitch configuration. (a) FSCW with $z = 1$ and $n_c = 2$. (b) FSCW with $z = 1$ and $n_c = 4$.

whereas analogous to standard stator, the equivalent coil-pitch angle is:

$$\alpha_\tau = \frac{2\pi}{Q_S}$$

4.3.1 MMF Approach

Considering a winding with $z = 1$ and $n_c = 4$, it is assumed that:

- the first subzone consists of n_{c1} coils with N_{w1} turns per coil and $\alpha_{\tau 1}$ coil-pitch angle
- the second subzone consists of n_{c2} coils with N_{w2} turns per coil and $\alpha_{\tau 2}$ coil-pitch angle.

The MMF distribution for the corresponding winding subzones is plotted in Figure 4.15. As in previous cases, the resulting MMF is found through the superposition method. The analytical expressions are identical in form to (4.32)–(4.34), with the only difference being in the respective coil-pitch factors:

$$^{\nu}\xi_{p,1} = \sin\left(\nu\frac{\alpha_{\tau 1}}{2}\right) \tag{4.37}$$

$$^{\nu}\xi_{p,2} = \sin\left(\nu\frac{\alpha_{\tau 2}}{2}\right) \tag{4.38}$$

If the number of zones per phase is two, the analytical model is similar to Equation (4.35), with the additional consideration of the zone factors.

To systematically study the effect of coil-pitch across different winding types, the coil-pitch parameters are expressed as a design variable:

$$\alpha_{\tau 1} = (1 - k_{\alpha,\tau})\frac{2\pi}{Q_S}$$

$$\alpha_{\tau 2} = (1 + k_{\alpha,\tau})\frac{2\pi}{Q_S}$$

where the coil-pitch variation factor $(k_{\alpha,\tau})$ is defined as:

$$-50\%\frac{2\pi}{Q_S} < k_{\alpha,\tau} < 50\%\frac{2\pi}{Q_S}$$

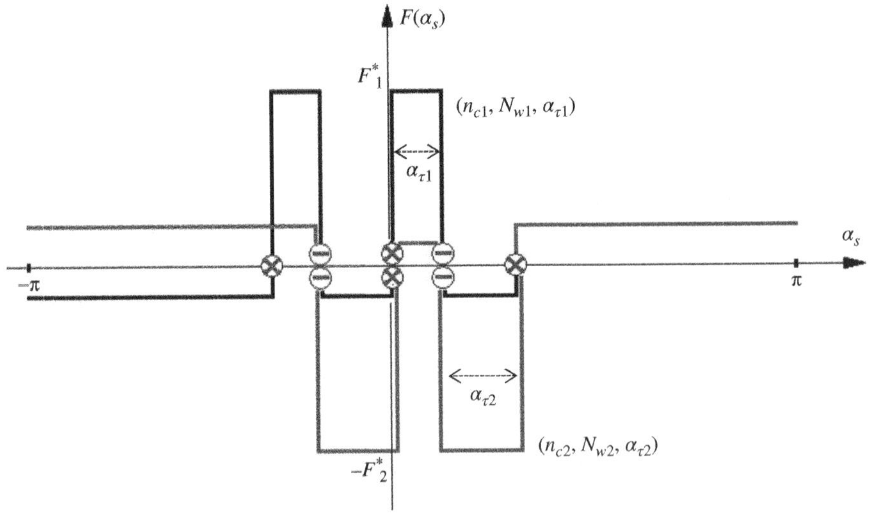

Figure 4.15 MMF function for an FSCW with $z = 1$, $n_c = 4$, and with unequal coil-pitches.

4.3.2 Coil-pitch Effect

For a standard winding with $z = 1$ and $n_c = 3$, the following assumptions are made:

- the two outer coils have a coil-pitch of $\alpha_{\tau 1}$
- the middle coil has a coil-pitch of $\alpha_{\tau 2}$.

For $k_{\alpha,\tau} = 0.3$, the MMF distributions corresponding to the standard winding and the configuration with unequal coil-pitch are shown in Figure 4.16a. Since the 1st and 2nd harmonics are subharmonic components, and the 4th harmonic represents the working harmonic, their winding factors are plotted in Figure 4.16b as a function of the variable coil-pitch parameter. When $k_{\alpha,\tau} = 0$, the configuration is equal to a conventional winding, and the resulting winding factors coincide with those of the standard winding. However, by appropriately selecting the coil-pitch relationship, certain harmonic components can be selectively attenuated or completely eliminated.

As a 2nd example, an FSCW with $z = 2$ and $n_c = 2$ is considered. In this case, it is assumed that:

- the first coil of the zone has a coil-pitch $\alpha_{\tau 1}$
- the second coil of the zone has a coil-pitch $\alpha_{\tau 2}$.

The MMF distribution for both the standard winding and the configuration with variable coil-pitch (with $k_{\alpha,\tau} = 0.3$) is compared in Figure 4.17a. Since the 1st harmonic is a subharmonic

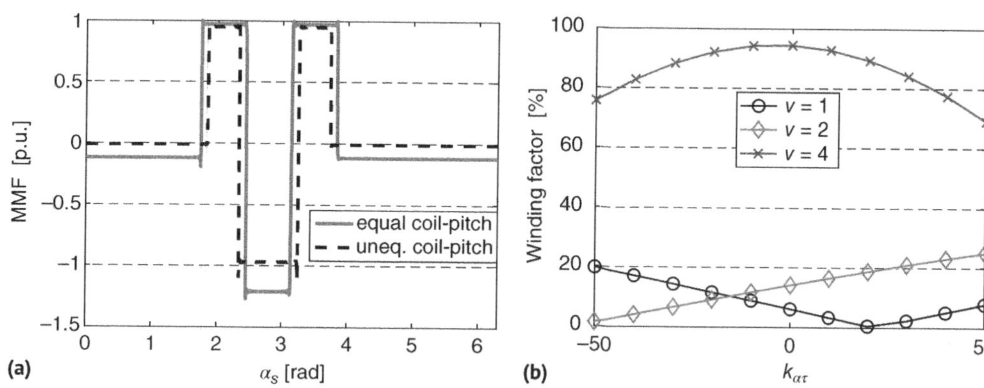

Figure 4.16 (a) Comparison of MMF distribution for an FSCW with $z = 1$ and $n_c = 3$. (b) Coil-pitch effect on the MMF harmonics for the winding with an unequal coil-pitch configuration.

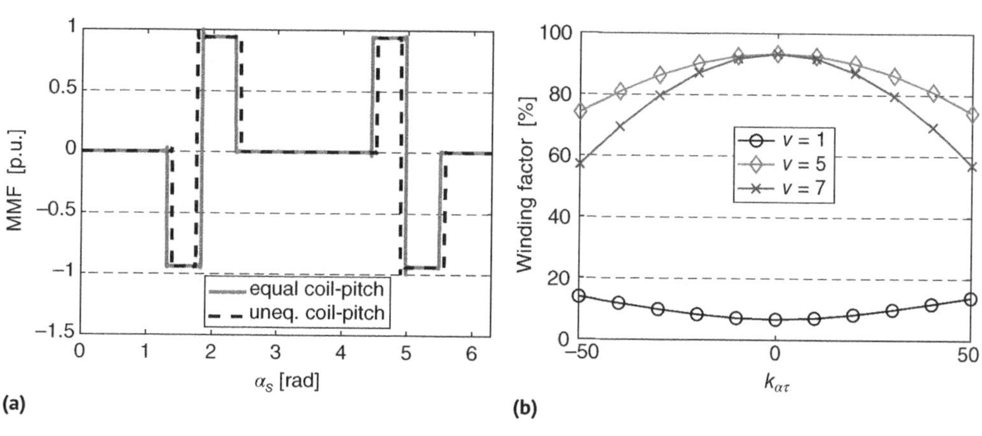

Figure 4.17 (a) Comparison of MMF distribution for an FSCW with $z = 2$ and $n_c = 2$. (b) Coil-pitch effect on the MMF harmonics for the winding with an unequal coil-pitch configuration.

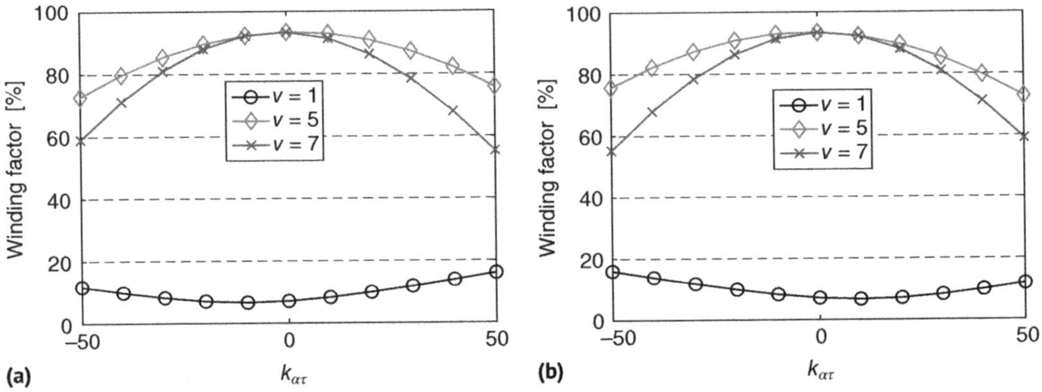

Figure 4.18 Coil-pitch effect on the MMF harmonics for the winding with various turns per coil. (a) Example-1 with $N_{w1} = 1, N_{w2} = 1.2$. (b) Example-2 with $N_{w1} = 1.2, N_{w2} = 1$.

component and the 5th or 7th harmonic is typically used as the working harmonic, the coil-pitch effect on these harmonic components is investigated in Figure 4.17b. The results indicate that optimal winding characteristics occur at $k_{\alpha,\tau} = 0$, thus the equal coil-pitch configuration remains the preferred choice in this case.

Finally, the coil-pitch variation technique can be combined with the unequal-turn concept (i.e. $N_{w1} \neq N_{w2}$). An example winding incorporating both methods is investigated in Figure 4.18, demonstrating that additional harmonic optimization potential is obtained when both design freedoms are used simultaneously.

4.4 Exemplary Windings with Special Coil Configurations

Two common FSCW topologies are examined below using various coil configuration techniques:

- 12-slot/10-pole FSCW; optimized using the "asymmetric coil" method
- 9-slot/8-pole FSCW; optimized using the "unequal-turn per coil" and the "unequal coil-pitch" method.

Example 4.1 12-slot/10-pole FSCW with asymmetric coils

The winding layout of the 12-slot/10-pole DL-FSCW with $z = 2$ and $n_c = 2$, and with asymmetric coils, is illustrated in Figure 4.19a. Here, N_{w11} represents the number of turns per coil-side located in slots shared by different phases, and N_{w12} is the number of turns in the opposite coil-side, placed in slots carrying coils of the same phase. Selecting a turn ratio of $N_{w11}/N_{w12} = 7/6$, the resulting MMF characteristics for both the standard and asymmetric coils are compared in Figure 4.19b. The results indicate that the subharmonic component is completely eliminated through an appropriate selection of the turns ratio, while the working harmonic (5th or 7th harmonic) remains essentially unchanged. Consequently, this approach provides an effective means of mitigating parasitic harmonics without degrading the torque capability of the machine.

Example 4.2 9-slot/8-pole FSCW with unequal turns per coil

The second example considers the well-known 9-slot/8-pole DL-FSCW with $z = 1$ and $n_c = 3$, which has been modified to include coils with unequal turns, as illustrated in Figure 4.20a. In this design, the two outer coils contain N_{w1} turns, the middle coil has N_{w2} turns, and the turn ratio is

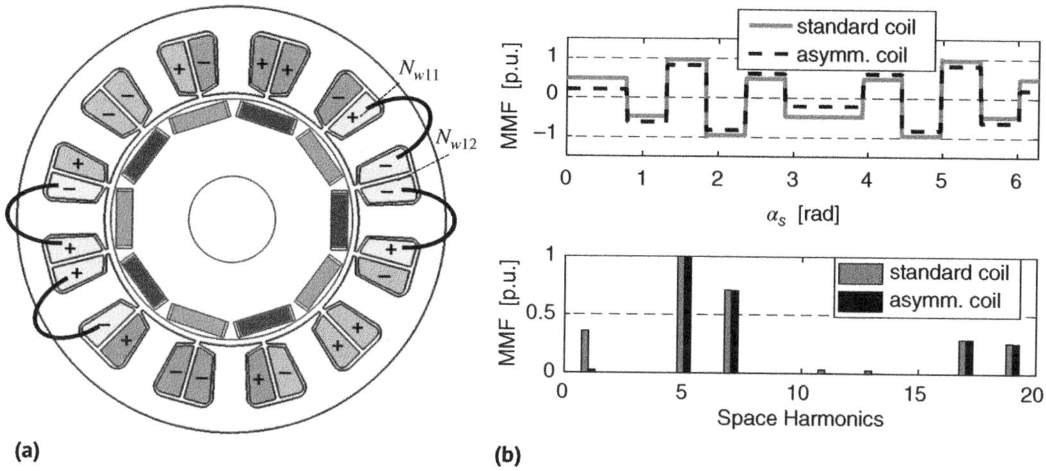

Figure 4.19 DL-FSCW with $z = 2$, $n_c = 2$, $Q_S = 12$, and with asymmetric coils. (a) Machine design with the winding layout. (b) Three-phase spatial MMF distribution and harmonic contents for $N_{w11} = 7$, $N_{w12} = 6$, and comparison with the standard winding.

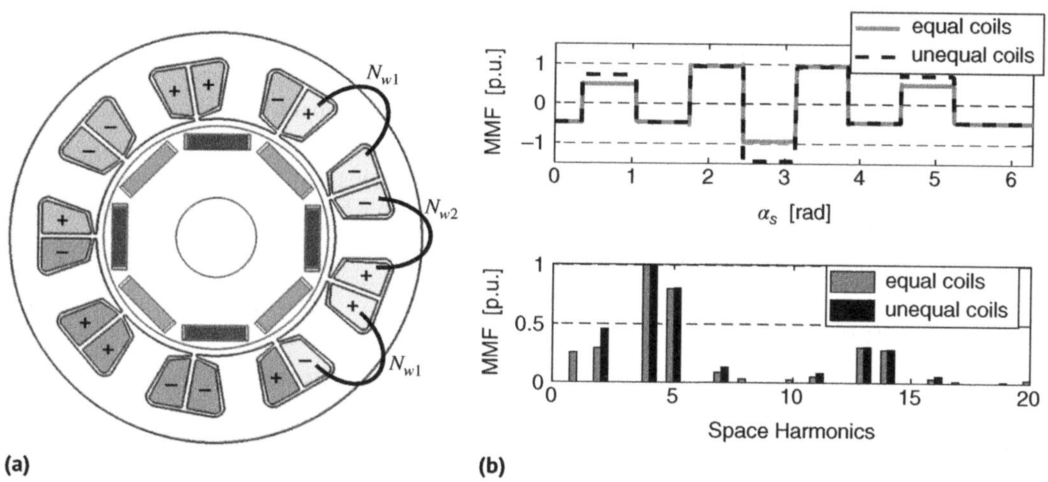

Figure 4.20 DL-FSCW with $z = 1$, $n_c = 3$, $Q_S = 9$, and with unequal turns per coil. (a) Machine design with the winding layout. (b) Three-phase spatial MMF distribution and harmonic contents for $N_{w1} = 0.65 \cdot N_{w2}$, and comparison with the standard winding.

selected to be $N_{w1} = 0.65N_{w2}$. A comparison of the MMF characteristics for the modified winding and the corresponding standard configuration is provided in Figure 4.20b. The results show that the 1st subharmonic is completely eliminated, whereas the operating harmonic remains unaffected, thereby maintaining the output torque capability of the machine. However, since the 2nd subharmonic component is amplified, additional optimization steps are required here to achieve full subharmonic mitigation.

Example 4.3 9-slot/8-pole FSCW with unequal coil-pitch

In the final example, the variable coil-pitch technique is applied to the same 9-slot/8-pole winding. The resulting modified winding layout is illustrated in Figure 4.21a. For the purpose of comparison,

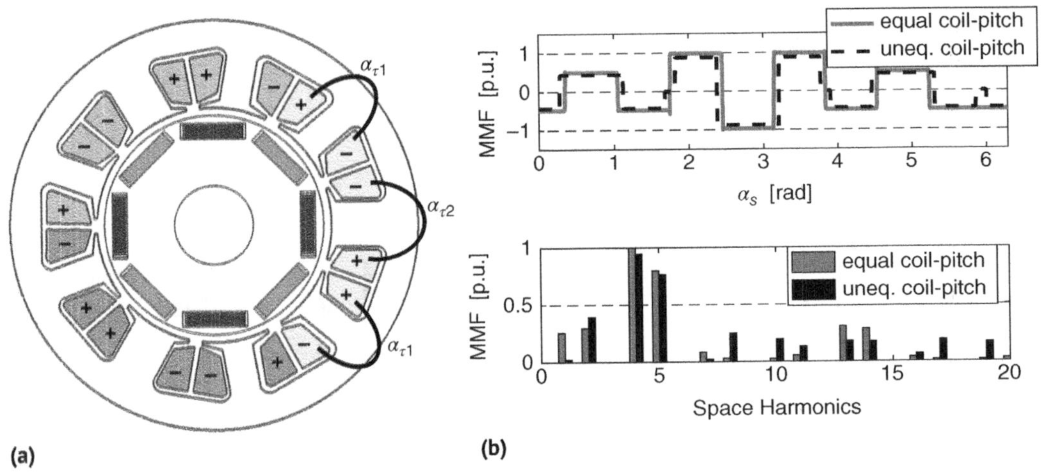

Figure 4.21 DL-FSCW with $z = 1$, $n_c = 3$, $Q_S = 9$, and with different coil pitches. (a) Machine design with the winding layout. (b) Three-phase spatial MMF distribution and harmonic contents for $k_{\alpha,\tau} = 0.2$, and comparison with the standard winding.

a coil-pitch variation factor of $k_{\alpha,\tau} = 0.2$ is selected. The corresponding MMF characteristics, presented in Figure 4.21b, indicate that the 1st subharmonic can once again be effectively suppressed. However, in contrast to the "unequal-turn per coil" approach, the modified coil-pitch produces additional side effects. In particular, the amplitude of the desired operating harmonic is slightly reduced, while the 2nd and several higher-order harmonics show a noticeable increase. These changes in the MMF spectrum result in a less favorable harmonic profile and ultimately in a degradation of the overall electromagnetic performance of the machine compared to the standard winding.

4.4.1 Summary of the Results

Each winding modification offers unique advantages in reducing unwanted harmonics, although they also introduce certain trade-offs. The essential effects of the three methods may be summarized as follows.

Method	**Effect on subharmonics**	**Effect on working harmonic**
Asymmetric coils	Completely eliminate	Unaffected
Unequal turns per coil	Completely eliminate	Unaffected, but amplifies a 2nd subharmonic
Unequal coil-pitch	Completely eliminate	Slightly reduced, but amplifies several MMF harmonics

5

Multiphase Fractional Slot Concentrated Windings

In traditional alternating current (AC) machines, stator windings typically consist of three-phase windings, which are spatially shifted by $2\pi/3$ electrical degrees and supplied with a three-phase AC with the same $2\pi/3$ phase shift. Since their invention, these windings have been widely used and continuously developed, leading to various topologies, including both distributed and concentrated windings. In multiphase windings, however, the number of phases is higher than three $(m > 3)$, which are shifted in space and time by $2\pi/m$ electrical degrees. Although multiphase drives were first introduced in the late 1960s by researchers such as Ward and Härer [55], Nelson and Krause [56], Jahns [57], Klingshirn [58], and Abbas et al. [59], interest in multiphase AC machines has grown significantly since the early 21st century, driven by advancements in the field of electric and hybrid vehicles, electric ship propulsion, and aerospace applications. The primary objectives of these developments were to achieve high reliability and improved fault tolerance, power splitting, lower torque pulsations, better noise performances, and modularity for flexible and scalable machine designs. These benefits have been extensively studied in Haylock (1998) [60], Parsa and Toliyat [61], Mitcham et al. [62], Levi [63], Aslan et al. [64], and Bojoi et al. [65].

Among the various multiphase winding topologies, the multi-three-phase solution is one of the most preferred. This category includes the six-phase and the nine-phase windings, which are commonly used in PMSMs, as reported by Barcaro et al. [66] and Burkhardt et al. [67]. Additionally, five-phase and seven-phase windings have also gained popularity in applications requiring fault tolerance and efficiency improvements [68–70]. Moreover, fractional slot concentrated windings (FSCWs) can also be implemented in multiphase configurations using a simple setup with a single coil per phase, allowing for an arbitrary number of phases. This approach enhances the operation flexibility in electric machines, enabling the electronic pole switching, multi-pole operation mode, more redundancy, and noise reduction. As reported by Dajaku [71], these winding configurations also enable topologies with high winding factor for working harmonics, making them particularly suitable for machine applications with high torque density.

5.1 Multiphase FSCW Topologies

Multiphase windings can be generally classified into two main categories: windings with a prime number of phases and those with an even or composite-odd number of phases. For machines with a prime number of phases (e.g. $m = 3, 5, 7, 11, ...$), the successive phase shift is $2\pi/m$. In contrast, machines with an even or a composite-odd number of phases (e.g. $m = 6, 9, 12, ...$) can be designed with k_{ws} sets of windings, where each set consists of m_0 phases and has a $2\pi/m_0$ spatial shift within the set. Moreover, if the winding sets are three phase $(m_0 = 3)$, the configuration is referred to as a multiple three-phase winding.

Depending on the spatial displacement between the winding sets (i.e. the φ_m shift angle), multiphase windings can be classified into three distinct configurations:

- dual m_0-phase ($\varphi_m = 0°$),
- symmetrical m-phase ($\varphi_m = 2\pi/m$), and
- asymmetrical m-phase ($\varphi_m = \pi/m$).

These classifications have been widely studied by Nelson and Krause [56] and Klingshirn [58]. As an illustration, Figure 5.1 presents the phase vectors for a six-phase winding arranged in different configurations.

5.1.1 Conversion of a Three-phase FSCW into a Six-phase Winding

The transformation between different winding configurations can be easily achieved by rearranging coil connections, modifying inverter supply leads in the machine's terminal box, and reconfiguring neutral point connections. For example, a common three-phase FSCW with 12 stator slots ($Q_s = 12$), 2 zones per phase ($z = 2$), and 2 adjacent coils per zone ($n_c = 2$) is shown in Figure 5.2.

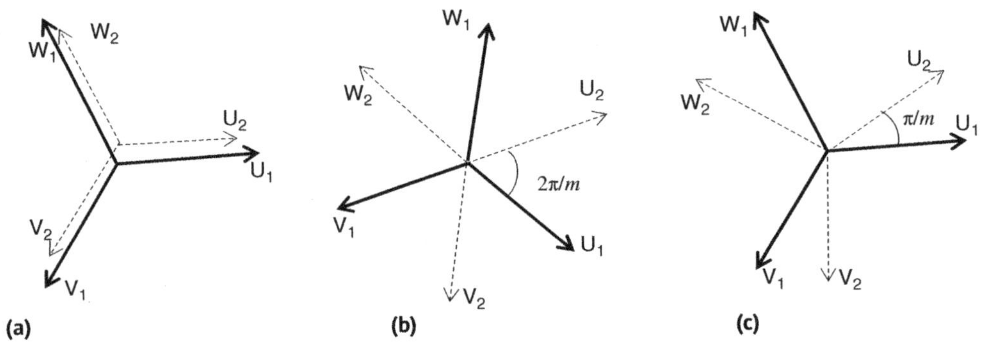

Figure 5.1 Phase vectors for different six-phase winding arrangements. (a) Dual three-phase. (b) Symmetric six-phase. (c) Asymmetric six-phase.

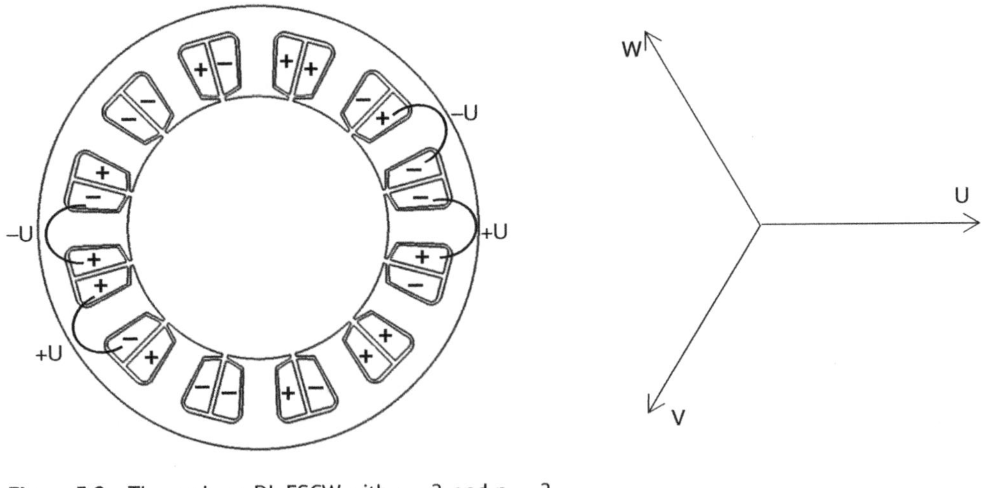

Figure 5.2 Three-phase DL-FSCW with $z = 2$, and $n_c = 2$.

By rearranging the coil-groups into two different phases, this winding can be converted into a six-phase winding consisting of two three-phase sets, as illustrated in Figure 5.3. The first set of phases is denoted as U_1, V_1, and W_1, while the second set is labeled with U_2, V_2, and W_2.

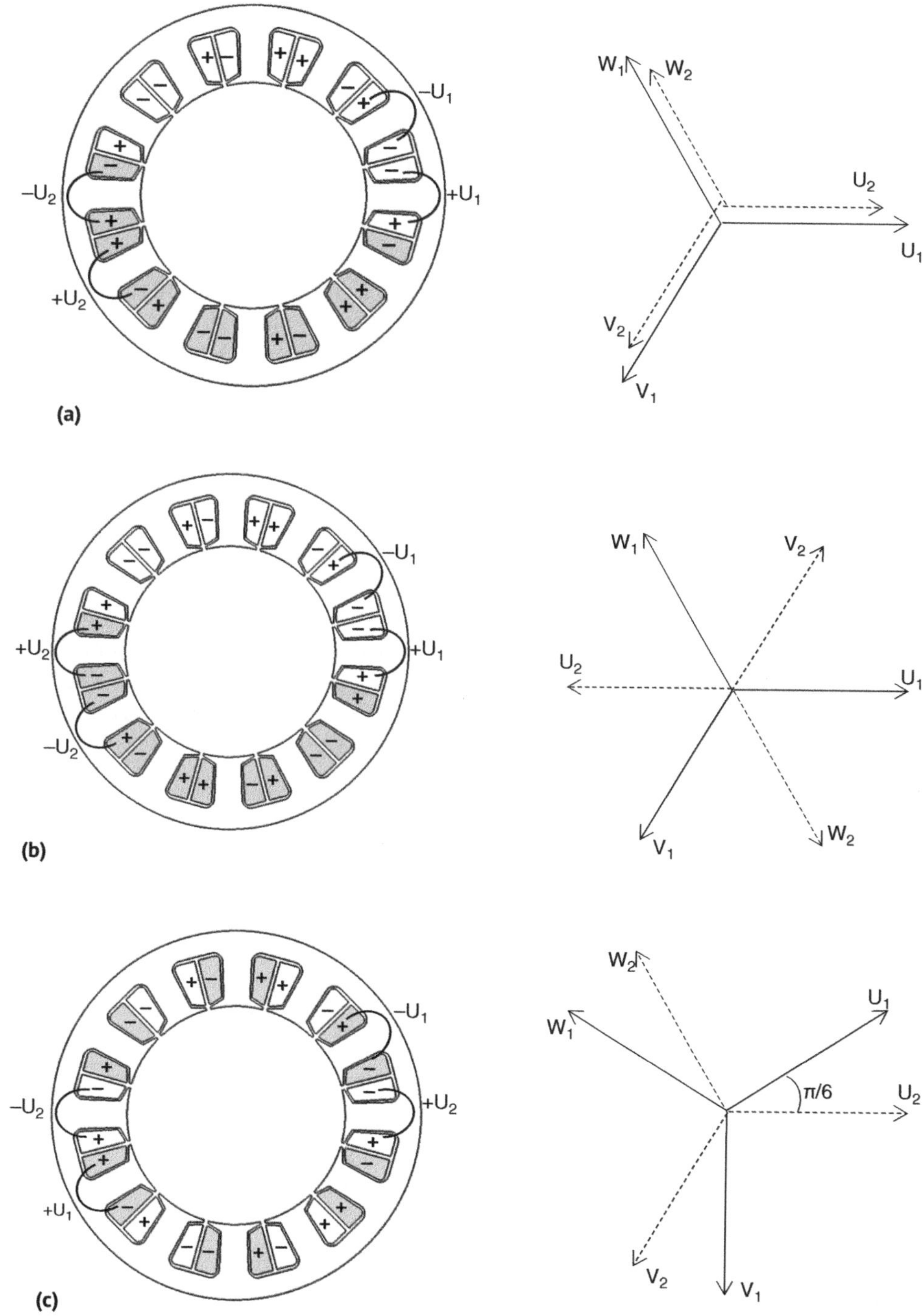

Figure 5.3 (a) Dual three-phase DL-FSCW with $z = 1$ and $n_c = 2$. (b) Symmetrical six-phase DL-FSCW with $z = 1$ and $n_c = 2$. (c) Asymmetric six-phase DL-FSCW with $z = 2$ and $n_c = 1$.

Example 5.1 Dual three-phase winding
The initial winding is divided into two distinct phases, each with one zone per phase and two adjacent coils per zone ($z = 1$ and $n_c = 2$), as illustrated in Figure 5.3a. In order to maintain the MMF characteristics of the original three-phase winding, the coil polarity remains unchanged, and both winding sets are supplied with identical three-phase current systems, resulting in a dual three-phase winding configuration.

Example 5.2 Symmetric six-phase winding
Here, the coil polarity of the second winding set is reversed (Figure 5.3b). To maintain the same MMF distribution, an opposite-phase current excitation with a 180° phase shift is applied, leading to a symmetric six-phase winding.

Example 5.3 Asymmetric six-phase winding
In this case, the original winding is split into two sets with $z = 2$ and $n_c = 1$, while the coil polarity remains unchanged. The current systems are configured with a 30° phase shift, resulting in an asymmetric six-phase winding. Unlike the previous configurations, the MMF characteristics of this asymmetrical solution differ from the original three-phase reference winding, as will be explored in Section 5.2.

5.1.2 Other Multiphase FSCW Configurations

Beyond the six-phase examples, Figures 5.4–5.6 illustrate additional multiphase FSCW configurations that can be realized as either single- or double-layer windings. Figure 5.4 presents a five-phase SL-FSCW (U, V, W, X, Y) employing two zones per phase ($z = 2$) and one coil per zone ($n_c = 1$). Maintaining the same stator slot number as in the previous example, the corresponding double-layer version of this five-phase winding is shown in Figure 5.5. A further variant, consisting of 10 phases (U_1–U_{10}), with each phase formed by a single coil, is shown in Figure 5.6. These examples illustrate that the design process for a multiphase FSCW follows

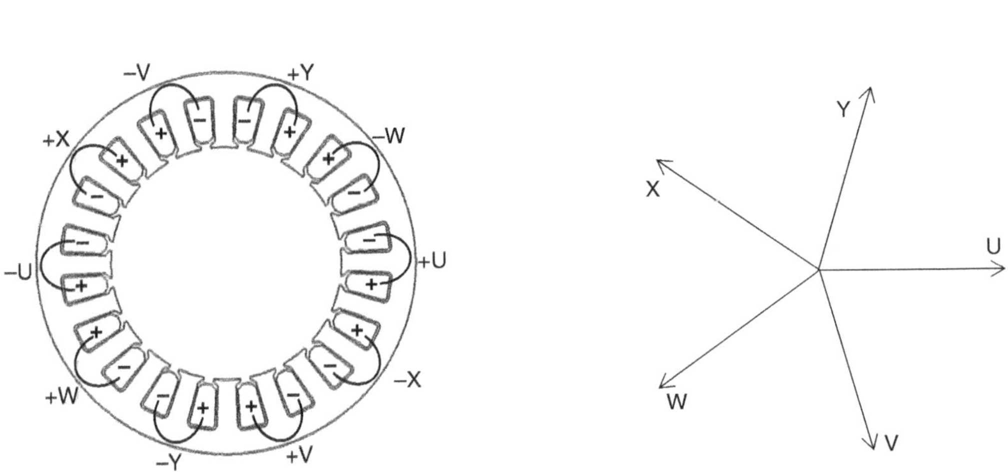

Figure 5.4 Five-phase SL-FSCW with $z = 2$ and $n_c = 1$.

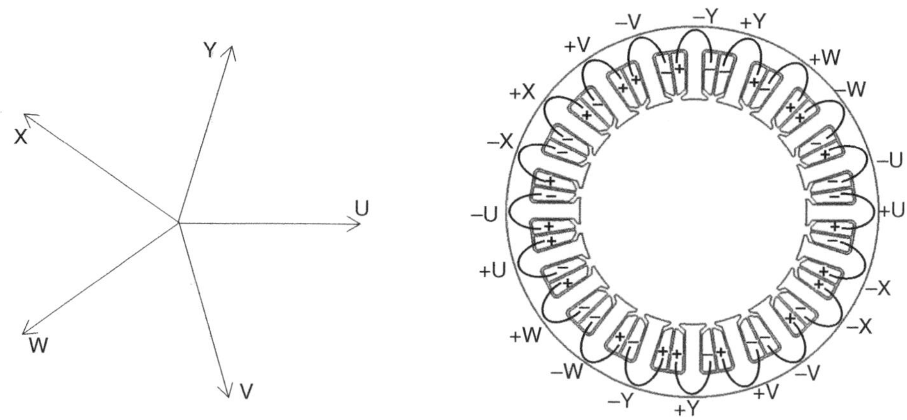

Figure 5.5 Five-phase DL-FSCW with $z = 2$ and $n_c = 2$.

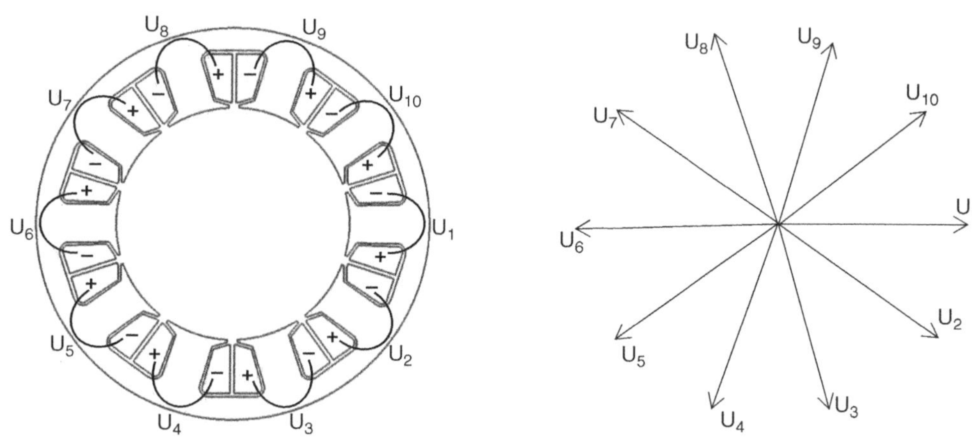

Figure 5.6 10-phase DL-FSCW with $z = 1$ and $n_c = 1$.

that of the conventional three-phase case. Essentially, it involves selecting and combining key winding parameters, including:

- number of phases: $m = 3$ or $m > 3$,
- number of zones: $z = 1$ or $z = 2$,
- number of coils per zone: $n_c = 1$ or $n_c > 1$,
- coils polarity: same or opposite,
- number of layers: single or double layer,
- phase arrangement: dual, symmetric, or asymmetric.

Moreover, the multiphase concept can be applied to various FSCWs with specific zone and coil arrangements introduced in Chapters 3 and 4. As a result, the number of feasible multiphase FSCW topologies becomes extremely large. Nevertheless, since the underlying design principles do not differ fundamentally from those of three-phase windings, the following sections will focus on a selected subset of representative topologies.

5.2 Multi-three-phase FSCWs

The multi-three-phase winding is widely used in various multiphase machine applications due to its similarity to the common three-phase machine topologies. In this design, the stator winding consists of two or more three-phase winding sets, each maintaining a $2\pi/3$ phase shift within the set. As noted earlier, the spatial offset between these sets determines whether the stator winding is arranged symmetrically, asymmetrically, or as a dual three-phase configuration. Among these, the asymmetric configuration produces a different MMF distribution compared to the standard three-phase reference winding. Due to its unique properties, two representative examples are discussed below.

5.2.1 Asymmetric Six-phase FSCW

As already discussed, the standard 12-slot/10-pole three-phase FSCW of Figure 5.2 can be converted into an asymmetric six-phase winding by dividing the original winding into two three-phase winding sets shifted in space by $\pi/6$. Each set is energized by an independent three-phase inverter, and the current systems are phase-shifted by $\pi/6$. The resulting arrangement is illustrated in Figure 5.3c.

For the reference three-phase case in Figure 5.2, the winding parameters are $Q_S = 12$, $z = 2$, and $n_c = 2$. The corresponding MMF function is given by Equation (5.1), while the associated winding factor parameters are listed in Equation (5.2).

$$F_{3\text{ph}}(\alpha_S, t) = \frac{mzn_cN_w\hat{i}}{\pi}\sum_{\nu}\frac{{}^{\nu}\xi_p\,{}^{\nu}\xi_d\,{}^{\nu}\xi_z}{\nu}\sin(\omega t - \nu\alpha_S) \qquad (5.1)$$
$$\nu = mg + 1 \quad g = 0, \pm 1, \pm 2, \pm 3, \ldots$$

$$^{\nu}\xi_p = \sin\left(\nu\frac{\pi}{Q_S}\right), \quad {}^{\nu}\xi_z = \sin\left(\nu\frac{\pi}{2}\right), \quad {}^{\nu}\xi_d = \frac{\sin\left(\nu n_c\frac{\pi}{Q_S} + n_c\frac{\pi}{2}\right)}{n_c\cdot\cos\left(\nu\frac{\pi}{Q_S}\right)} \qquad (5.2)$$

In the asymmetric six-phase arrangement, the original winding is divided into two three-phase sets, with each phase being split into two sub-phases (e.g. phase U is split into U_1 and U_2), and with the new winding parameters being $z = 2$ and $n_{c0} = 1$. The corresponding MMF waveforms are shown in Figure 5.7. Assuming that both winding sets have the same number of turns per coil (i.e. $N_{w1} = N_{w2} = N_w$) and are supplied with the same current phase amplitude $(\hat{i}_1 = \hat{i}_2 = \hat{i})$, the MMF distribution generated by each three-phase set can be expressed as,

$$\begin{aligned} F_1(\alpha_S, t) &= \frac{m_0 z n_{c0} N_w \hat{i}}{\pi}\sum_{\nu}\frac{{}^{\nu}\xi_p\,{}^{\nu}\xi_{d,nc0}\,{}^{\nu}\xi_z}{\nu}\sin(\omega t - \nu\alpha_S) \\ F_2(\alpha_S, t) &= -\frac{m_0 z n_{c0} N_w \hat{i}}{\pi}\sum_{\nu}\frac{{}^{\nu}\xi_p\,{}^{\nu}\xi_{d,nc0}\,{}^{\nu}\xi_z}{\nu}\sin\left[\left(\omega t - \frac{\pi}{6}\right) - \nu\left(\alpha_S - \frac{\pi}{6}\right)\right] \end{aligned} \qquad (5.3)$$

where m_0 is the number of phases per set (i.e. $m_0 = 3$), and n_{c0} denotes the number of coils per zone in each set.

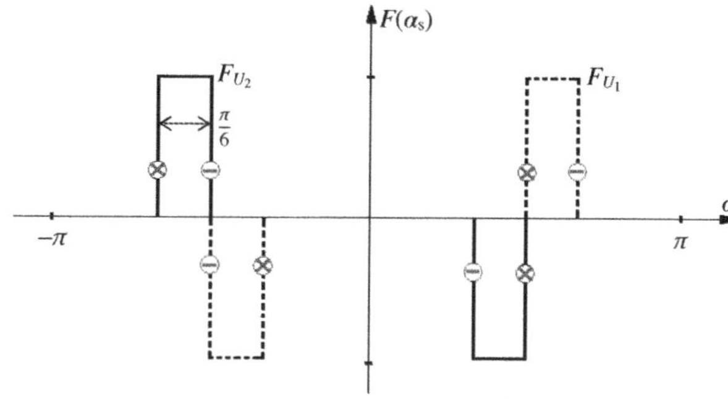

Figure 5.7 MMF distribution of phase U_1 in solid line and U_2 in dashed line.

For the case $n_c = n_{c0} = 1$, Equation (5.2) yields the maximum distribution factor for each individual winding sets.

$$^{\nu}\xi_{d,n_{c0}}\Big|_{n_{c0}=1} = 1$$

Further, using (5.3), the total air-gap MMF generated by both winding sets becomes

$$\begin{aligned} F(\alpha_S, t) &= F_1(\alpha_S, t) + F_2(\alpha_S, t) \\ &= \frac{m_0 z N_w \hat{i}}{\pi} \sum_{\nu} \frac{^{\nu}\xi_p\,^{\nu}\xi_z}{\nu} \left\{ \sin(\omega t - \nu\alpha_S) - \sin\left[\left(\omega t - \frac{\pi}{6}\right) - \nu\left(\alpha_S - \frac{\pi}{6}\right)\right]\right\} \\ &= \frac{2 m_0 z N_w \hat{i}}{\pi} \sum_{\nu} \frac{^{\nu}\xi_w}{\nu} \cos\left(\omega t - \nu\alpha_S - (\nu - 1)\frac{\pi}{12}\right) \end{aligned} \tag{5.4}$$

The resulting winding factor for the asymmetric six-phase configuration is therefore

$$^{\nu}\xi_w = {}^{\nu}\xi_p\,^{\nu}\xi_z \cdot \sin\left((\nu - 1)\frac{\pi}{12}\right) \tag{5.5}$$

Depending on the harmonic order, the *sine* term in the above relation leads to two distinct values. Specifically,

$$\sin\left((\nu - 1)\frac{\pi}{12}\right) = \begin{cases} 0, & \nu = +1, -11, +13 \ldots \\ 1, & \nu = -5, +7, -17 \ldots \end{cases}$$

As a result, this configuration inherently suppresses the 1st MMF harmonic as well as certain higher-order harmonics, notably the 11th and 13th. On the other hand, due to an increased distribution factor, the working harmonic ($\nu = 5$) along with the 7th, 17th, and 19th harmonics show an increase of approximately 3.3% points. A comparison of the MMF characteristics for the original three-phase winding and the asymmetrical six-phase FSCW is provided in Figure 5.8 and Table 5.1.

This example demonstrates that the dual three-phase technique improves the MMF performance while maintaining a relatively simple winding structure. The same approach is applicable to any three-phase FSCW with one or two zones per phase and an even number of coils per zone. Moreover, when the parameter n_c is an integer multiple of three, the winding can be converted into a multiphase winding using the $k_{ws} \times$ three − phase transformation concept (where k_{ws} is an integer number), which will be explored in the next example.

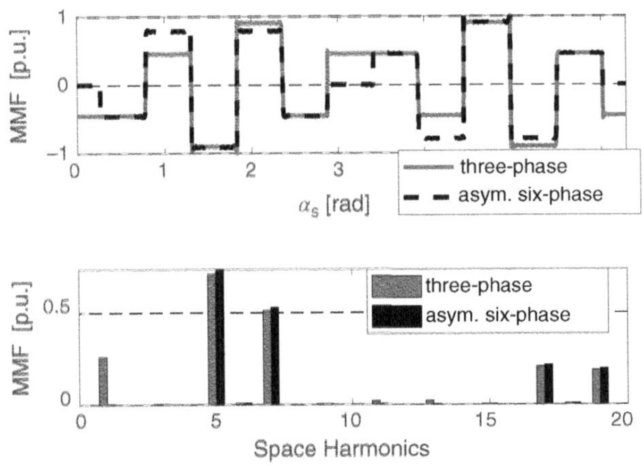

Figure 5.8 MMF comparison for the three-phase and the asymmetric six-phase 12-slot/10-pole FSCW.

Table 5.1 Winding factors for the three-phase and asymmetric six-phase winding.

Harmonic order	Three-phase (%)	Dual three-phase (%)
1	6.7	0
5	93.3	96.6
7	93.3	96.6

5.2.2 Asymmetric Nine-phase FSCW

To examine the asymmetric case, the 9-slot/8-pole FSCW is considered. In its standard layout, each phase comprises a single zone with three adjacent coils connected in series, as illustrated in Figure 5.9. For this topology, the key winding parameters are $Q_S = 9$, $z = 1$, and $n_c = 3$. The resulting MMF function is expressed in Equation (5.6), with the associated coil-pitch and distribution factors defined in Equation (5.2).

$$F_{3\text{ph}}(\alpha_S, t) = \frac{mn_cN_w\hat{i}}{\pi} \sum_{\nu} \frac{{}^{\nu}\xi_p\, {}^{\nu}\xi_d}{\nu} \cos(\omega t - \nu\alpha_S)$$
$$\nu = mg + 1 \quad g = 0, \pm1, \pm2, \pm3, \ldots \tag{5.6}$$

In the asymmetric nine-phase arrangement, the original three-coil phase winding is split into three separate phases, each consisting of a single coil per zone, i.e. $n_{c0} = 1$. This transformation yields a stator comprising three independent *three-phase* winding systems, as shown in Figure 5.10 (index 1–3 denotes the corresponding winding sets). The winding sets are spatially shifted by $2\pi/9$ electrical radians and are supplied by three-phase independent inverters, each shifted in time by $\pi/9$.

The MMF distribution for one of the split phases, specifically phase-U, is presented in Figure 5.11. Based on the new winding parameters (m_0 and n_{c0}), the corresponding MMF functions are defined in Equations (5.7)–(5.9). Since $n_{c0} = 1$, the distribution factor ξ_d becomes maximal $(\xi_d = 1)$ and therefore does not explicitly appear in the MMF equations.

$$F_1(\alpha_S, t) = \frac{m_0N_w\hat{i}}{\pi} \sum_{\nu} \frac{{}^{\nu}\xi_p}{\nu} \cos(\omega t - \nu\alpha_S) \tag{5.7}$$

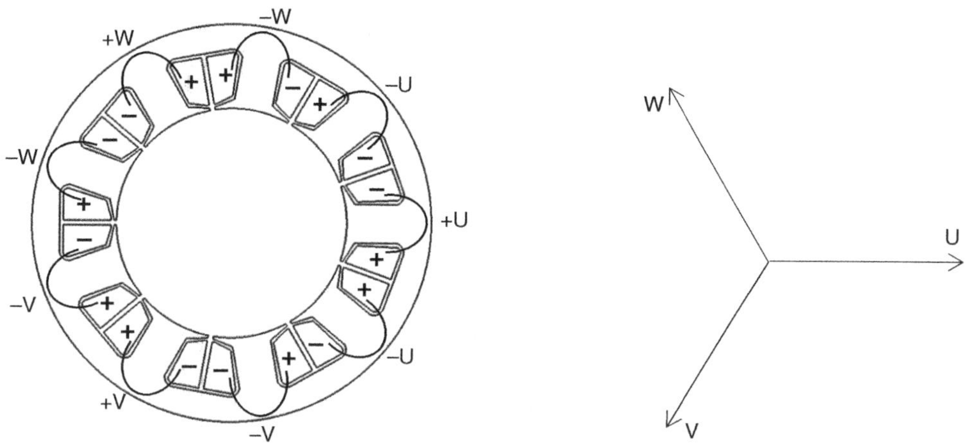

Figure 5.9 Three-phase DL-FSCW with $z = 1$ and $n_c = 3$.

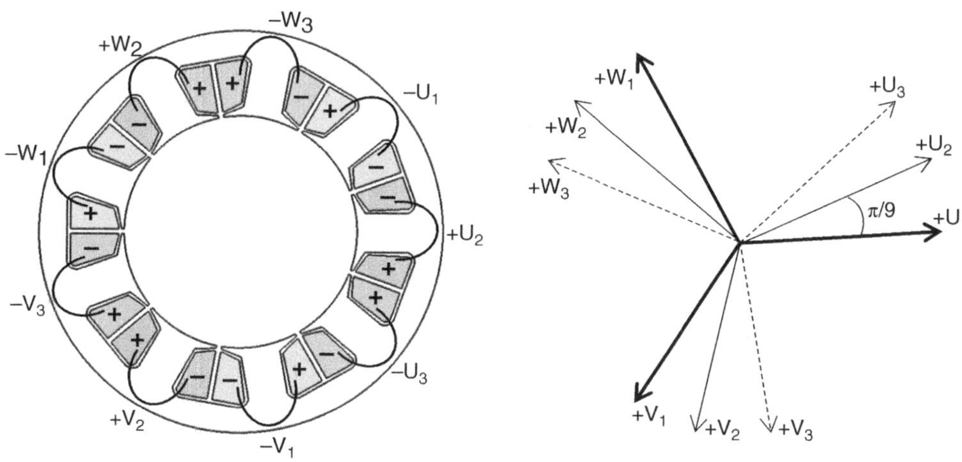

Figure 5.10 Asymmetric nine-phase DL-FSCW with $z = 1$ and $n_c = 1$.

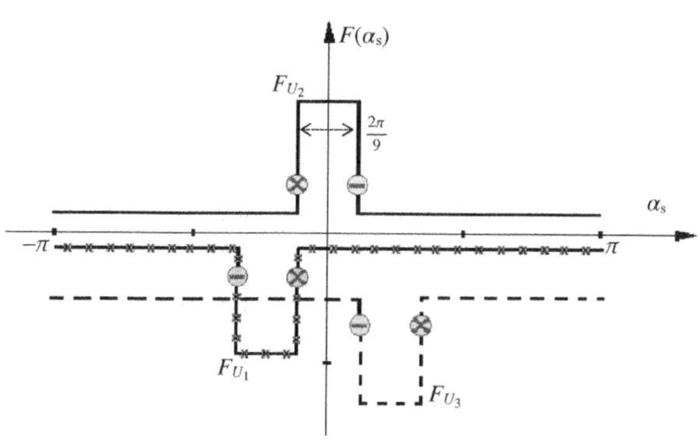

Figure 5.11 MMF distribution of phase U_1 in x-marked line, U_2 in solid-line and U_3 in dashed line.

$$F_2(\alpha_S,t) = -\frac{m_0 N_w \hat{i}}{\pi}\sum_{\nu}\frac{{}^{\nu}\xi_p}{\nu}\cos\left[\left(\omega t + \frac{\pi}{9}\right) - \nu\left(\alpha_S - \frac{2\pi}{9}\right)\right] \tag{5.8}$$

$$F_3(\alpha_S,t) = \frac{m_0 N_w \hat{i}}{\pi}\sum_{\nu}\frac{{}^{\nu}\xi_p}{\nu}\cos\left[\left(\omega t + 2\frac{\pi}{9}\right) - \nu\left(\alpha_S - 2\frac{2\pi}{9}\right)\right] \tag{5.9}$$

From Equations (5.7) to (5.9), the resulting air-gap MMF is

$$\begin{aligned} F_1(\alpha_S,t) &= F_1(\alpha_S,t) + F_2(\alpha_S,t) + F_3(\alpha_S,t) \\ &= \frac{m_0 N_w \hat{i}}{\pi}\sum_{\nu}\frac{{}^{\nu}\xi_p}{\nu}\left\{\cos\left(\omega t - \nu\alpha_S\right) - \cos\left[\left(\omega t + \frac{\pi}{9}\right) - \nu\left(\alpha_S - \frac{2\pi}{9}\right)\right]\right. \\ &\qquad \left. + \cos\left[\left(\omega t + 2\frac{\pi}{9}\right) - \nu\left(\alpha_S - 2\frac{2\pi}{9}\right)\right]\right\} \\ &= \frac{m_0 N_w \hat{i}}{\pi}\sum_{\nu}\frac{{}^{\nu}\xi_p}{\nu}\sum_{h=0}^{k_{ws}-1}\cos\left[\left(\omega t + h\frac{\pi}{9}\right) - \nu\left(\alpha_S - h\frac{2\pi}{9}\right) + h\pi\right] \end{aligned} \tag{5.10}$$

The sum of *cosine* term in Equation (5.10) can be expressed as,

$$\begin{aligned} &\sum_{h=0}^{k_{ws}-1}\cos\left(\left(\omega t - \nu\alpha_S\right) + (2\nu+1)h\frac{\pi}{9} + h\pi\right) \\ &= e^{j(\omega t - \nu\alpha_S)}\sum_{h=0}^{k_{ws}-1} e^{jh\left((2\nu+1)\frac{\pi}{9} + \pi\right)} \\ &= \frac{\sin\left(k_{ws}\left((2\nu+1)\frac{\pi}{18} + \frac{\pi}{2}\right)\right)}{\sin\left((2\nu+1)\frac{\pi}{18} + \frac{\pi}{2}\right)} e^{j\left(\omega t - \nu\alpha_S + (k_{ws}-1)\left((2\nu+1)\frac{\pi}{18} + \frac{\pi}{2}\right)\right)} \\ &= \frac{\cos\left((2\nu+1)\frac{\pi}{6}\right)}{\cos\left((2\nu+1)\frac{\pi}{18}\right)} e^{j\left(\omega t - \nu\alpha_S - (2\nu+1)\frac{\pi}{9} + \pi\right)}, \quad k_{ws} = 3 \end{aligned} \tag{5.11}$$

Substituting Equations (5.11) into (5.10) yields the final MMF expression

$$\begin{aligned} F(\alpha_S,t) &= \frac{3 m_0 N_w \hat{i}}{\pi}\sum_{\nu}\frac{{}^{\nu}\xi_p}{\nu}\frac{\cos\left(3(2\nu+1)\frac{\pi}{18}\right)}{3\cdot\cos\left((2\nu+1)\frac{\pi}{18}\right)} \\ &\quad \cos\left(\omega t - \nu\alpha_S - (2\nu+1)\frac{\pi}{9} + \pi\right) \end{aligned} \tag{5.12}$$

The trigonometric term in Equation (5.12) may be expressed as

$$\frac{\cos\left(3(2\nu+1)\frac{\pi}{18}\right)}{3\cdot\cos\left((2\nu+1)\frac{\pi}{18}\right)} = \frac{4}{3}\cos^2\left((2\nu+1)\frac{\pi}{18}\right) - 1$$

Since $(2\nu + 1)$ is always an odd integer, it follows that

$$(2\nu + 1) = \pm 3k, \quad k = 1, 3, 5, \ldots$$

This leads to two distinct angular values for the cosine term, i.e.

$$\left((2\nu+1)\frac{\pi}{18}\right)=\begin{cases}\pm k\dfrac{\pi}{6}, & k=1,5,7...\\ \pm\left(\dfrac{k}{3}\right)\dfrac{\pi}{2}, & k=3,9,...\end{cases}$$

Thus, depending on the harmonic order ν, the trigonometric function reduces to

$$\left|\frac{4}{3}\cos^2\left((2\nu+1)\frac{\pi}{18}\right)-1\right|=\begin{cases}1, & \text{for } \nu=\dfrac{\pm 3k-1}{2}, \quad k=3,9,\ldots\\ 0, & \text{for } \nu=\dfrac{\pm 3k-1}{2}, \quad k=1,5,7,\ldots\end{cases}$$

Consequently, the air-gap MMF function is simplified to

$$\begin{aligned}F(\alpha_S,t) &= \frac{3m_0N_w\hat{i}}{\pi}\sum_{\nu}\frac{{}^{\nu}\xi_p}{\nu}\cos\left(\omega t-\nu\alpha_S-(2\nu+1)\frac{\pi}{9}+\pi\right)\\ \nu &= \frac{\pm 3k-1}{2}, \quad k=3,9,15\ldots\end{aligned} \tag{5.13}$$

From Equation (5.13), it follows that the resulting MMF distribution contains only specific harmonics, namely:

$$\nu=+4\text{th}, -5\text{th}, +13\text{th}, \ldots$$

Moreover, due to the maximal distribution factor, the corresponding winding factor of these harmonics is higher than in the reference three-phase winding. This enhancement is evident in the MMF characteristics shown in Figure 5.12 and Table 5.2. The results clearly indicate that the asymmetric multiphase technique eliminates the MMF subharmonic components while increasing the magnitude of the working harmonic by approximately 4.2%. This improvement leads to higher torque density and reduced rotor losses compared to the conventional three-phase design.

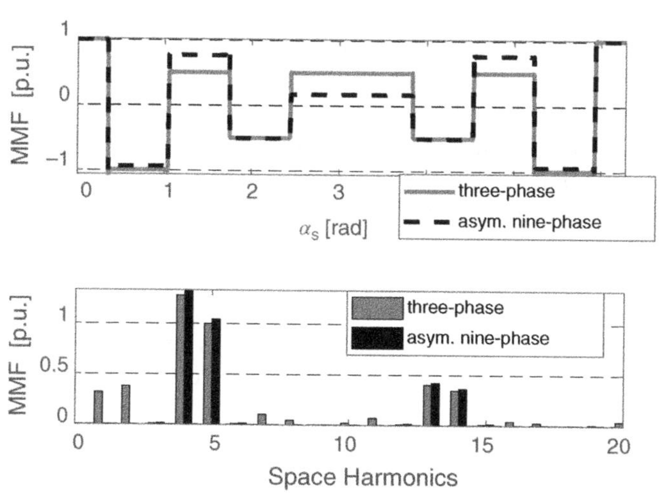

Figure 5.12 MMF comparison for the three-phase and the asymmetric nine-phase 9-slot/8-pole FSCW.

Table 5.2 Winding factors for both three-phase and asymmetric nine-phase winding.

Harmonic order	Three phase (%)	Asymmetric nine-phase (%)
1	6.0	0
2	13.9	0
4	94.5	98.5
5	94.5	98.5

5.2.3 Concluding Remarks

The most important advantages and features of the asymmetric multiphase winding solution can be summarized as follows:

- Winding layout:
  - The number of coils per zone is reduced by the factor k_{ws} (number of winding sets).
- Winding factor:
  - Reducing the number of coils per zone increase the distribution factor, yielding in a higher winding factor for the working harmonic.
- Suppression of MMF harmonics:
  - All MMF subharmonic components are fully eliminates.
  - Higher-order MMF harmonics, except the slot-related harmonics (i.e. $v_{p1,p2} + jQ_S$, where $j = 1, 2, 3, \ldots$), are also suppressed.
- Performance improvements:

  The asymmetric multiphase FSCW solution leads to enhanced machine performance due to:
  - an increased winding factor for the working harmonic, and
  - the lower MMF harmonic contents and the absence of subharmonic components.

5.3 FSCWs with an Arbitrary Number of Phases

Beyond the winding configurations discussed earlier, multiphase FSCWs can also be extended to an arbitrary number of phases using a simple winding arrangement, as reported in Dajaku [71]. An example is the 11-phase FSCW shown in Figure 5.13, where each stator tooth carries its own coil, and each coil is directly supplied by an individual power module. In this manner, every coil operates as an individual phase. This design not only simplifies the winding process but also enables precise and independent control of each stator tooth, thereby enhancing the overall machine performance.

In the studied example, the number of phases m is directly related to the number of stator teeth or slots Q_S. Specifically, for a double-layer winding layout, the number of phases is equal with the number of stator teeth, whereas for a single-layer winding it is half of the Q_S, as defined in (5.14). This relationship implies that the number of phases in a winding system can be defined or adjusted by selecting the desired number of stator modules.

$$m = \begin{cases} Q_S/2, & \text{single-layer FSCW} \\ Q_S, & \text{double-layer FSCW} \end{cases} \tag{5.14}$$

A notable advantage of this winding configuration is its capability to generate variable MMF waveforms when supplied by an appropriate modulated phase current. This feature is highly

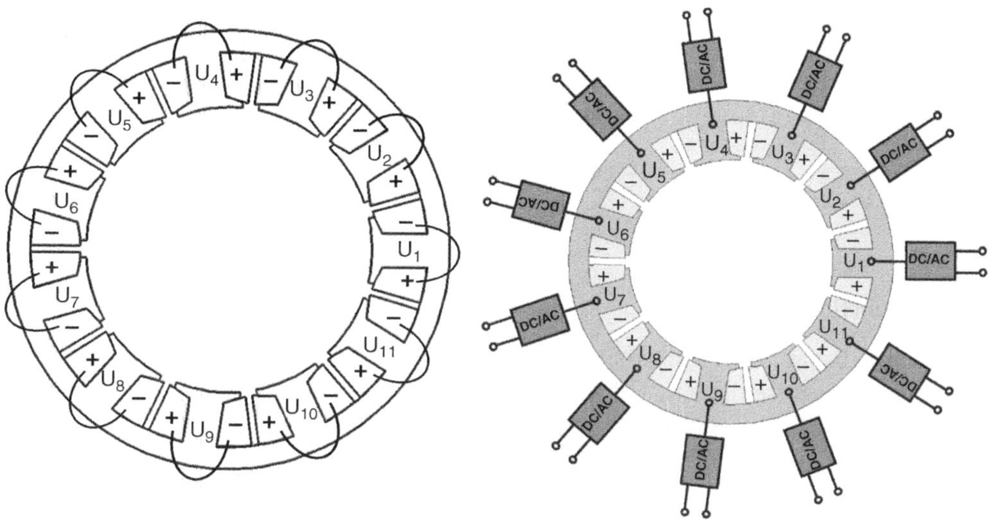

Figure 5.13 Multiphase concept for the 11-phase DL-FSCW.

advantageous in applications that require independent control of individual harmonic components of the air-gap flux density. The generalized expression for the phase excitation that enables this behavior is given in Equation (5.15).

$$i_j(t) = \hat{i} \cdot \cos\left(\omega t - \nu_0 \cdot (j-1)\frac{2\pi}{m}\right) \tag{5.15}$$

Here, j denotes the phase number, and ν_0 is a control parameter that defines the MMF waveform or the injected MMF harmonic in the machine air-gap. This parameter allows for a wide range of MMF waveforms, providing greater flexibility in both design and operation. The available range of ν_0 is directly related to the number of phases m, as defined in Equation (5.16).

$$\nu_0 = \begin{cases} 1, 2, \ldots, \dfrac{m+1}{2} - 1, & m - \text{odd} \\ 1, 2, \ldots, \dfrac{m}{2} - 1, & m - \text{even} \end{cases} \tag{5.16}$$

For example:

- with a three-phase winding $(m = 3)$, only one MMF waveform can be generated $(\nu_0 = 1)$,
- with a five-phase winding $(m = 5)$, two different MMF waveforms become available $(\nu_0 = 1 \text{ and } \nu_0 = 2)$,
- and as m increases, additional MMF harmonic orders can be generated.

To further enhance the winding factor for the working harmonics, a special coil configuration with unequal turns per coil side, described in Chapter 4, can be applied. An example of such a multiphase FSCW with asymmetric coils is illustrated in Figure 5.14. Here, the number of turns on the first coil-side is denoted as N_{w11}, while on the second coil-side as N_{w12}, with the relationship being as follows

$$N_{w11} = N_{w12} + 1.$$

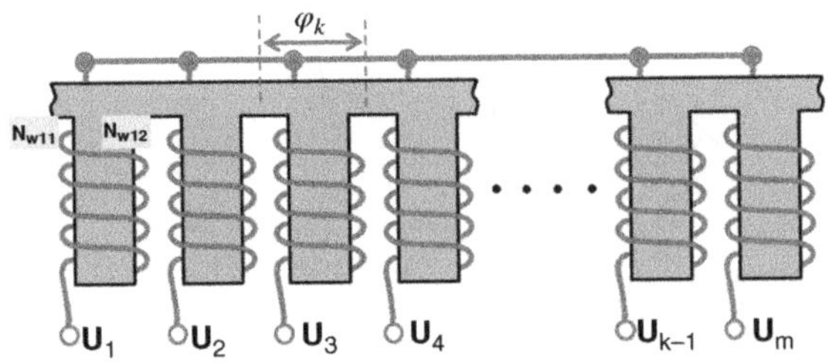

Figure 5.14 Multiphase FSCW with asymmetric coils.

5.3.1 MMF Function Analysis

The MMF analysis below focuses on the specific coil with unequal turns per coil-side, representing the most general case. By setting $\Delta n_w = 0$, the formulation reduces directly to standard symmetric coils. A detailed analysis of this configuration is provided in Chapter 4, particularly in Equations (4.6)–(4.8). The MMF function of the phase-j is given by

$$F_j(t, \alpha_S) = \frac{2N_{w,\mathrm{eq}} i_j(t)}{\pi} \sum_{\nu} \frac{{}^{\nu}\xi_{p,\mathrm{eq}}}{\nu} \sin(\nu\alpha_S + \varphi_{1,2}) \tag{5.17}$$

with the equivalent coil-pitch factor defined as

$${}^{\nu}\xi_{p,\mathrm{eq}} = \frac{1}{N_{w,\mathrm{eq}}} \sqrt{\left(N_{w1}{}^{\nu}\xi_p\right)^2 + \left(\frac{\Delta n_w}{2}\right)^2 + N_{w1}\Delta n_w {}^{\nu}\xi_p \cos\left(\nu\frac{\pi}{Q_S} + \frac{\pi}{2}\right)} \tag{5.18}$$

Here, ${}^{\nu}\xi_p$ denotes the coil-pitch factor of the common coil, $N_{w,\mathrm{eq}}$ is the equivalent number of turns of the asymmetric coil, and Δn_w is the turn difference between coil sides ($\Delta n_w = 1$ in this case).

$${}^{\nu}\xi_p = \sin\left(\nu\frac{\pi}{Q_S}\right) \tag{5.19}$$

$$N_{w,\mathrm{eq}} = \frac{2N_w - \Delta n_w}{2} = N_w - 1/2, \quad \Delta n_w = 1, \quad N_w = N_{w11} \tag{5.20}$$

According to the winding arrangement in Figure 5.14 and the excitation currents in Equation (5.15), the multiphase FSCW is characterized by

- m-phases, each shifted spatially by φ_k, and
- m-phase current systems, with a phase shift of $\nu_0.\varphi_k$,

where

$$\varphi_k = \frac{2\pi}{m}$$

Summing the MMF contributions of all phase yields

$$\begin{aligned} F(t, \alpha_S) &= \sum_{j=1}^{m} F_j(t, \alpha_S) = F_1(t, \alpha_S) + F_2(t, \alpha_S - \varphi_\kappa) + \dots \\ &\quad + F_m(t, \alpha_S - (m-1)\,\varphi_\kappa) \\ &= \frac{m}{2}\frac{2N_{w,\mathrm{eq}} i(t)}{\pi} \sum_{\nu} \frac{1}{\nu} {}^{\nu}\xi_{p,\mathrm{eq}} \cos\left(\nu\alpha_S + (\nu_0 - \nu)\frac{2\pi}{m} + \varphi_{1,2}\right) \end{aligned} \tag{5.21}$$

with

$$\nu = m \cdot g + \nu_0, \quad g = 0, \pm 1, \pm 2, \pm 3, \ldots \tag{5.22}$$

and

$$|\nu| = |m \cdot g + \nu_0| \geq \nu_0 \tag{5.23}$$

Equations (5.22) and (5.23) clearly show that the MMF harmonic contents for this winding category are affected by:

- the parameter ν_0, which plays a crucial role in determining the MMF waveform,
- the number of phases (m), which has a direct effect on the harmonic content and the available MMF waveforms.

Furthermore, the resulting MMF spectrum contains both even and odd harmonics, where:

- the first harmonics in the MMF spectrum is defined by ν_0 parameter,
- whereas the working harmonic, i.e. the one with the highest winding factor, is defined by the winding layout, as outlined below.

5.3.1.1 Single-layer Winding Layout

From Equations (5.14) and (5.19), the coil-pitch factor becomes

$$^{\nu}\xi_p = \sin\left(\nu \frac{\pi}{2m}\right) \tag{5.24}$$

and the winding factors for harmonic order ν_0 and $\nu > \nu_0$ are:

$$^{\nu}\xi_p(\nu_0) = \sin\left(\nu_0 \frac{\pi}{2m}\right) \tag{5.25}$$

$$^{\nu}\xi_p(\nu)\Big|_{\nu>\nu_0} = \sin\left((mg + \nu_0)\frac{\pi}{2m}\right) = \begin{cases} \pm\cos\left(\nu_0 \frac{\pi}{2m}\right), & g = \text{odd} \\ \pm\sin\left(\nu_0 \frac{\pi}{2m}\right), & g = \text{even} \end{cases} \tag{5.26}$$

According to (5.16), the trigonometric relationship in (5.26) always satisfy the condition

$$\left|\cos\left(\nu_0 \frac{\pi}{2m}\right)\right| > \left|\sin\left(\nu_0 \frac{\pi}{2m}\right)\right| \tag{5.27}$$

Therefore, the following conclusions can be made:

- higher-order harmonics $(\nu > \nu_0)$ have higher winding factor,
- the operating wave is a harmonic of order $\nu > \nu_0$ and is defined as

  $$\nu_p = m - \nu_0,$$

- the harmonic ν_0 is a subharmonic component, and
- the maximum winding factor is achieved for the harmonic order $\nu = m - 1$.

5.3.1.2 Double-layer Winding Layout

From Equations (5.14) and (5.19), the respective coil-pitch factors are

$$^{\nu}\xi_p(\nu_0) = \sin\left(\nu_0\frac{\pi}{m}\right) \tag{5.28}$$

$$^{\nu}\xi_p(\nu)\Big|_{\nu>\nu 0} = \sin\left((mg - \nu_0)\frac{\pi}{m}\right) = \pm\sin\left(\nu_0\frac{\pi}{m}\right) \tag{5.29}$$

From above, the key conclusions are:

- the winding factors for harmonic orders ν_0 and $\nu > \nu_0$ are equal,
- therefore the operating wave is defined by $\nu_0(\nu_p = \nu_0)$,
- as a result, this winding type doesn't contain any subharmonic component, and
- the highest winding factor is achieved for the maximal available ν_0, therefore the operating wave is defined as: $\nu_p = \max(\nu_0)$.

5.3.2 Operation Modes

A multiphase FSCW in which each coil is supplied independently and the phase currents are controlled via the ν_0 parameter offers substantially greater operational flexibility than traditional three-phase and multiphase FSCWs.

Key advantages include:

a. Generation of different pole-pair numbers: The same winding can generate different operating harmonics simply by adjusting the excitation harmonic, enabling greater flexibility in design and operation.

b. Multi pole-pair operation: Multiple operating harmonics can be generated simultaneously. This capability is particularly advantageous for applications requiring complex or combined field excitations.

This innovative approach significantly improves machine adaptability, harmonic control, and performance optimization and therefore highly beneficial for advanced electric motor applications.

5.3.2.1 FSCW with Variable Pole-pair Number

Equation (5.21) provides the general MMF expression for a particular multiphase FSCW with an arbitrary number of phases, providing crucial insight into its behavior under different excitation conditions. The key parameter is ν_0, which determines the harmonic order induced in the MMF function, as demonstrated in Figure 5.15a–c. Using the 11-phase FSCW of Figure 5.13 as an example, the excitation cases $\nu_0 = 1, 3,$ and 5 are examined. The results show that the MMF waveform becomes increasingly sinusoidal as ν_0 decreases. On the other hand, the harmonic content increases linearly with ν_0 as described in (5.31).

$$-\nu_0\frac{\nu_{\max} + 1}{m} < g < \nu_0\frac{\nu_{\max} - 1}{m} \tag{5.30}$$

$$h_C \sim 2\nu_0\frac{\nu_{\max}}{m} \tag{5.31}$$

Here, $\nu_{\max}$ denotes the highest MMF harmonic order considered in the evaluation of the harmonic content (h_C). The behavior of h_C with ν_0 for the investigated 11-phase winding is

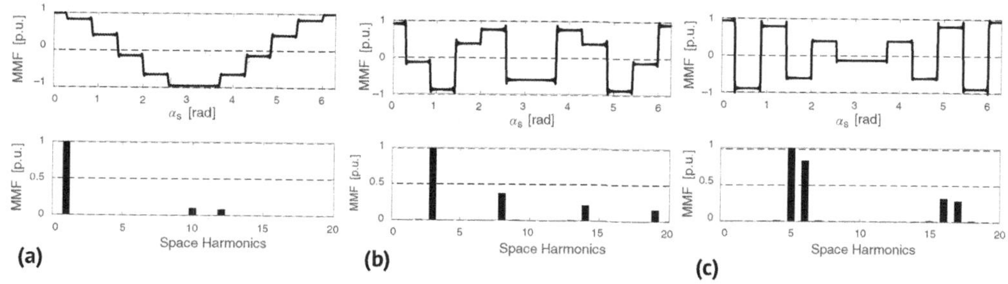

Figure 5.15 MMF characteristics for the 11-phase DL-FSCW. (a) MMF for $v_0 = 1$. (b) MMF for $v_0 = 3$. (c) MMF for $v_0 = 5$.

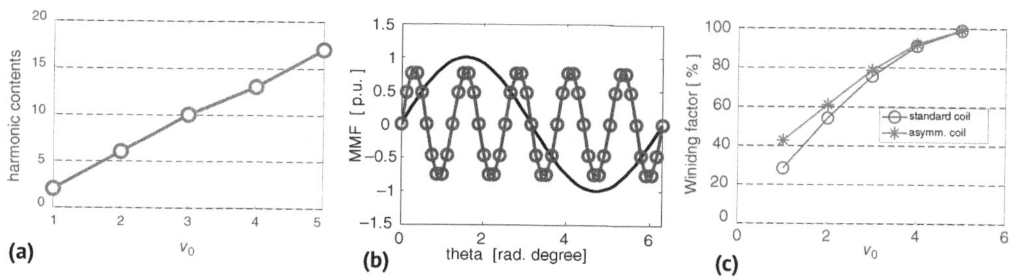

Figure 5.16 Winding characteristics for the DL-FSCW with $m = 11$. (a) Harmonic contents vs. v_0 parameter. (b) Multi pole-pairs operation $p = 1$ and $p = 5$. (c) Winding factors vs. v_0 parameter for the conventional and the asymmetric coil configuration.

shown in Figure 5.16a. For each excitation mode ν_0, the harmonic content is evaluated over the range from zero up to $\nu_{max} = 100$.

5.3.2.2 Multi-pole Operation mode

Another important capability of this winding configuration is the simultaneous generation of multiple pole-pair numbers. This is achieved by exciting the winding with multiple harmonic currents, as defined in Equation (5.32), where index-k identifies the injected harmonic current. The resulting MMF function under this excitation mode is given by Equation (5.33).

$$i_j(t) = \sum_{k=1}^{k_{max}} \hat{i}_k \cdot \cos\left(\omega_k t - \nu_{0,k} \cdot (j-1)\frac{2\pi}{m}\right) \tag{5.32}$$

$$F(t,\alpha_S) = \frac{m}{2}\frac{N_{w,eq}}{\pi}\sum_k\sum_\nu \frac{\hat{i}_k}{\nu_k}\,{}^{\nu}\xi_{p,eq}\cos\left(\omega_k t - \nu_k\alpha_S - (\nu_{0,k} - \nu_k)\frac{2\pi}{m}\right) \tag{5.33}$$

where,

$\nu_k = mg + \nu_{0,k}$ and $g = 0, \pm1, \pm2, \pm2, \ldots$

Using again the 11-phase FSCW example, Figure 5.16b demonstrates the simultaneous generation of two distinct MMF waves corresponding to $\nu_0 = 1$ and $\nu_0 = 5$ harmonics. Only the fundamental components are shown for clarity. As mentioned earlier, this configuration allows full independent control of the amplitude, frequency, and the rotation direction of various induced MMF components. Therefore, this design is highly versatile and well-suited for advanced machine concepts, including self-exciting synchronous machine (SESM) or multi pole-pair asynchronous machines (ASM), as studied in Dajaku and Gerling [72].

5.3.3 Multiphase FSCWs with Various Slot/Pole Combination

Unlike the conventional three-phase machines, where the stator slot number is always a multiple of three (e.g. $Q_S = 3, 6, 9, 12\ldots$), the FSCWs with an arbitrary number of phases introduced here enable greater flexibility in the stator design. For example, instead of a 12-slot stator, the electric machine may employ a non-traditional stator configuration, such as 10- or 11-slot stators, paired with the corresponding 10- or 11-phase windings. In addition, both single-layer and double-layer realizations are feasible, as illustrated in Figure 5.17.

This expanded design flexibility enhances the optimization potential concerning the cogging torque and torque ripples, power density, and noise and vibration issues. In addition, multiphase FSCWs can achieve very high winding factors for the working harmonics. Representative values for various SL- and DL-FSCW configurations are summarized in Tables 5.3 and 5.4, where the highest winding factors, reaching up to 99.6%, are highlighted to emphasize their significance.

One notable drawback of this winding configuration is that the winding factor for the injected working harmonic decreases when the number of pole-pairs is smaller than the number of phases (i.e. $\nu_0 \ll m$). In such cases, an asymmetric coil design, where the two coil sides carry an unequal number of turns, can be employed to enhance the winding factor. This effect can be observed in the results presented in Figure 5.16c for the case with $N_{w11} = 2$ and $N_{w12} = 1$. However, the benefit of adding an extra turn to one coil-side is modest unless the total number of turns per coil is small, a situation commonly encountered in low-voltage machines.

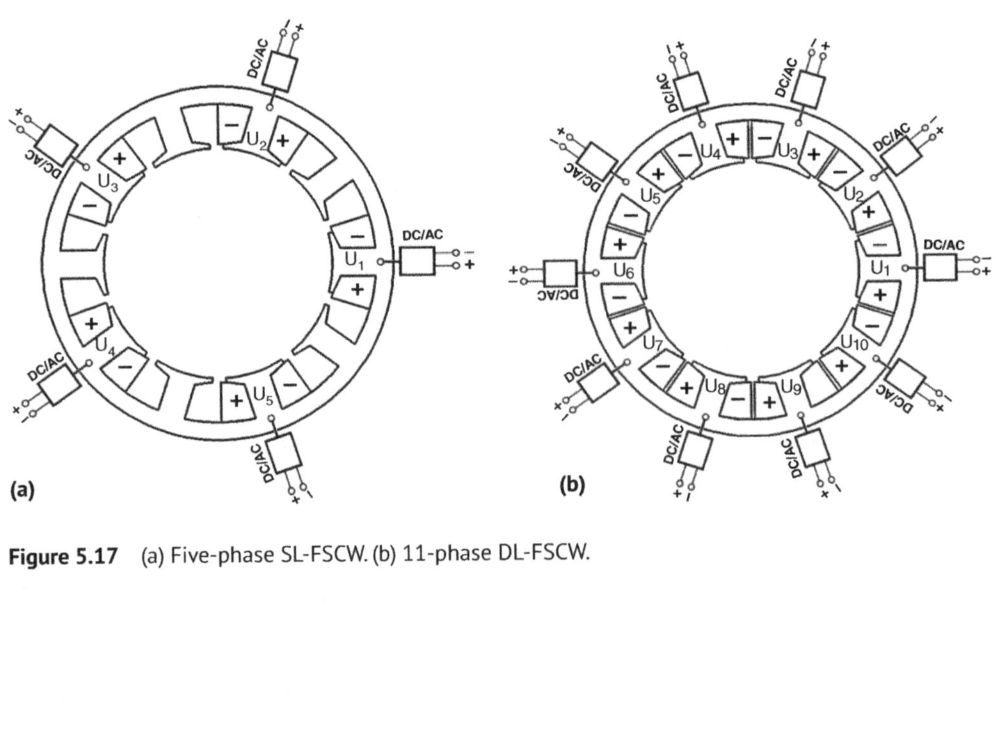

Figure 5.17 (a) Five-phase SL-FSCW. (b) 11-phase DL-FSCW.

Table 5.3 Winding factors (in %) for different multiphase DL-FSCWs with $m = Q_S$ (Bold > 80%; italics > 90%; bold italics > 95%; underline > 97%).

$2p/Q_S$	2	4	6	8	10	12	14	16	18	20
4	70.7	—	70.7	—	70.7	—	70.7	—	70.7	—
5	58.8	***95.1***	***95.1***	58.8	—	58.8	***95.1***	***95.1***	58.8	—
6	50	**86.6**	—	**86.6**	50	—	50	**86.6**	—	**86.6**
7	43.4	78.2	97.5	97.5	—	43.4	78.2	43.4	78.2	97.5
8	38.3	70.7	*92.5*	—	*92.5*	70.73	38.3	—	38.3	70.73
9	34.2	64.3	**86.6**	98.5	98.5	**86.6**	64.3	34.2	—	—
10	30.9	58.8	**80.9**	***95.2***	—	***95.2***	**80.9**	58.8	30.9	—
11	28.2	54.1	75.6	*91*	**99.1**	99.1	91	75.6	54.1	28.2
12	25.8	50	70.6	***86.7***	***96.7***	—	***96.7***	**86.7**	70.6	50
13	23.9	46.4	66.3	**82.3**	**93.5**	99.4	99.4	*93.5*	**82.3**	66.3
14	22.25	43.4	62.4	78.2	*90.15*	97.15	—	97.15	*90.15*	78.2
15	20.8	40.67	58.8	74.3	**86.6**	***95.2***	99.6	99.6	***95.2***	**86.6**
16	19.5	38.3	55.6	70.73	**83.2**	***92.4***	98.2	—	98.2	92.4
17	18.37	36.12	52.64	67.3	79.8	**89.5**	***96.2***	99.6	99.6	***96.2***
18	17.4	34.2	50	64.3	76.6	**86.6**	*94.4*	98.6	—	98.6

Table 5.4 Winding factors (in %) for different multiphase SL-FSCWs with $m = Q_S/2$ (Bold > 80%; italics > 90%; bold italics > 95%; underline > 97%).

$2p/Q_S$	2	4	6	8	10	12	14	16	18	20
8	19.5	38.27	55.5	—	**83.3**	*92.38*	98.1	—	98.1	*92.38*
10	30.90	57.78	**80.90**	***95.10***	—	***95.10***	**80.90**	58.78	30.90	—
12	25.88	50	—	**86.60**	*96.59*	—	***96.59***	**86.60**	—	50
14	22.25	43.38	62.34	78.18	*90.09*	97.49	—	97.49	*90.09*	78.18
16	19.5	38.26	55.55	—	**83.14**	*92.39*	98.08	—	98.08	*92.38*
18	17.36	34.20	50.00	64.27	76.60	**86.60**	*93.96*	98.48	—	98.48
20	15.64	30.90	45.39	58.77	—	**80.90**	**89.10**	***95.10***	98.76	—
22	14.23	28.17	41.54	54.06	65.48	65.48	**84.12**	*90.96*	***95.94***	98.98

5.4 Representative Multiphase FSCWs

As already mentioned, multiphase FSCWs with an arbitrary number of phases offer a simple physical structure and at the same time enable a variety of operation modes. The number of phases is defined by Equation (5.14), whereas the phase current function, which is essential for variable MMF excitation, is defined in Equations (5.15) and (5.32). The key parameter to predict the MMF waveform and its harmonic contents is ν_0. To demonstrate its impact, various example windings are analyzed below, focusing exclusively on the standard coils with equal turns on both sides.

5.4.1 Multiphase Double-layer FSCWs

Example 5.4 Five-phase DL-FSCW
For a five-phase double-layer winding, the required number of stator slots is $Q_S = 5$. Following the winding layout of Figure 5.17b, the example winding consists of one coil per phase, and with all phases having the same coil polarity. According to Equation (5.16), this configuration is able to produce two pole-pairs or MMF distributions. For the excitation modes $\nu_0 = 1$ and $\nu_0 = 2$, the corresponding MMF characteristics are shown in Figure 5.18. Since the working harmonic of the double-layer configuration is directly defined by ν_0, the corresponding operating waves in this case are the 1st or the 2nd harmonic, whose winding factors increases with ν_0.

Example 5.5 Seven-phase DL-FSCW
With $Q_S = 7$, the seven-phase double-layer winding is analogous in structure to the previous case. According to Equation (5.16), this configuration is able to produce three pole-pair MMF distributions, i.e. the possible excitation modes are $\nu_0 = 1$, $\nu_0 = 2$, and $\nu_0 = 3$, as shown in Figure 5.19. Compared with the five-phase design, the seven-phase winding produces a more sinusoidal MMF waveform for $\nu_0 = 1$ and achieves a higher winding factor for the highest operating harmonic, i.e. for $\nu_p = 3$.

Example 5.6 10-phase DL-FSCW
The 10-phase double-layer winding, which is illustrated in Figure 5.17b, allows four operation modes ($\nu_0 = 1$ to $\nu_0 = 4$), as shown in Figure 5.20. This configuration yields an MMF waveform that closely approximates a sinusoidal function when excited at $\nu_0 = 1$. In addition, Equations (5.32)

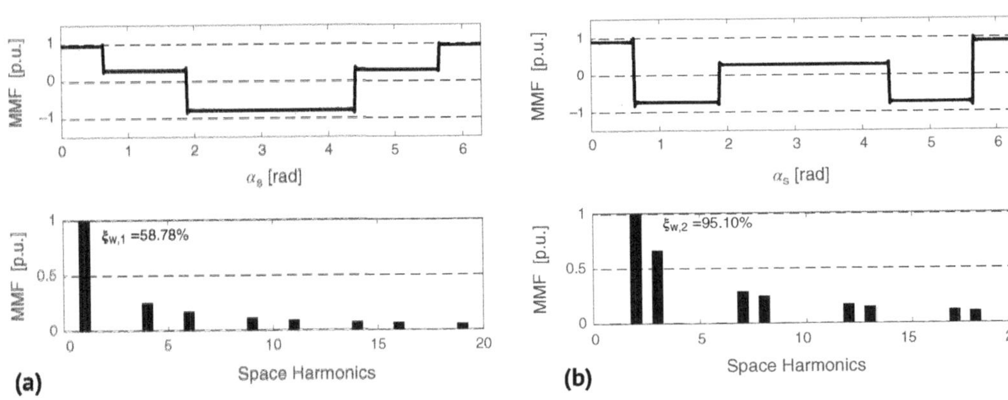

Figure 5.18 MMF distribution and harmonic contents for the DL-FSCW with $Q_s = 5$ and $m = 5$. (a) Excitation mode $\nu_0 = 1$. (b) Excitation mode $\nu_0 = 2$.

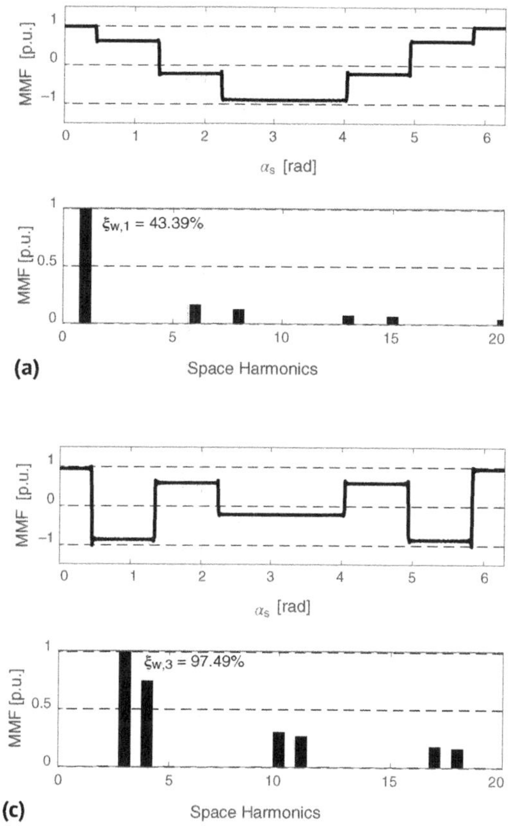

Figure 5.19 MMF distribution and harmonic contents for the DL-FSCW with $Q_s = 7$ and $m = 7$. (a) Excitation mode $v_0 = 1$. (b) Excitation mode $v_0 = 2$. (c) Excitation mode $v_0 = 3$.

and (5.33) show that multiple MMF waves may be generated simultaneously with independently controlled amplitude, rotation speed, and rotation direction. This is demonstrated in Figure 5.20d, where the 10-phase winding is simultaneously excited with $v_{0,1} = 1$ and $v_{0,2} = 2$. In this scenario, the current amplitudes of the first and second harmonics are chosen such that $I_1 = 0.5 \cdot I_2$.

5.4.2 Multiphase Single-layer FSCWs

Example 5.7 Five-phase SL-FSCW

The layout for the five-phase single-layer winding is presented in Figure 5.17a. According to Equations (5.14) and (5.16), this configuration requires 10 stator slots ($Q_S = 10$) and can generate two MMF waveforms, as demonstrated in Figure 5.21 for $v_0 = 1$ and $v_0 = 2$. In contrast to double-layer windings, where v_0 directly defines the operating wave, v_0 in single-layer configuration represents a subharmonic component, while the operating wave is defined as $v_p = m - v_0$. Thus, the working harmonics for this example are the 3rd or the 4th, with winding factors decreasing as v_0 increases.

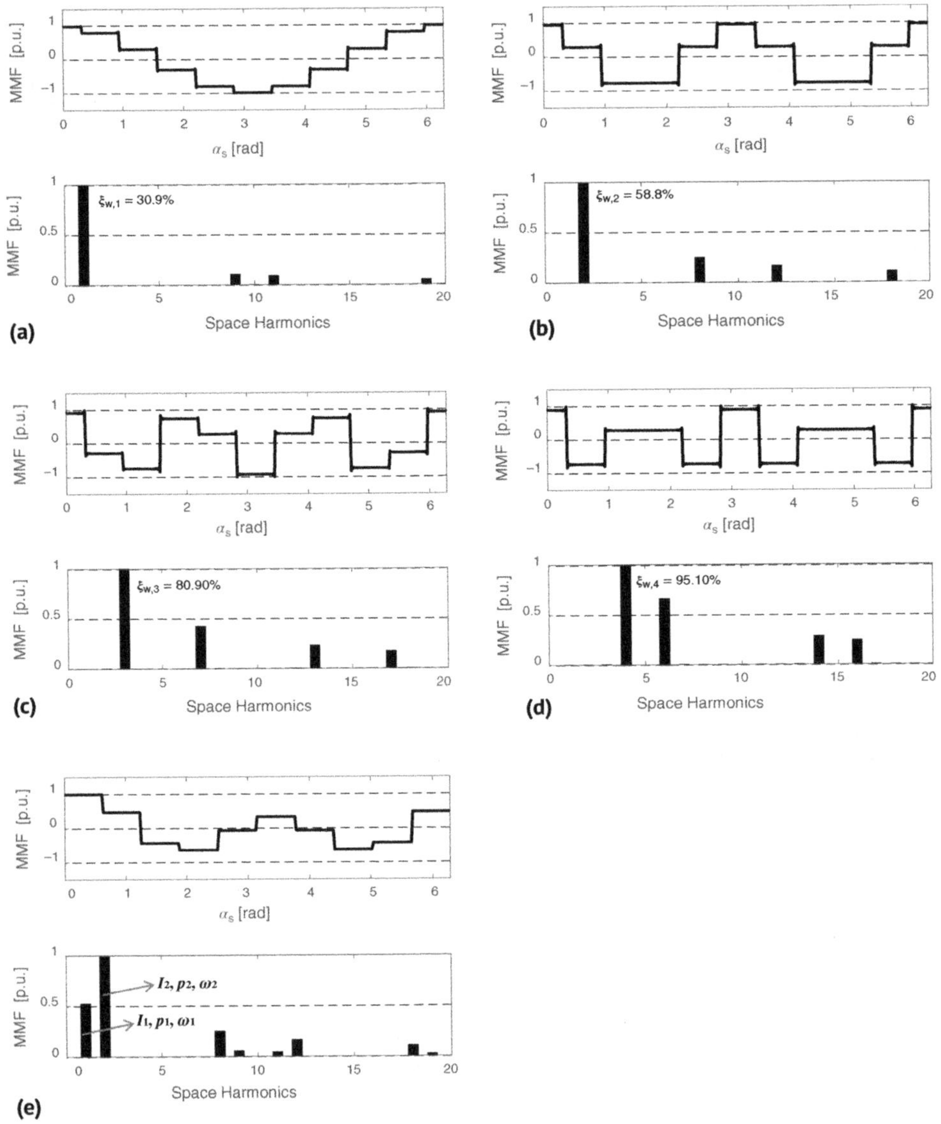

Figure 5.20 MMF distribution and harmonic contents for the DL-FSCW with $Q_S = 10$ and $m = 10$. (a) Excitation mode $\nu_0 = 1$. (b) Excitation mode $\nu_0 = 2$. (c) Excitation mode $\nu_0 = 3$, (d) Excitation mode $\nu_0 = 4$. (e) Multi pole-pair operation mode with $\nu_{0,1} = 1$ and $\nu_{0,2} = 2$.

Example 5.8 Seven-phase SL-FSCW

The winding layout of the seven-phase SL-FSCW is similar to that of the five-phase design, but it features 7 stator coils and 14 stator slots $(Q_S = 14)$. According to Equation (5.16), this topology can generate three MMF waveforms corresponding to $\nu_0 = 1$, $\nu_0 = 2$, and $\nu_0 = 3$, as shown in Figure 5.22. The corresponding working harmonics are the 6th, 5th, and 4th, respectively. As in the previous example, the highest winding factor is associated with highest harmonic order, in this case $\nu_p = 6$.

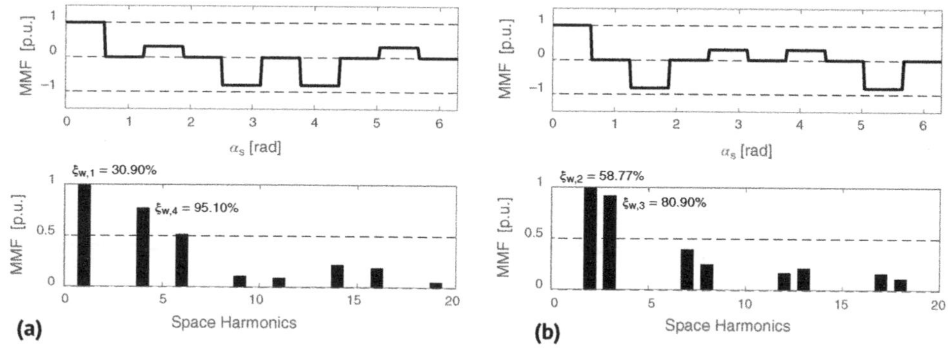

Figure 5.21 MMF distribution and harmonic contents for the SL-FSCW with $Q_s = 10$ and $m = 5$. (a) Excitation mode $v_0 = 1$. (b) Excitation mode $v_0 = 2$.

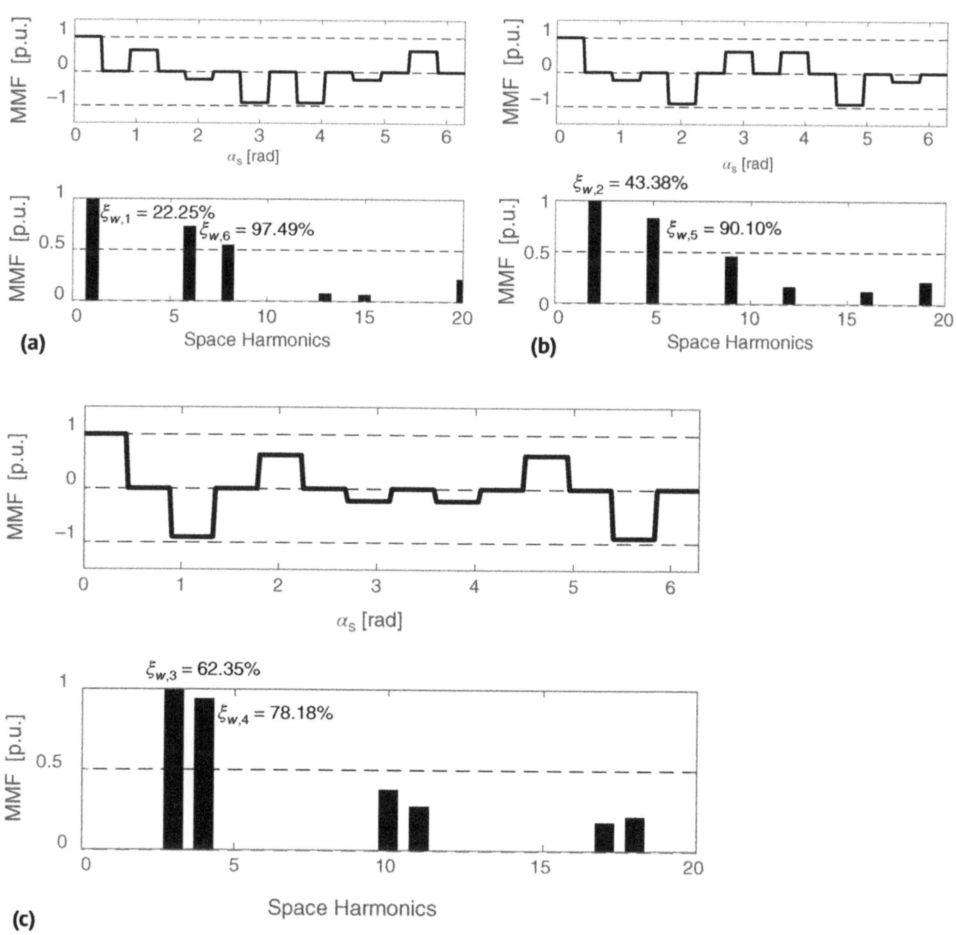

Figure 5.22 MMF distribution and harmonic contents for the SL-FSCW with $Q_s = 14$ and $m = 7$. (a) Excitation mode $v_0 = 1$. (b) Excitation mode $v_0 = 2$, (c) Excitation mode $v_0 = 3$.

Example 5.9 11-phase SL-FSCW

The final example is the 11-phase SL-FSCW, which consist of 11 coils and 22 stator slots ($Q_S = 22$). With this configuration five distinct MMF waveforms may be produced (shown in Figure 5.23), with the highest winding factor being 98.98% for the 10th MMF harmonic.

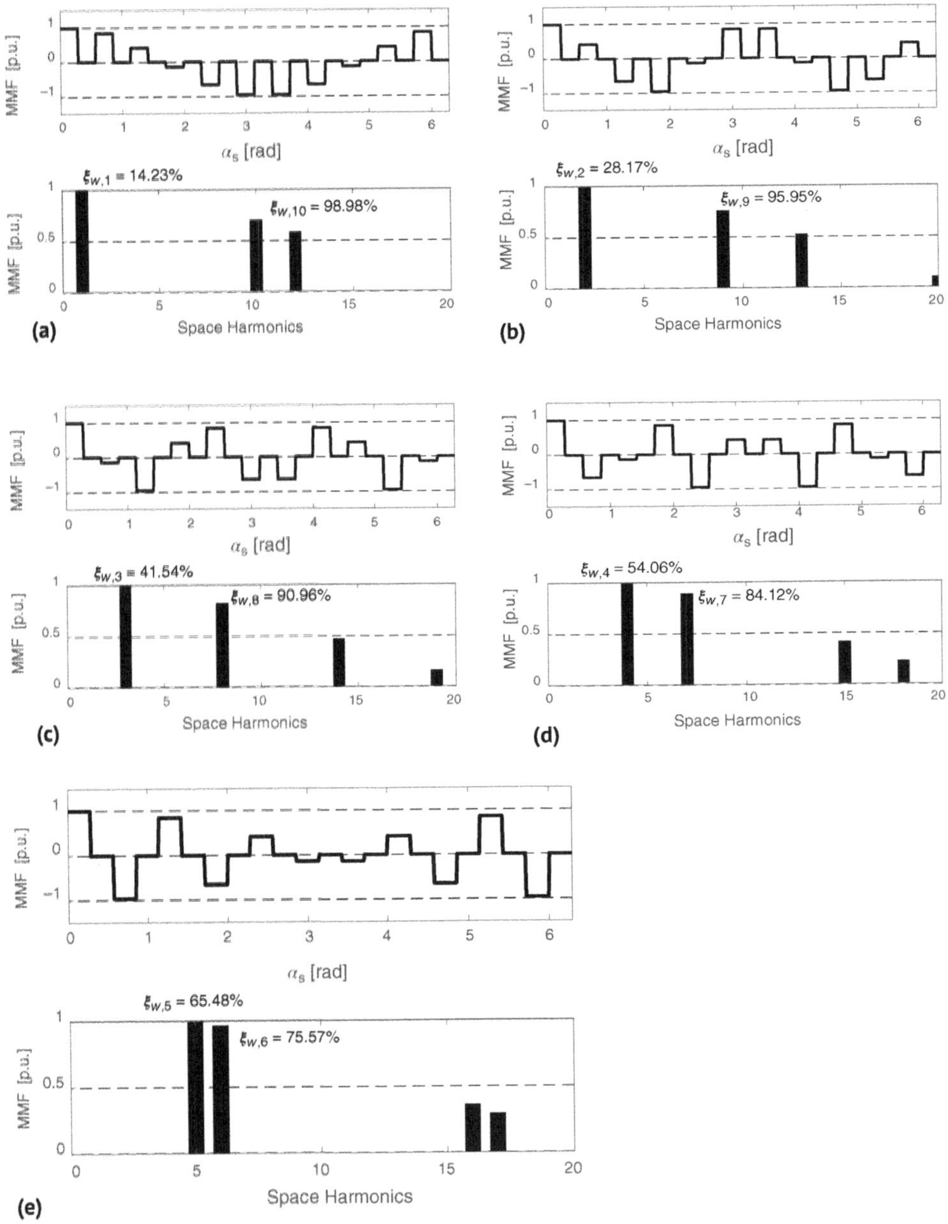

Figure 5.23 MMF distribution and harmonic contents for the SL-FSCW with $Q_s = 22$ and $m = 11$. (a) Excitation mode $\nu_0 = 1$. (b) Excitation mode $\nu_0 = 2$. (c) Excitation mode $\nu_0 = 3$. (d) Excitation mode $\nu_0 = 4$. (e) Excitation mode $\nu_0 = 5$.

5.4.3 Concluding Remarks

Based on the MMF results for different SL and DL multiphase FSCWs, the following conclusions can be drawn.

A. Multiphase DL-FSCWs
- Capable of generating different pole-pair numbers.
- Can operate simultaneously in multi-pole pair mode.
- Operating harmonic: $\nu_p = \nu_0$.
- No subharmonic components.
- Winding factor for the operating wave increases with ν_0.

B. Multiphase SL-FSCWs
- Capable of generating different pole-pair numbers.
- Can operate simultaneously in multi-pole pair mode.
- Operating harmonic: $\nu_p = m - \nu_0$.
- Includes subharmonic components: $\nu = \nu_0$.
- Winding factor for the operating wave decreases with increasing ν_0.

6

Fractional Slot Concentrated Windings with Flux-barrier Stator

As discussed in the previous chapters, electric machines with fractional slot concentrated windings (FSCWs) offer a wide range of slots/pole combinations. From a manufacturing perspective, the stator coils can be wound either on all teeth or on alternate teeth, as illustrated in Figure 6.1 for a 12-slots/14-poles permanent magnet synchronous motor (PMSM). Despite their many advantages, a key drawback of concentrated windings is their inherently non-sinusoidal MMF distribution, which contains numerous low- and high-order spatial harmonics. This behavior is clearly evident in the magnetomotive force (MMF) characteristics presented in Table 6.1 and Figure 6.2, where the 1st, 5th, 7th, 17th, and 19th harmonics dominate in both configurations. Since the 5th and 7th harmonics have relatively large winding factors, one of these components is typically selected as the working (or operating) harmonic. This selection leads to the well-known 12-slot/10-pole or the 12-slot/14-pole PMSMs. In the case of the 14-pole machine, for example, torque production is primarily achieved through the interaction between the rotor permanent magnet field and the 7th harmonic of the stator winding MMF. The remaining harmonic components, particularly the 1st, 5th, and higher-order harmonics, do not contribute to the generation of useful torque. Instead, they cause a number of undesirable electromagnetic effects, including:

- localized magnetic saturation in both the stator and rotor cores,
- increased iron losses in the stator and rotor,
- additional eddy current losses in the permanent magnets, and
- noise and vibration problems.

The negative effects of MMF harmonics in FSCWs have been extensively examined in the literature, as reported, for example, by Bianchi and Fornasiero [73], Zhu et al. [74], Yamazaki and Fukushima [75], Krotsch and Piepenbreier [76].

To address issues related to sub- and higher harmonic components, several novel techniques have been proposed and investigated in recent years. In general, these optimization methods can be classified into two principal categories:

I. Modification of the stator core structure: It has been shown that the application of magnetic flux-barriers in certain locations of the stator core can effectively suppress the subharmonics of the winding MMF. By strategically positioning flux gaps within the tooth region or along the stator yoke, the flux distribution within the stator core and thus the air-gap flux density harmonics can be controlled.

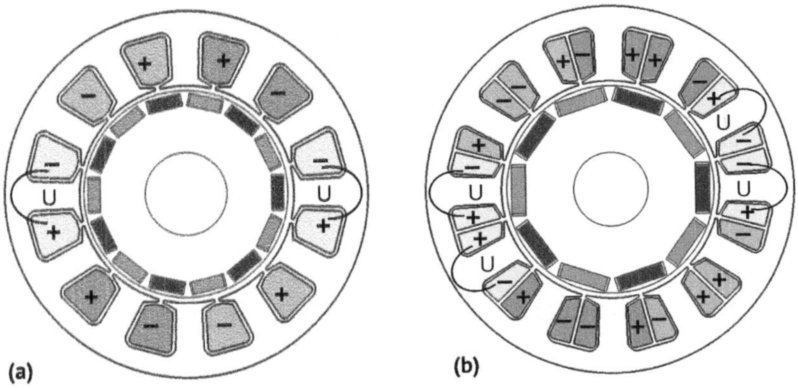

Figure 6.1 The winding layout for a 12-slot stator with concentrated winding and a conventional stator core. (a) Single-layer winding. (b) Double-layer winding.

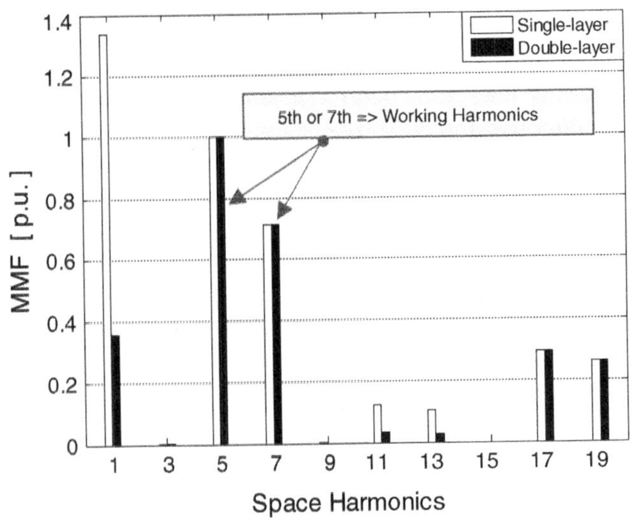

Figure 6.2 Comparison of MMF winding harmonics for the 12-slot SL- and DL-FSCW (MMF amplitudes normalized to the 5th harmonic).

Table 6.1 Winding factors of the 12-slot FSCW.

ξ_w^{ν}	1st	5th	7th
SL-FSCW	25.8%	96.6%	96.6%
DL-FSCW	6.7%	93.3%	93.3%

II. Modification of the winding layout: An alternative and equally promising approach involves optimizing the winding arrangement itself through specific coil configurations or combinations of multiple winding sets. This approach has led to numerous novel winding configurations, some of which are capable of simultaneously suppressing both subharmonic and critical higher-order MMF harmonics.

Among the various techniques available for improving FSCW performance, the flux-barrier approach stands out due to its conceptual simplicity and high effectiveness. This method is particularly well suited for single-layer-FSCWs (SL-FSCWs), which are typically characterized by high MMF harmonic content and relatively large subharmonic amplitudes. When applied to this winding category, the flux-barrier technique results in a substantial reduction of MMF subharmonics, making it more efficient and powerful than any other concentrated winding type used so far. Beyond the subharmonic suppression, this method offers numerous additional advantages, including the increase of the working harmonic component, the possibility of implementing direct stator cooling, and increased flexibility in slot-shape optimization, all of which contribute to improved manufacturability and overall machine performance.

This chapter provides a comprehensive investigation of the flux-barrier technique and its application to a range of winding topologies. By examining different stator core configurations with flux-barriers positioned at specific locations, the resulting effect on the MMF harmonic content, particularly on the subharmonic components reduction are systematically evaluated. Through a detailed analysis and comparison of various FSCWs in combination with flux-barrier stators, this chapter provides valuable insights into the design and optimization of flux-barrier PMSMs.

6.1 Stator Flux-barrier Techniques

As already mentioned, FSCWs are characterized by a rich presence of sub- and higher-order MMF harmonics, which leads to undesirable effects in electric machines. In particular, the subharmonic components in SL windings have higher amplitudes compared to double-layer windings with the same stator core. For example, as summarized in Table 6.1, the winding factor for the MMF subharmonic in a SL winding is nearly four times greater than that of the equivalent DL configuration. As a consequence, despite their relatively stronger working harmonics (i.e. 5th and 7th), SL-FSCW have been limited in many applications in the past.

Figure 6.3 shows the magnetic field distribution under a single-phase excitation for the selected example winding that clearly highlights the presence of a strong subharmonic component. Although this type of winding operates with a higher number of poles, e.g. a 10 or 14-pole rotor, the generated magnetic flux distribution resembles that of a two-pole machine. This behavior indicates that the subharmonic component dominates the entire stator and rotor core. Moreover, since it rotates six times faster than the working magnetic field (i.e. the 5th or the 7th harmonic), this

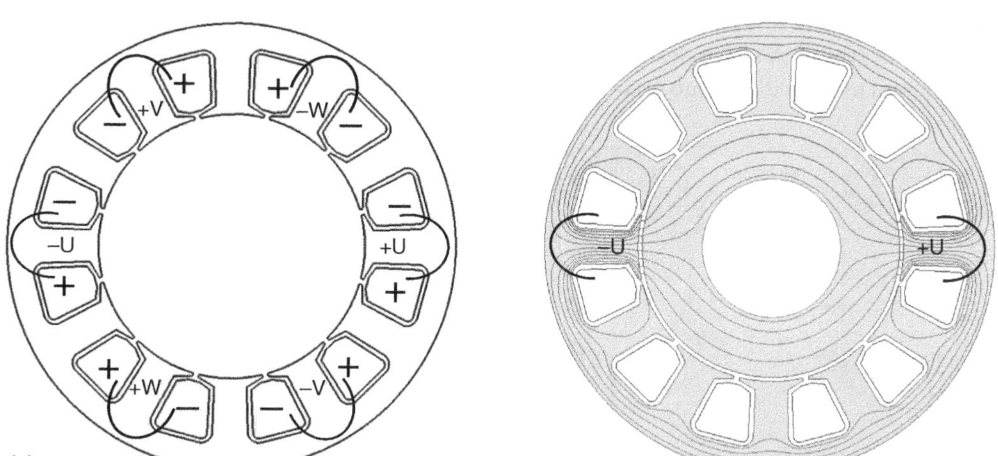

(a) (b)

Figure 6.3 (a) SL-FSCW with $z = 2$ and $n_c = 1$. (b) Flux density distribution under one-phase excitation.

leads to a substantial increase in iron losses in both the stator and rotor core, as well as eddy current losses in the permanent magnets. In addition, the strong subharmonic field increases the degree of magnetic saturation, which ultimately reduces the torque-per-ampere capability of the machine.

6.1.1 Flux-barrier Effect

To overcome the challenges associated by MMF subharmonics and to enable the practical use of these winding types, an optimization method based on the magnetic flux-barriers was initially introduced by Dajaku and Gerling [77, 78]. Since then, this technique has been widely investigated and further developed to address various issues in electric machines. Reported benefits include:

- improved torque density and efficiency in various FSCWs and DWs [79–88],
- improved thermal performance through enhanced cooling capability [89, 90],
- and reduced demagnetization risk in PMSMs [91].

The fundamental principle of the flux-barrier technique is to strategically insert flux gaps in areas of the stator magnetic circuit, either in the slot-adjacent regions of the stator-yoke or within the stator-teeth, as illustrated in Figure 6.4. In both cases, the flux-barriers are arranged alternately into the adjacent teeth or slot-yokes. Their main objective is to selectively increase the magnetic resistance (reluctance) associated with the first-order MMF harmonic, while leaving the reluctance seen by the working harmonic largely unaffected. This targeted modification in the magnetic circuit leads to a substantial reduction of the air-gap flux density components corresponding to MMF subharmonics when compared with a conventional stator design, as shown in Figure 6.5. In other words, the two-pole magnetic field is significantly weakened or entirely eliminated in machines employing magnetic flux gaps in the stator core.

6.1.2 Realization

From a manufacturing perspective, the flux-barrier stator can be realized using either a complete stator core structure or a modular design, both incorporating predefined flux-barrier regions. A fundamental requirement is that these regions must be made of a non-magnetic material to effectively prevent unwanted magnetic flux components. The use of non-conductive material is also preferred to avoid possible parasitic eddy current losses caused by stray flux components.

Figure 6.6 illustrates several representative stator configurations incorporating flux-barriers within the tooth region. A stator design with separate modular sections is shown in Figure 6.6a, which is referred to as an E-core modular stator in Zhou et al. [90]. Another construction, depicted

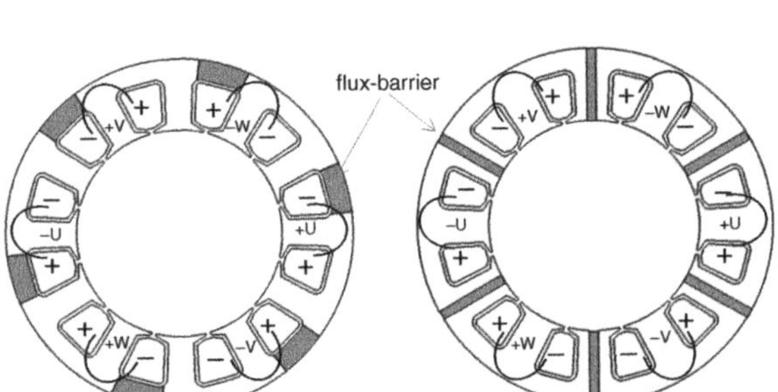

Figure 6.4 SL-FSCW with $z = 2$ and $n_c = 1$. (a) Stator with flux-barriers in stator yoke. (b) Stator with flux-barriers in stator teeth.

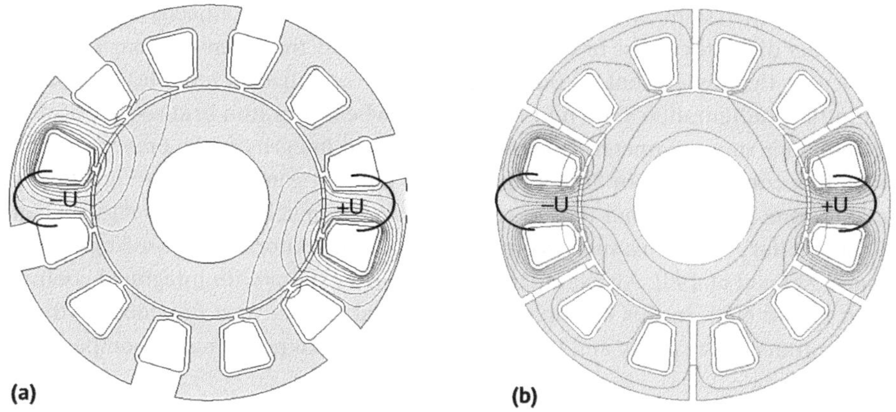

Figure 6.5 (a) SL-FSCW with $z = 2$ and $n_c = 1$, and under one-phase excitation conditions. (b) Flux density distribution with flux-barriers in stator yoke.

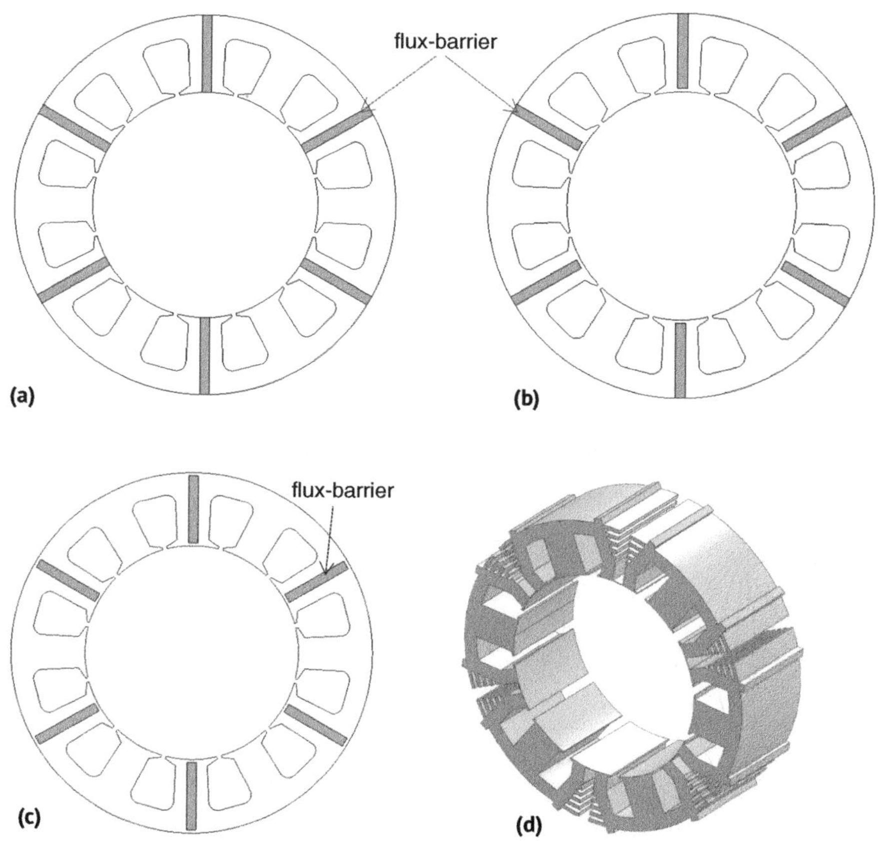

Figure 6.6 Stator cross section for different flux-barrier configurations. (a) Modular design. (b) Single sheet design with iron bridges in the stator inner diameter. (c) Single sheet design with iron bridges on both the inner and outer stator diameters. (d) Stator with integrated cooling fins in flux-barrier regions.

in Figure 6.6b, features a unified stator sheet in which the stator modules are connected by thin iron bridges in the stator inner diameter. These bridges must be sufficiently thin to ensure magnetic saturation at low flux levels, thereby preventing flux coupling between adjacent modules. A similar concept is applied in the configuration shown in Figure 6.6c, where thin iron bridges are located at both the inner and outer stator diameters to ensure mechanical integrity while preventing the unwanted magnetic flux components. In all configurations, the flux gap areas are magnetically inactive and therefore provide additional usable space within the machine. These regions can therefore be used for auxiliary functions, such as direct cooling of the stator, as proposed by Nollau and Gerling [89], and Zhou et al. [90]. An example of a flux-barrier stator with integrated cooling fins is illustrated in Figure 6.6d. In this example, the additional surface area provided by these fins improves the heat transfer capability and leads to superior thermal performance compared to standard stators.

6.2 FSCWs with Stator-yoke Flux-barriers

A simplified representation of an electric machine with Q_S stator slots, a solid rotor, and a single excitation coil wound around one tooth are presented in Figure 6.7. The Figure also illustrates the resulting magnetic flux lines produced when the coil is excited with a current. Based on the flux distribution, the spatial MMF distribution can be approximated by a square pulse with a width of $2\pi/Q_S$, as depicted in Figure 6.7b. According to (2.14), the corresponding MMF function is given by

$$F_{\text{Coil}}(t, a_S) = \frac{2N_w i(t)}{\pi} \sum_{\nu} \frac{1}{\nu} {}^{\nu}\xi_p \cdot \cos(\nu a_S) \tag{6.1}$$

where the coil-pitch factor, previously defined in Equation (2.15) is

$${}^{\nu}\xi_p = \sin\left(\nu \frac{\pi}{Q_S}\right)$$

Figure 6.8 illustrates the modified stator configuration with flux-barriers in the yoke region. In this arrangement, slot-shaped gaps are embedded within the yoke area of alternating stator slots. The stator tooth carrying the excitation coil is labeled as region "1," the adjacent tooth without flux-barriers in between is labeled as region "2," and the remaining stator circumference is labeled as stator region "3." Assuming a rectangular flux-gap geometry, the dominant geometric parameters

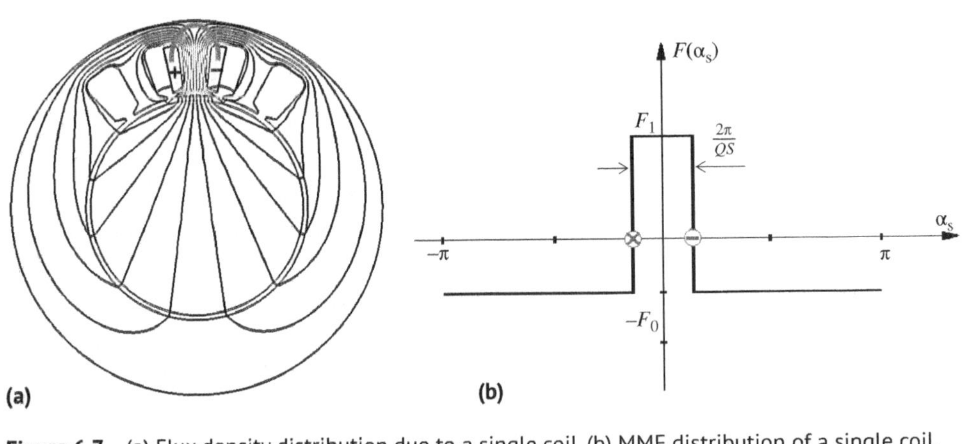

Figure 6.7 (a) Flux density distribution due to a single coil. (b) MMF distribution of a single coil.

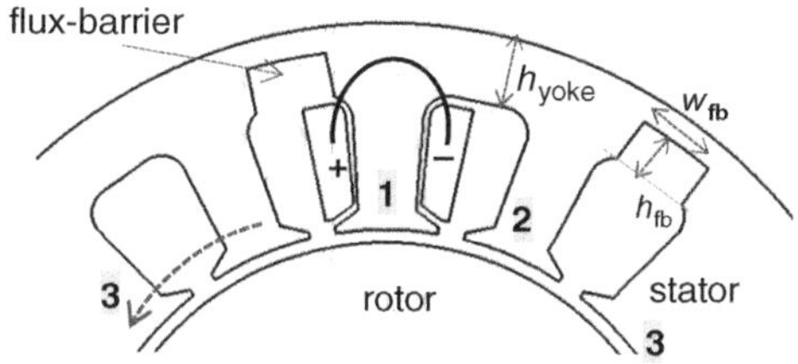

Figure 6.8 Cross section of an electric machine with flux-barriers in the stator-yoke.

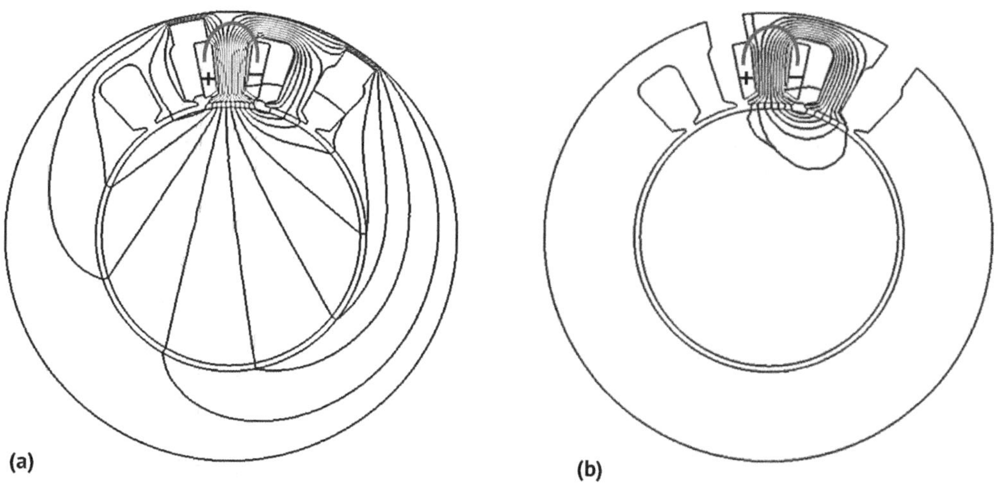

Figure 6.9 Flux density distribution due to a single excitation coil. (a) Flux-barrier stator with $h_{fb} < h_{yoke}$. (b) Flux-barrier stator with $h_{fb} = h_{yoke}$.

are the barrier depth (h_{fb}) and width (w_{fb}), which directly determine the ability of the flux-barriers to modify the MMF waveform and its specific harmonics. To evaluate their impact, two limited cases are considered.

1. *Partial yoke flux-barrier* (Figure 6.9a): In this scenario, the barrier depth is smaller than the stator yoke thickness $(h_{fb} < h_{yoke})$, but sufficiently large to increase the local magnetic resistance. As a result, the flux density in stator region "3" decreases compared to a conventional stator, while the amount of the reduced flux is transferred (concentrated) in the adjacent tooth "2."
2. *Full yoke flux-barrier* (Figure 6.9b): Here, the barrier depth is equal to the yoke thickness $(h_{fb} = h_{yoke})$, thereby completely interrupting the flux path in the stator yoke. Under this condition, the entire coil flux is forced to pass through the adjacent stator tooth "2," resulting in zero flux in the stator region "3." As a result, the presence of flux-barriers causes a complete flux concentration effect in stator tooth "2."

Based on the flux density behaviors described above and shown in Figure 6.9, the resulting MMF distribution, including the flux-barrier effect, is approximated by the dashed line in Figure 6.10. For the conventional stator, the MMF function parameters F_0 and F_1 are defined in Equation (2.12). On the other hand, for the flux-barrier stator, the modified MMF magnitudes associated with stator regions 1, 2, and 3 are denoted as F^*_{11}, F^*_{12}, and F^*_0, respectively. These quantities are determined based on the flux density behavior with the barrier depth, as described below.

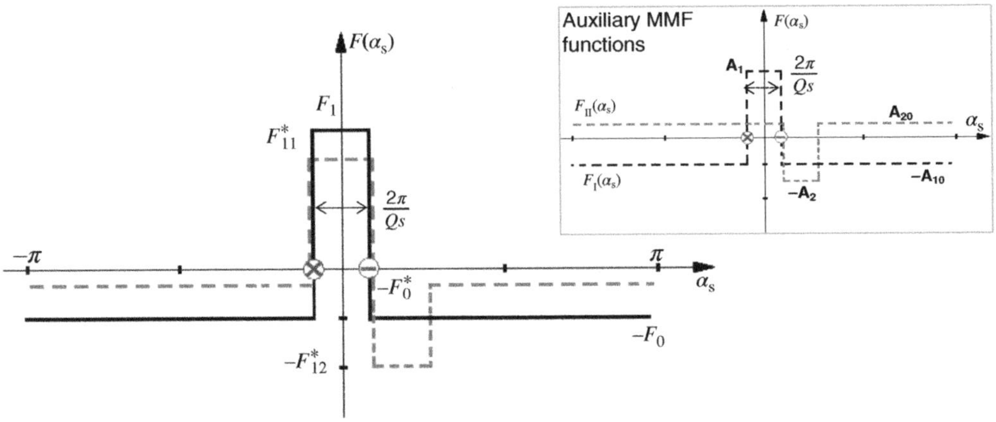

Figure 6.10 Flux-barrier effect on the MMF distribution of a single coil. Solid line for the conventional stator, dashed line for the flux-barrier stator.

Stator region "3": As the barrier depth increases, the flux density in this region decreases and disappears completely when the yoke is fully interrupted. Accordingly, the MMF function magnitude $\left(F_0^*\right)$ decreases linearly from its initial value F_0 to zero as h_{fb} increases from 0 to h_{yoke}, yielding

$$F_0^* = \left(1 - \frac{h_{\text{fb}}}{h_{\text{yoke}}}\right) F_0 = \left(1 - \frac{h_{\text{fb}}}{h_{\text{yoke}}}\right) \frac{1}{Q_S} N_w \tag{6.2}$$

Stator tooth "1": In order to keep the total MMF excitation of the winding equal to that of the original design, the increase in the MMF magnitude F_{12}^* at tooth "2," caused by flux concentration, must be compensated by a corresponding reduction in the MMF contribution F_{11}^* of the excitation coil on tooth "1." This condition is expressed as

$$F_{11}^* = \begin{cases} F_1 = \dfrac{2\pi - a_\tau}{2\pi} N_w, & h_{\text{fb}} = 0 \\[2ex] \dfrac{1}{2} N_w, & h_{\text{fb}} = h_{\text{yoke}} \end{cases} \tag{6.3}$$

from which, the F_{11}^* for the tooth region "1" is defined as

$$F_{11}^* = \frac{N_w}{2} \left\{ \frac{\pi - a_\tau}{\pi} \left(1 - \frac{h_{\text{fb}}}{h_{\text{yoke}}}\right) + 1 \right\} \tag{6.4}$$

Stator tooth "2": In tooth "2," the local increase in flux density due to the flux-barrier effect results in an enhanced MMF magnitude, which can be written as

$$F_{12}^* = \begin{cases} F_0 = \dfrac{\alpha_\tau}{2\pi} N_w, & h_{\text{fb}} = 0 \\[2ex] \dfrac{1}{2} N_w, & h_{\text{fb}} = h_{\text{yoke}} \end{cases} \tag{6.5}$$

leading to the following expression for the variation of F^*_{12} with the barrier depth

$$F^*_{12} = \frac{N_w}{2}\left\{\frac{\pi - \alpha_\tau}{\pi}\left(\frac{h_{\mathrm{fb}}}{h_{\mathrm{yoke}}} - 1\right) + 1\right\} \tag{6.6}$$

6.2.1 Equivalent MMF Function

To derive the analytical expression for the modified MMF distribution resulting from the flux-barrier effect, a superposition approach is employed. The total MMF is expressed as the sum of two auxiliary MMF functions F_I and F_{II}, spatially shifted by $2\pi/Q_S$, as illustrated in Figure 6.10 (top right). According to Equation (6.1), the corresponding MMF functions are

$$F_I(t, \alpha_S) = \frac{2\widehat{F}_I}{\pi} i(t) \sum_\nu \frac{{}^\nu\xi_p}{\nu} \cdot \cos(\nu\alpha_S) \tag{6.7}$$

$$F_{II}(t, \alpha_S) = \frac{2\widehat{F}_{II}}{\pi} i(t) \sum_\nu \frac{{}^\nu\xi_p}{\nu} \cdot \cos(\nu(\alpha_S - 2\pi/Q_S) + \pi) \tag{6.8}$$

where their respective magnitudes are given by,

$$\begin{aligned} \widehat{F}_I &= \mathrm{A}_1 + \mathrm{A}_{10} \\ \widehat{F}_{II} &= \mathrm{A}_2 + \mathrm{A}_{20} \end{aligned} \tag{6.9}$$

According to the modelling approach shown in Figure 6.10, the auxiliary functions must satisfy the following conditions

$$\begin{aligned} \mathrm{A}_1 + \mathrm{A}_{20} &= F^*_{11} \\ \mathrm{A}_2 + \mathrm{A}_{10} &= F^*_{12} \end{aligned} \tag{6.10}$$

and

$$\begin{aligned} \mathrm{A}_{10} &= \frac{\mathrm{A}_1}{Q_S - 1} \\ \mathrm{A}_{20} &= \frac{\mathrm{A}_2}{Q_S - 1} \end{aligned} \tag{6.11}$$

Solving this system yields to

$$\begin{aligned} \mathrm{A}_1 &= F^*_{11} - \left(F^*_{12} - \frac{F^*_{11}}{Q_S - 1}\right)\bigg/\left(Q_S - 1 - \frac{1}{Q_S - 1}\right) \\ &= \frac{N_w}{2}\left(1 - \frac{\alpha_\tau}{2\pi}\right)\left(2 - \frac{h_{\mathrm{fb}}}{h_{\mathrm{yoke}}}\right) \end{aligned} \tag{6.12}$$

$$\begin{aligned} \mathrm{A}_2 &= \left(F^*_{12} - \frac{F^*_{11}}{Q_S - 1}\right)\bigg/\left(1 - \frac{1}{(Q_S - 1)^2}\right) \\ &= \frac{N_w}{2}\left(1 - \frac{\alpha_\tau}{2\pi}\right)\left(\frac{h_{\mathrm{fb}}}{h_{\mathrm{yoke}}}\right) \end{aligned} \tag{6.13}$$

Substituting Equations (6.11)–(6.13) into (6.9) results in

$$\begin{aligned}\widehat{F}_I &= A_1 + A_{10} = A_1\frac{Q_S}{Q_S-1}\\ &= \frac{N_w}{2}\left(2-\frac{h_{fb}}{h_{yoke}}\right)\end{aligned} \tag{6.14}$$

$$\begin{aligned}\widehat{F}_{II} &= A_2 + A_{20} = A_2\frac{Q_S}{Q_S-1}\\ &= \frac{N_w}{2}\frac{h_{fb}}{h_{yoke}}\end{aligned} \tag{6.15}$$

Finally, using Equations (6.7) and (6.8), the modified coil MMF function accounting the stator-yoke flux-barrier effect is obtained as

$$\begin{aligned}F_{fb}(t,a_S) &= \frac{2i(t)}{\pi}\left\{\sum_\nu \frac{\widehat{F}_I}{\nu}{}^\nu\xi_p \cos(\nu a_S) + \sum_\nu \frac{\widehat{F}_{II}}{\nu}{}^\nu\xi_p \cos(\nu(a_S - 2\pi/Q_S)+\pi)\right\}\\ &= \frac{2(\widehat{F}_I+\widehat{F}_{II})i(t)}{\pi}\sum_\nu \frac{{}^\nu\xi_{p,fby}}{\nu}\sin\left(\nu a_S + {}^\nu\varphi_{1,2}\right)\end{aligned} \tag{6.16}$$

with

$${}^\nu\xi_{p,fby} = \frac{|{}^\nu\xi_p|}{\widehat{F}_I+\widehat{F}_{II}}\sqrt{(\widehat{F}_I)^2 + (\widehat{F}_{II})^2 - 2\widehat{F}_I\widehat{F}_{II}\cos\left(\nu\frac{2\pi}{Q_S}\right)} \tag{6.17}$$

and

$$\varphi_{1,2} = a\tan\left(\frac{\widehat{F}_I\sin(\varphi_1)+\widehat{F}_{II}\sin(\varphi_2)}{\widehat{F}_I\cos(\varphi_1)+\widehat{F}_{II}\cos(\varphi_2)}\right) = a\tan\left(\frac{\widehat{F}_I - \widehat{F}_{II}\cos\left(\nu\frac{2\pi}{Q_S}\right)}{\widehat{F}_{II}\sin\left(\nu\frac{2\pi}{Q_S}\right)}\right) \tag{6.18}$$

Substituting (6.14) and (6.15) into the above relations yields explicit formulae for the coil MMF function, the associated phase shift, and the equivalent coil-pitch factor.

Modified coil MMF function

$$F_{fb}(t,a_S) = \frac{2N_w i(t)}{\pi}\sum_\nu \frac{{}^\nu\xi_{p,fby}}{\nu}\sin\left(\nu a_S + {}^\nu\varphi_{1,2}\right) \tag{6.19}$$

Phase shift

$${}^\nu\varphi_{1,2} = a\tan\left(\frac{2\frac{h_{yoke}}{h_{fb}} - 1 - \cos\left(\nu\frac{2\pi}{Q_S}\right)}{\sin\left(\nu\frac{2\pi}{Q_S}\right)}\right) \tag{6.20}$$

Equivalent coil-pitch factor

$${}^\nu\xi_{p,fby} = |{}^\nu\xi_p|\cdot{}^\nu\Gamma_{fby} \tag{6.21}$$

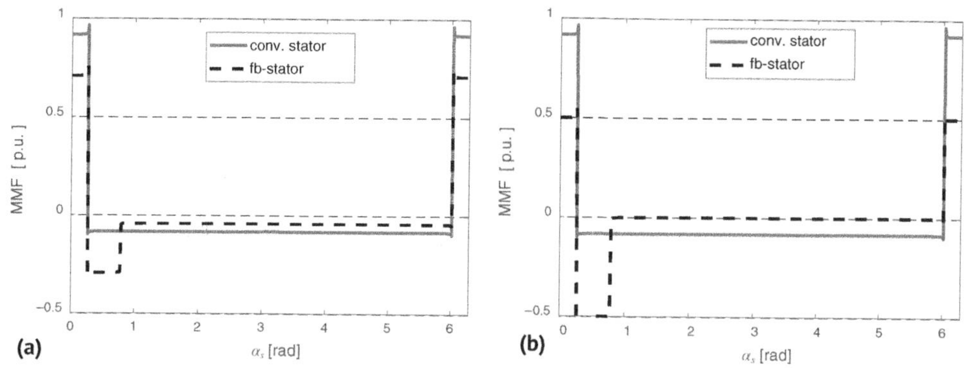

Figure 6.11 Flux-barrier effect on the MMF distribution of a single coil. (a) Flux-barrier stator with $h_{fb} = 0.5h_{yoke}$. (b) Flux-barrier stator with $h_{fb} = h_{yoke}$.

$$^{\nu}\Gamma_{fby} = \frac{1}{2}\sqrt{\left(2 - \frac{h_{fb}}{h_{yoke}}\right)^2 + \left(\frac{h_{fb}}{h_{yoke}}\right)^2 - 2\left(2 - \frac{h_{fb}}{h_{yoke}}\right)\frac{h_{fb}}{h_{yoke}}\cos\left(\nu\frac{2\pi}{Q_S}\right)} \quad (6.22)$$

Equations (6.21) and (6.22) explicitly show that the presence of stator-yoke flux-barriers directly affects the winding coil-pitch factor. This property can therefore be used to selectively influence or suppress specific MMF harmonics, which provides an additional degree of freedom for machine performance optimization.

To demonstrate this effect, two representative examples are shown in Figure 6.11. The studied topologies employ a stator with $Q_S = 12$ slots, a single excitation coil, and barrier depth equal to 50% and 100% of the stator yoke thickness. Compared to a conventional stator, the presence of stator-yoke flux-barriers produces a significant modification of the MMF waveform. In the limiting case where $h_{fb} = h_{yoke}$, the resulting MMF of a single coil becomes similar to that of a concentrated double-layer winding with two adjacent coils of opposite polarity, as studied in Section 2.2.2 and illustrated in Figure 2.14. In the following plots, the legend label "fb-stator" denotes the modified stator design including flux-barriers in the stator yoke.

6.2.2 SL-FSCWs with Stator-yoke Flux-barrier

As described in Chapter 2, the single-layer concentrated winding can be realized using either a single-zone or two-zone configuration. Within each zone, the coil arrangement may be implemented with either the same or opposite polarity. In order to achieve higher winding factors for the operating harmonic, configurations with coil arrangements of same polarity (e.g. +U+U+U,...) show better performance. Therefore, the following analysis is focused exclusively on the implementation of flux-barrier technique to this winding category.

6.2.2.1 SL-FSCW with $z = 1$ and $n_c \geq 1$

Following the methodology introduced previously, the auxiliary functions for a single-layer concentrated winding with $z = 1$ and $n_c \geq 1$ are obtained from Equation (2.23) and expressed in (6.23) and (6.24). Here, the distribution factor $^{\nu}\xi_d$ is defined in (2.24), while the parameters $\hat{F}_I$ and $\hat{F}_{II}$ have already been determined in (6.14) and (6.15).

$$F_{j,I}(t,a_S) = \frac{2\widehat{F}_I}{\pi} i_j(t) \sum_{\nu} \frac{{}^{\nu}\xi_p\, {}^{\nu}\xi_d}{\nu} \cdot\cos\left(\nu\left(a_S - (n_c - 1)\frac{2\pi}{Q_S}\right)\right) \tag{6.23}$$

$$F_{j,II}(t,a_S) = \frac{2\widehat{F}_{II}}{\pi} i_j(t) \sum_{\nu} \frac{{}^{\nu}\xi_p\, {}^{\nu}\xi_d}{\nu} \cdot\cos\left(\nu\left(a_S - n_c\frac{2\pi}{Q_S}\right) - \pi\right) \tag{6.24}$$

Using the above equations, the resulting MMF distribution, including the stator-yoke flux-barrier effect, is

$$\begin{aligned} F_{j,\mathrm{fb}}(t,a_S) &= F_{j,I}(t,a_S) + F_{j,I}(t,a_S) \\ &= \frac{2n_c N_w}{\pi} i_j(t) \sum_{\nu} \frac{{}^{\nu}\xi_{w,\mathrm{fby}}}{\nu} \cdot\cos\left(\nu a_S + {}^{\nu}\varphi_{1,2}\right) \end{aligned} \tag{6.25}$$

The resulting equivalent winding factor is the product of the coil-pitch, distribution, and the flux-barrier factor,

$${}^{\nu}\xi_{w,\mathrm{fby}} = \left|{}^{\nu}\xi_p\, {}^{\nu}\xi_d\right| \cdot {}^{\nu}\Gamma_{\mathrm{fby}} \tag{6.26}$$

Meanwhile, the phase shift angle is defined as

$${}^{\nu}\varphi_{1,2} = a\tan\left(\frac{\left(2 - \frac{h_{\mathrm{fb}}}{h_{\mathrm{yoke}}}\right)\sin\left(\nu(n_c - 1)\frac{2\pi}{Q_S}\right) - \frac{h_{\mathrm{fb}}}{h_{\mathrm{yoke}}}\sin\left(\nu n_c\frac{2\pi}{Q_S}\right)}{\left(2 - \frac{h_{\mathrm{fb}}}{h_{\mathrm{yoke}}}\right)\cos\left(\nu(n_c - 1)\frac{2\pi}{Q_S}\right) - \frac{h_{\mathrm{fb}}}{h_{\mathrm{yoke}}}\cos\left(\nu n_c\frac{2\pi}{Q_S}\right)}\right) \tag{6.27}$$

Example 6.1 SL-FSCW with $z = 1$, $n_c = 1$, and stator-yoke flux-barriers
The first example considers the winding configuration shown in Figure 6.12a, with $z = 1$, $n_c = 1$, and with flux-barriers on the stator yoke. When the barriers completely remove the yoke area adjacent to alternating stator slots, their impact on the MMF characteristics is examined in Figures 6.12b and 6.13a. Further investigations in Figure 6.13b show the influence of barrier depth on the dominant MMF winding harmonics, particularly the 1st and the 2nd harmonics. The results indicate that the applied technique significantly affects the harmonic content of the winding, with pronounced reduction of the 1st subharmonic component. The most effective reduction of subharmonics is achieved when the barrier depth equals the yoke height ($h_{\mathrm{fb}} = h_{\mathrm{yoke}}$), corresponding to the complete removal of the targeted yoke section. At the same time, the operating harmonic is reduced by around 13%.

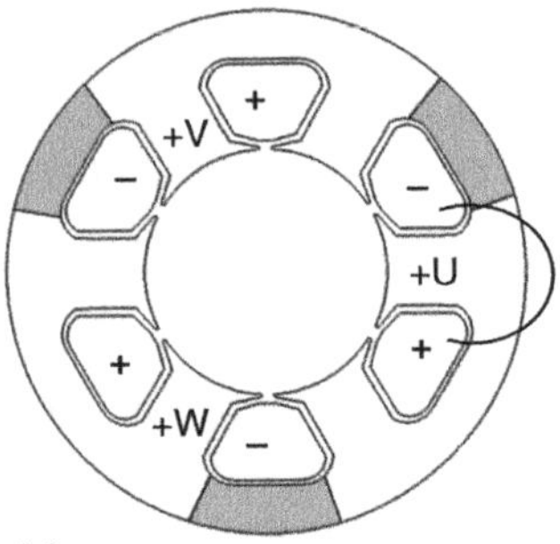

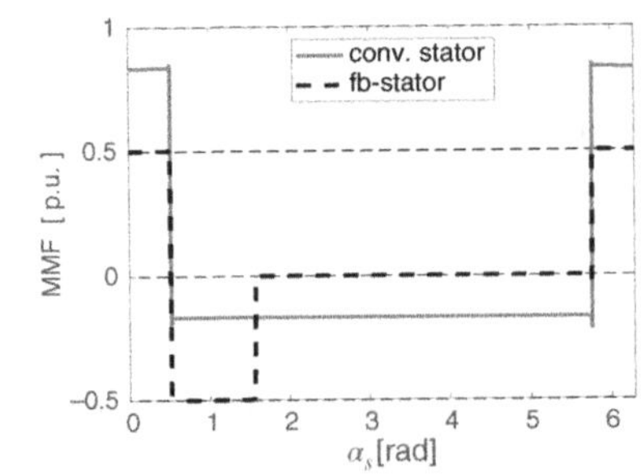

Figure 6.12 (a) Three-phase SL-FSCW with $z = 1$, $n_c = 1$, and flux-barrier stator. (b) MMF distribution of phase-j for both the conventional stator and the flux-barrier stator with $h_{\mathrm{fb}} = h_{\mathrm{yoke}}$.

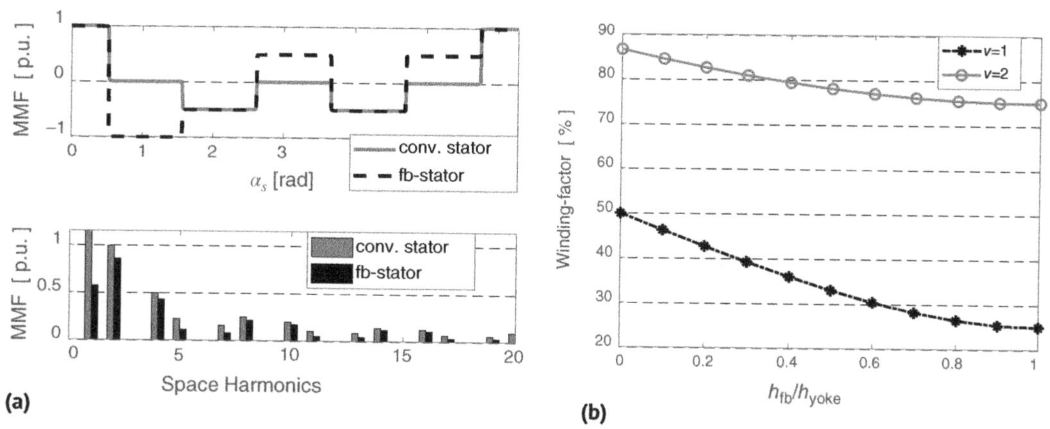

Figure 6.13 Three-phase SL-FSCW with $z = 1$ and $n_c = 1$. (a) Comparison of the MMF characteristics between the conventional stator and the flux-barrier stator with $h_{fb} = h_{yoke}$. (b) Flux-barrier effect on the MMF harmonic winding factors.

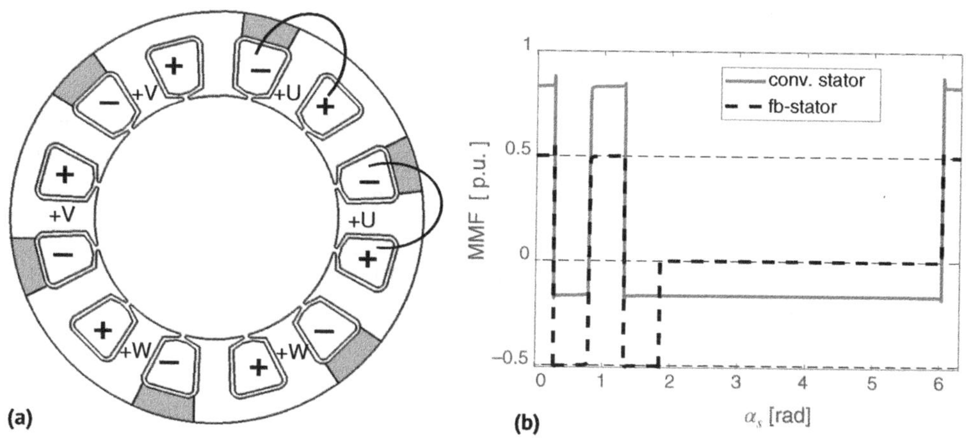

Figure 6.14 (a) Three-phase SL-FSCW with $z = 1$, $n_c = 2$, and flux-barrier stator. (b) MMF distribution of phase-j for both the conventional stator and the flux-barrier stator with $h_{fb} = h_{yoke}$.

Example 6.2 SL-FSCW with $z = 1$, $n_c = 2$ and stator-yoke flux-barriers

A second example, shown in Figure 6.14, considers a winding with $z = 1$ and $n_c = 2$. In this configuration, the 5th MMF harmonic serves as a working harmonic, while the 1st, 2nd, and 4th harmonics behave as subharmonics, which makes this winding type unsuitable for many applications. Nevertheless, as shown in Figure 6.15, the applications of stator-yoke flux-barriers significantly improve the MMF characteristics by effectively suppressing the undesirable harmonic components.

Conclusion: The impact of flux-barriers on the winding factors of various single-layer concentrated windings with same-polarity coil configurations is summarized in Table 6.2. For clarity, results for conventional stator are presented with a white background, whereas those for flux-barrier stator ($h_{fb} = h_{yoke}$) are highlighted in gray. Overall, the results demonstrate the effectiveness of this technique in mitigating MMF subharmonics in this winding group.

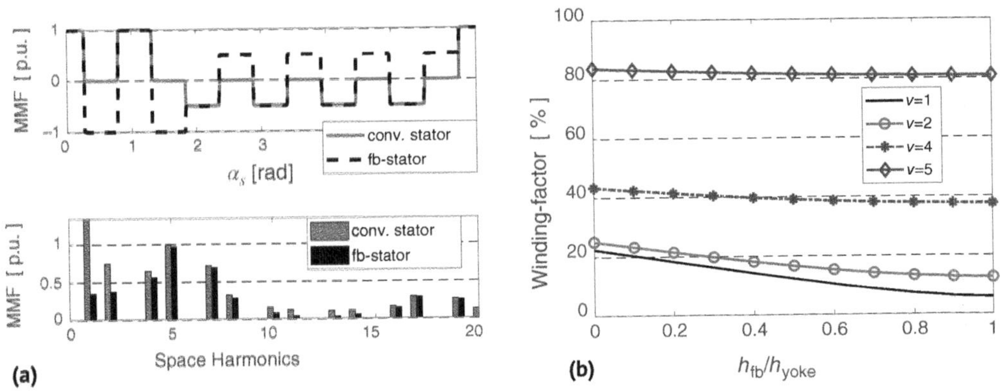

Figure 6.15 Three-phase SL-FSCW with $z = 1$ and $n_c = 2$. (a) Comparison of the MMF characteristics between the conventional stator and the flux-barrier stator with $h_{fb} = h_{yoke}$. (b) Flux-barrier effect on the MMF harmonic winding factors.

Table 6.2 Winding factors (in %) for the SL-FSCW with $z = 1$ and same-polarity coils. Comparison of the conventional stator (white row) and the flux-barrier stator with $h_{fb} = h_{yoke}$ (gray row).

n_c	Q_s	$\nu = 1$	2	3	4	5	6	7	8	9	10	11	12	13	14	15	16
1	6	50.0	**86.6**	—	**86.6**	50.0	—	50.0	**86.6**	—	**86.6**	50.0	—	50.0	**86.6**	—	**86.6**
		25.0	**75.0**	—	**75.0**	25.0	—	25.0	**75.0**	—	**75.0**	25.0	—	25.0	**75.0**	—	**75.0**
2	12	22.4	25.0	—	43.3	**83.7**	—	**83.7**	43.3	—	25.0	22.4	—	22.4	25.0	—	43.3
		5.80	12.5	—	37.5	**80.8**	—	**80.8**	37.5	—	12.5	5.80	—	5.80	12.5	—	37.5
3	18	14.7	15.4	—	18.8	22.5	—	42.2	**83.1**	—	**83.1**	42.2	—	22.5	18.8	—	15.4
		2.54	5.25	—	12.11	17.2	—	39.6	**81.8**	—	**81.8**	39.6	—	17.2	12.1	—	5.25

Note: Bold represents winding factors with highest values.

6.2.2.2 SL-FSCW with $z = 2$ and $n_c \geq 1$

The auxiliary functions for the two-zone single-layer concentrated windings, including the effect of the flux-barriers, are defined in Equations (6.28) and (6.29). Following the same analytical approach as before, the resulting MMF function is determined in Equation (6.30).

$$F_{j,I}(t, a_S) = \frac{2\widehat{F}_I}{\pi} i_j(t) \sum_{\nu} \frac{{}^{\nu}\xi_p \, {}^{\nu}\xi_d \, {}^{\nu}\xi_z}{\nu} \cdot \sin\left(\nu\left(a_S - (n_c - 1)\frac{2\pi}{Q_S}\right)\right) \tag{6.28}$$

$$F_{j,II}(t, a_S) = \frac{2\widehat{F}_{II}}{\pi} i_j(t) \sum_{\nu} \frac{{}^{\nu}\xi_p \, {}^{\nu}\xi_d \, {}^{\nu}\xi_z}{\nu} \cdot \sin\left(\nu\left(a_S - n_c\frac{2\pi}{Q_S}\right) - \pi\right) \tag{6.29}$$

$$F_{j,\text{fb}}(t, a_S) = \frac{2zn_cN_w}{\pi} i_j(t) \sum_{\nu} \frac{{}^{\nu}\xi_{w,\text{fb}}}{\nu} \cdot \sin\left(\nu a_S + {}^{\nu}\varphi_{1,2}\right) \tag{6.30}$$

The equivalent winding factor for this winding group is given by the product of coil-pitch, distribution, zone, and the flux-barrier factor, as expressed in Equation (6.31)

$$ {}^{\nu}\xi_{w,\text{fby}} = \left| {}^{\nu}\xi_p \, {}^{\nu}\xi_d \, {}^{\nu}\xi_z \right| \cdot {}^{\nu}\Gamma_{\text{fby}} \tag{6.31} $$

Example 6.3 Different two-zone SL-FSCWs with stator-yoke flux-barriers
Figures 6.16–6.20 provide a detailed analysis of the influence of flux-barriers on the MMF characteristics of various two-zone single-layer concentrated windings. These results highlight how the MMF distribution and specific harmonic components depends on the flux-barrier geometry, particularly the barrier depth. A comparative summary of winding factors, with and without flux-barriers, is presented in Table 6.3. As in the single-zone case, this technique is shown to be highly effective in reducing subharmonic components.

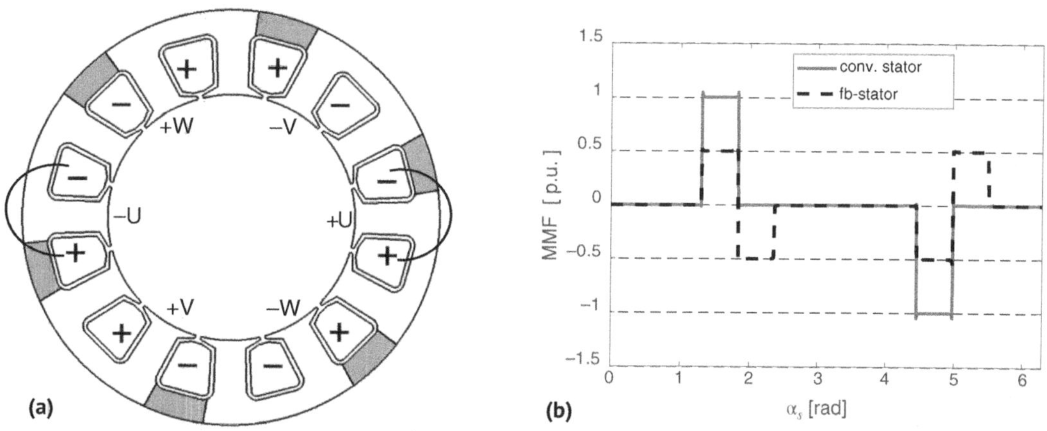

Figure 6.16 (a) Three-phase SL-FSCW with $z = 2$, $n_c = 1$, and flux-barrier stator. (b) MMF distribution of phase-j for both the conventional stator and the flux-barrier stator with $h_{\text{fb}} = h_{\text{yoke}}$.

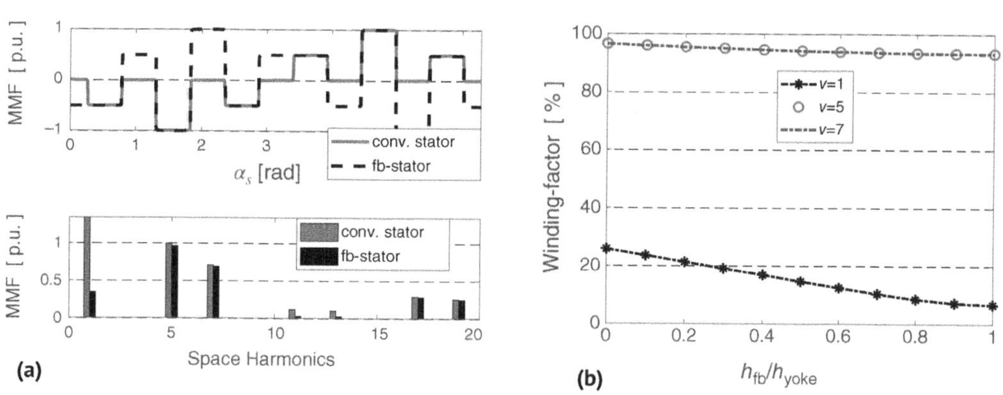

Figure 6.17 Three-phase SL-FSCW with $z = 2$ and $n_c = 1$. (a) Comparison of the MMF characteristics between the conventional stator and the flux-barrier stator with $h_{\text{fb}} = h_{\text{yoke}}$. (b) Flux-barrier effect on the MMF harmonic winding factors.

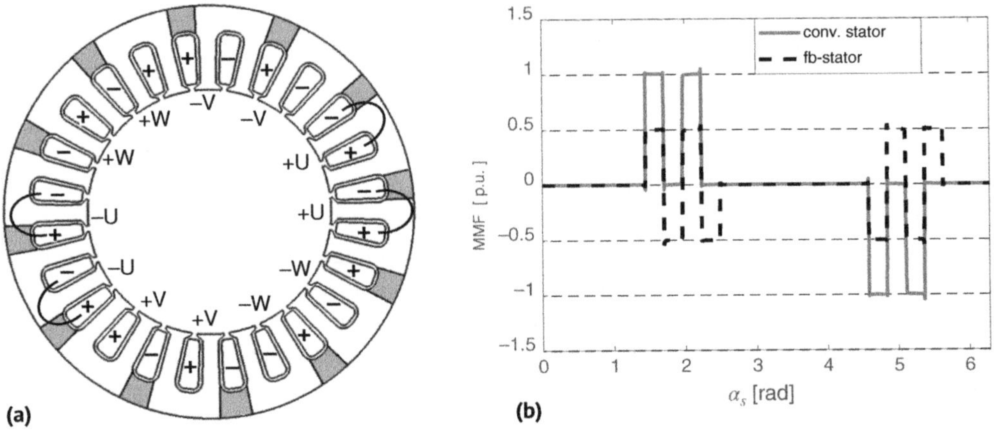

Figure 6.18 (a) Three-phase SL-FSCW with $z = 2$, $n_c = 2$, and flux-barrier stator. (b) MMF distribution of phase-j for both the conventional stator and the flux-barrier stator with $h_{fb} = h_{yoke}$.

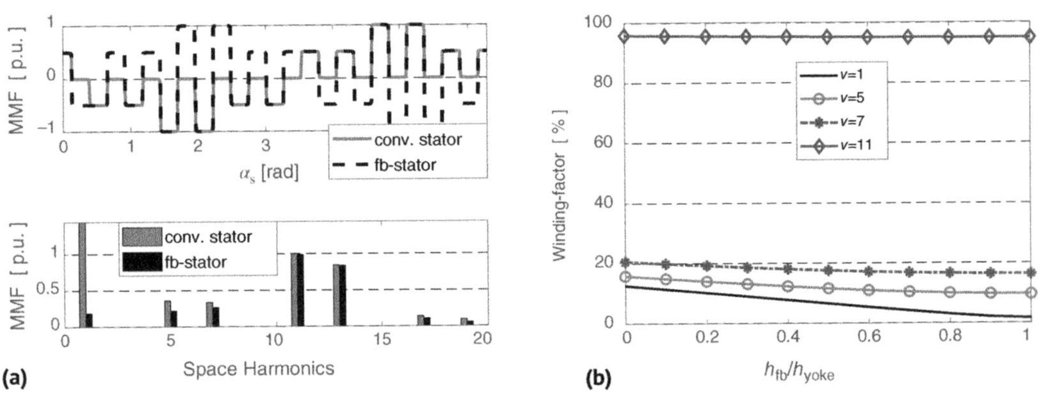

Figure 6.19 Three-phase SL-FSCW with $z = 2$ and $n_c = 2$. (a) Comparison of the MMF characteristics for the conventional stator and the flux-barrier stator with $h_{fb} = h_{yoke}$. (b) Flux-barrier effect on the MMF harmonic winding factors.

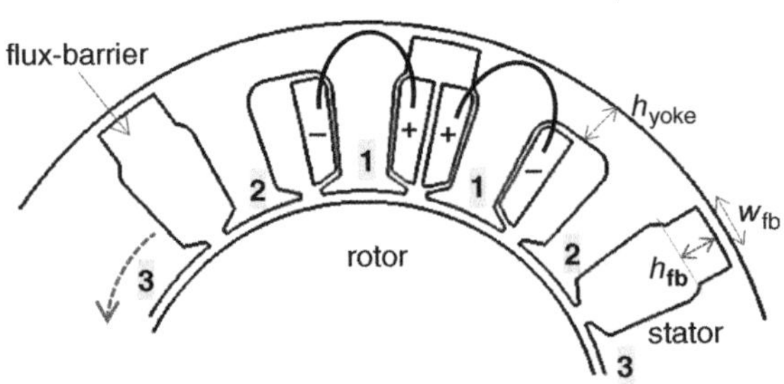

Figure 6.20 Cross section of an electric machine with flux-barriers in the stator yoke.

Table 6.3 Winding factor (in %) for the SL-FSCW with $z = 2$ and same-polarity coils. Comparison of the conventional stator (white row) and the flux-barrier stator with $h_{fb} = h_{yoke}$ (gray row).

n_c	Q_s	$v = 1$	2	3	4	5	6	7	8	9	10	11	12	13	14	15	16
1	12	25.9	—	—	—	**96.6**		**96.6**	—	—	—	25.9		25.9	—	—	—
		6.70	—	—	—	**93.3**		**93.3**	—	—	—	6.70	—	6.70	—	—	—
2	24	12.6	—	—	—	15.8		20.5	—	—	—	**95.8**	—	**95.8**	—	—	—
		1.64	—	—	—	9.60		16.3	—	—	—	**94.95**	—	**94.95**	—	—	—
3	36	8.4	—	—	—	9.2		10.2	—	—	—	14.5	—	19.7	—	—	—
		0.73	—	—	—	3.88		5.83	—	—	—	11.9	—	17.8	—	—	—

6.2.3 DL-FSCWs with Stator-yoke Flux-barrier

Figure 6.20 illustrates a schematic of a DL-FSCW with slot-shaped inserts in the stator-yoke. Unlike single-layer windings, where such modifications can be applied flexibly to every second slot-yoke region, their effectiveness in double-layer windings is restricted to specific locations. In particular, according to Dajaku and Gerling [77, 78], a positive effect is only achieved if flux-barriers are placed in yoke regions adjacent to stator slots containing two coil sides of the same phase. However, in double-layer winding topologies with a large number of coils per zone, several adjacent slots are usually assigned to the same phase. In such cases, the introduction of flux-barriers leads to pronounced non-uniform magnetic reluctances associated with these slots, which ultimately degrades the overall winding performance. As a result, this approach is particularly effective only for windings with two coils per zone ($n_c = 2$), and the following analysis is therefore limited to this class of windings.

In the modified stator depicted in Figure 6.20, the geometric parameters of the slot-shaped gaps are denoted by w_{fb} and h_{fb}. The stator teeth carrying the excitation coils are labeled with "1," adjacent teeth without flux gaps between them are designated as "2," and the remaining stator circumference is referred to as stator region "3."

To demonstrate the impact of flux-barriers on the magnetic flux distribution, Figure 6.21 compares two stator configurations equipped with a DL-FSCW with $n_c = 2$. In the conventional stator (Figure 6.21a), the generated magnetic flux is mainly concentrated in teeth labelled as "1." In the modified stator (Figure 6.21b) however, the increased magnetic resistance in the barrier regions forces the magnetic flux to redistribute and close mainly through the neighboring teeth "2." Consequently, this redistribution changes the MMF waveform of the winding, as approximated by the dashed curve in Figure 6.22 and affects the corresponding MMF function magnitudes as described in Equation (6.32).

$$F_{11}^* = \begin{cases} N_w, & h_{fb} = 0 \\ \frac{1}{2}N_w, & h_{fb} = h_{yoke} \end{cases}, \quad F_{12}^* = \begin{cases} 0, & h_{fb} = 0 \\ \frac{1}{2}N_w, & h_{fb} = h_{yoke} \end{cases} \tag{6.32}$$

The dependence of the modified coefficients F_{11}^* and F_{12}^* with respect to the barrier depth is described by Equations (6.33) and (6.34).

$$F_{11}^* = N_w\left(1 - \frac{h_{fb}}{2h_{yoke}}\right) \tag{6.33}$$

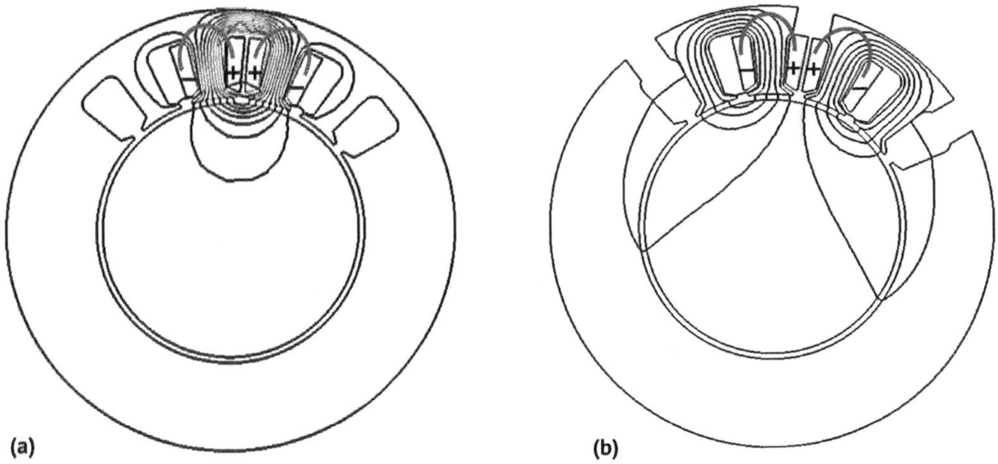

Figure 6.21 Flux density distribution due to two neighbor excitation coils. (a) Conventional stator. (b) Flux-barrier stator with $h_{fb} = h_{yoke}$.

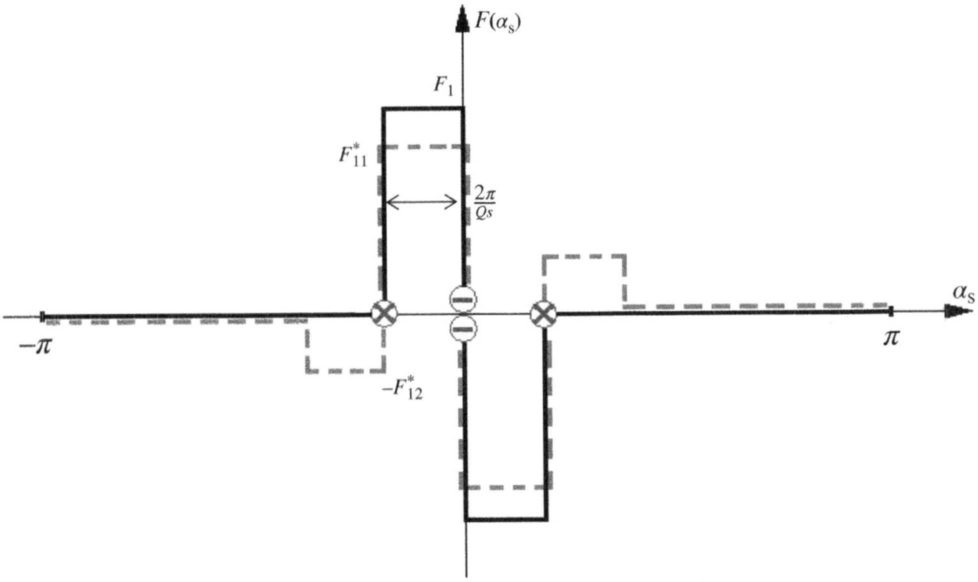

Figure 6.22 Flux-barrier effect on the MMF distribution of a DL-FSCW ($z = 1$ and $n_c = 2$); Solid line for the conventional stator, dashed line for the flux-barrier stator.

$$F^*_{12} = N_w \frac{h_{fb}}{2h_{yoke}} \tag{6.34}$$

It is important to note that, unlike single-layer windings, which produce asymmetric MMF distributions when a flux-barrier stator is used, the MMF of double-layer windings remains spatially symmetric (as shown in Figure 6.22), which simplifies the subsequent Fourier analysis. The resulting MMF series is given in Equation (6.35), with the corresponding coefficient defined in (6.36).

$$F_{\text{fb}}(t, a_S) = i(t) \cdot \sum_{\nu} {}^{\nu}\widehat{F}_{\text{FB}} \cdot \sin(\nu a_S) \tag{6.35}$$

$$\begin{aligned} {}^{\nu}\widehat{F}_{\text{FB}} &= \frac{2}{\nu\pi}\left\{ \left(F_{11}^{*} + F_{12}^{*}\right)\cos(\nu a_\tau) - F_{12}^{*}\cos(2\nu\alpha_\tau) - \Theta_{11}^{*} \right\} \\ &= \frac{2N_w}{\nu\pi}\left\{ \cos(\nu a_\tau) - \frac{h_{\text{fb}}}{2h_{\text{yoke}}}\cos(2\nu a_\tau) - \left(1 - \frac{h_{\text{fb}}}{2h_{\text{yoke}}}\right) \right\} \end{aligned} \tag{6.36}$$

From these expressions, the final MMF function and the equivalent winding factor are obtained in Equations (6.37) and (6.38), respectively.

$$F_{\text{fb}}(t, a_S) = \frac{4N_w i(t)}{\pi} \sum_{\nu} \frac{{}^{\nu}\xi_{w,\text{fby}}}{\nu} \sin(\nu a_S) \tag{6.37}$$

$${}^{\nu}\xi_{w,\text{fby}} = \frac{1}{2}\left(1 - \frac{h_{\text{fb}}}{2h_{\text{yoke}}} - \cos(\nu a_\tau) + \frac{h_{\text{fb}}}{2h_{\text{yoke}}}\cos(2\nu\alpha_\tau)\right) \tag{6.38}$$

6.2.3.1 DL-FSCW with $z = 1$ and $n_c = 2$

An example DL-FSCW with $z = 1$ and $n_c = 2$, and including stator-yoke flux-barriers is illustrated in Figure 6.23a. The corresponding MMF function is analogous to that given in Equation (6.37); its general analytical expression is provided in (6.39).

$$F_{j,\text{fb}}(t, a_S) = \frac{2n_c N_w}{\pi} i_j(t) \sum_{\nu} \frac{{}^{\nu}\xi_{w,\text{fby}}}{\nu} \cdot \sin(\nu a_S) \tag{6.39}$$

Example 6.4 DL-FSCW with $z = 1$, $n_c = 2$, and stator-yoke flux-barriers

Figures 6.23b and 6.24 show the MMF characteristics of the examined winding. As discussed in Chapter 2 (Table 2.6), the 2nd MMF harmonic serves here as the working harmonic, whereas the 1st harmonic represents the subharmonic component. Accordingly, these two harmonics are selected for detailed investigation in Figure 6.24b.

At an optimal barrier depth of $h_{\text{fb}} = 0.67h_{\text{yoke}}$, the subharmonic component is completely eliminated. The analysis also shows that, for this particular winding topology, the introduction of

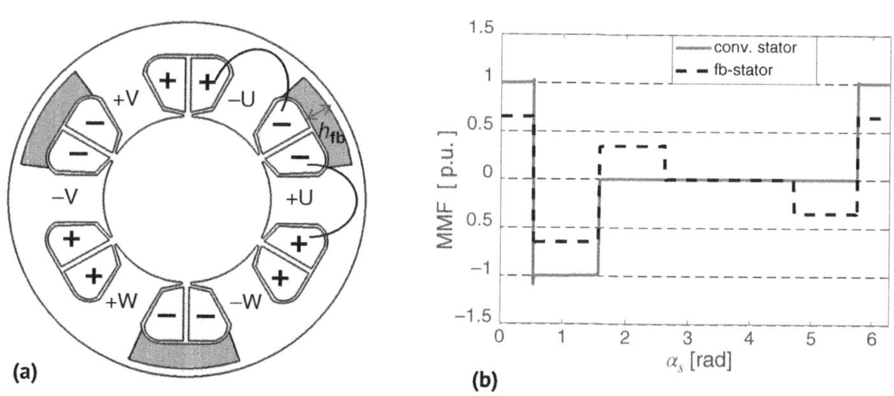

Figure 6.23 (a) Three-phase DL-FSCW with $z = 1$, $n_c = 2$, and flux-barrier stator. (b) MMF distribution of phase-j for both the conventional stator and the flux-barrier stator with $h_{\text{fb}} = 0.67h_{\text{yoke}}$.

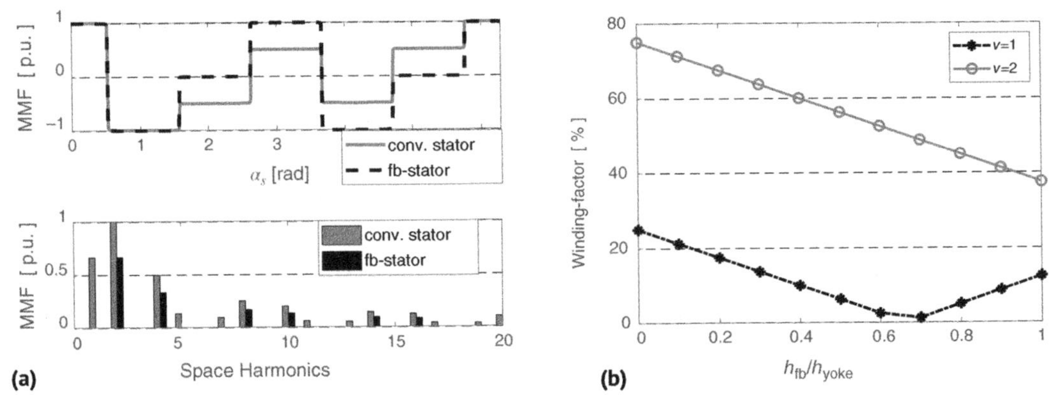

Figure 6.24 Three-phase DL-FSCW with $z = 1$ and $n_c = 2$. (a) Comparison of the MMF characteristics between the conventional stator and the flux-barrier stator with $h_{fb} = 0.67h_{yoke}$. (b) Flux-barrier effect on the MMF harmonic winding factors.

Table 6.4 Winding factor (in %) for the DL-FSCW with $z = 1$ and opposite polarity coils. Comparison of the conventional stator (white row) and the flux-barrier stator with $h_{fb} = 0.67h_{yoke}$ (gray row).

n_c	Q_s	$v = 1$	2	3	4	5	6	7	8	9	10	11	12	13	14	15	16
2	6	25.0	**75.0**	—	**75.0**	25.0	—	25.0	**75.0**	—	**75.0**	25.0	—	25.0	**75.0**	—	**75.0**
		0	**50.0**	—	**50.0**	0	—	0	**50.0**	—	**50.0**	0	—	0	**50.0**	—	**50.0**

flux-barriers adversely affects the working harmonic. Nevertheless, Figure 6.24a demonstrates a substantial reduction in the overall harmonic contents. The corresponding equivalent winding factors for the investigated winding are listed in Table 6.4.

6.2.3.2 DL-FSCW with $z = 2$ and $n_c = 2$

The DL-FSCW topology with $z = 2$, $n_c = 2$, and a stator-yoke flux-barrier configuration can be realized as illustrated in Figure 6.25a. The resulting MMF function and the corresponding winding factor are analogous to those of the previous example, with the additional inclusion of the winding zone factor in the formulation, i.e.

$$F_{j,\text{fb}}(t, \alpha_S) = \frac{2zn_cN_w}{\pi} i_j(t) \sum_{\nu} \frac{{}^{\nu}\xi_{w,\text{fby}}}{\nu} \cdot \cos(\nu\alpha_S) \tag{6.40}$$

$${}^{\nu}\xi_{w,\text{fby}} = \frac{{}^{\nu}\xi_z}{2}\left(1 - \frac{h_{\text{fb}}}{2h_{\text{yoke}}} - \cos(\nu\alpha_\tau) + \frac{h_{\text{fb}}}{2h_{\text{yoke}}}\cos(2\nu\alpha_\tau)\right) \tag{6.41}$$

Example 6.5 DL-FSCW with $z = 2$, $n_c = 2$, and stator-yoke flux-barriers
The MMF characteristics for the studied winding are presented in Figures 6.25b and 6.26. In conventional stator, this winding type is widely used in PMSM applications, with the 5th or the 7th harmonic serving as the operating harmonics and the 1st harmonic acting as a subharmonic component. As demonstrated in Figure 6.26, the use of stator-yoke flux-barriers considerably reduces the subharmonic component. In particular, at $h_{\text{fb}} = 0.53h_{\text{yoke}}$, the 1st harmonic is entirely eliminated. The equivalent winding factors for this case are summarized in Table 6.5.

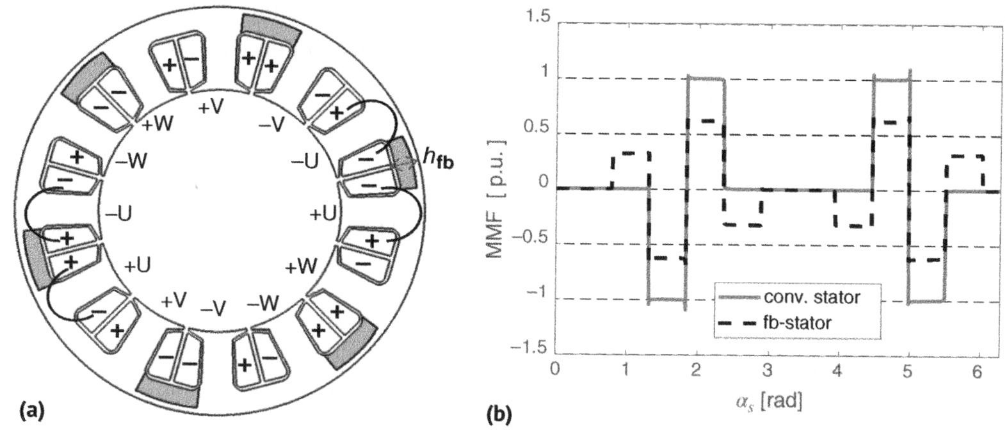

Figure 6.25 (a) Three-phase DL-FSCW with $z = 2$, $n_c = 2$, and flux-barrier stator. (b) MMF distribution of phase-j for both the conventional stator and the flux-barrier stator with $h_{fb} = 0.53h_{yoke}$.

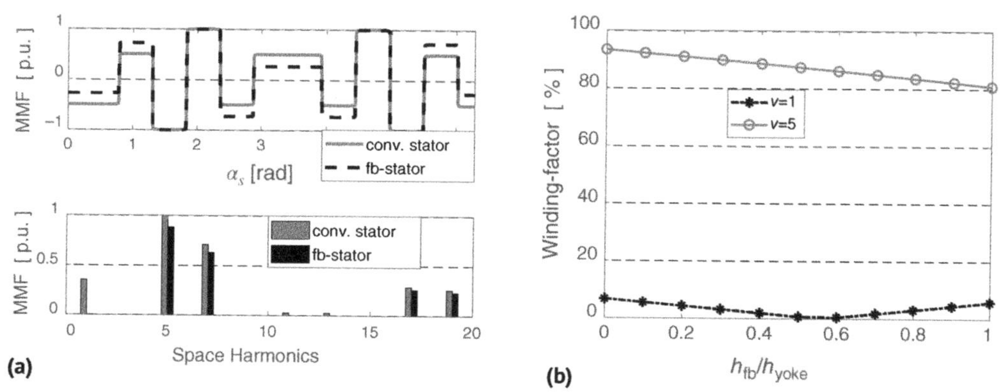

Figure 6.26 Three-phase DL-FSCW with $z = 2$ and $n_c = 2$. (a) Comparison of the MMF characteristics between the conventional stator and the flux-barrier stator with $h_{fb} = 0.53h_{yoke}$. (b) Flux-barrier effect on the MMF harmonic winding factors.

Table 6.5 Winding factor (in %) for the DL-FSCW with $z = 2$ and opposite polarity coils. Comparison of the conventional stator (white row) and the flux-barrier stator with $h_{fb} = 0.53h_{yoke}$ (gray row).

n_c	Q_s	$v = 1$	2	3	4	5	6	7	8	9	10	11	12	13	14	15	16
2	12	6.70	—	—	—	93.3		93.3	—	—	—	6.70		6.70	—	—	—
		0	—	—	—	86.7		86.7	—	—	—	0		0	—	—	—

6.2.4 Key Features and Potentials

The above investigations show that the stator-yoke flux-barrier technique is highly effective in mitigating the MMF winding subharmonics in both SL- and DL-FSCWs. However, its impact on the flux distribution and subharmonic reduction depends strongly on the winding type. In single-layer windings, the introduction of stator-yoke flux-barriers significantly changes the internal flux distribution, making it similar to that of an equivalent double-layer winding with a conventional stator. This phenomenon is referred to as the flux conservation effect induced by the flux-barriers.

The main conclusions can be summarized as follows:

SL-FSCWs:

- Impact: significant reduction of subharmonics;
- Application: all winding topologies;
- Transformation: converts them into the related DL-FSCWs with the same slot/pole combination.

DL-FSCWs:

- Impact: reduction of subharmonics;
- Application: limited to windings with $n_c = 2$;
- Transformation: no equivalent transformation to another FSCW type.

6.2.4.1 Flux Transformation Effect

The flux transformation effect in single-layer concentrated windings is examined in more detail in Figures 6.27 and 6.28. These figures highlight the substantial changes in flux distribution that occur when this technique is applied. In (a), the images show the flux density distribution for a conventional stator. When flux-barriers are applied to the stator yoke (middle images), the increased magnetic resistance in the barrier regions forces the magnetic flux to

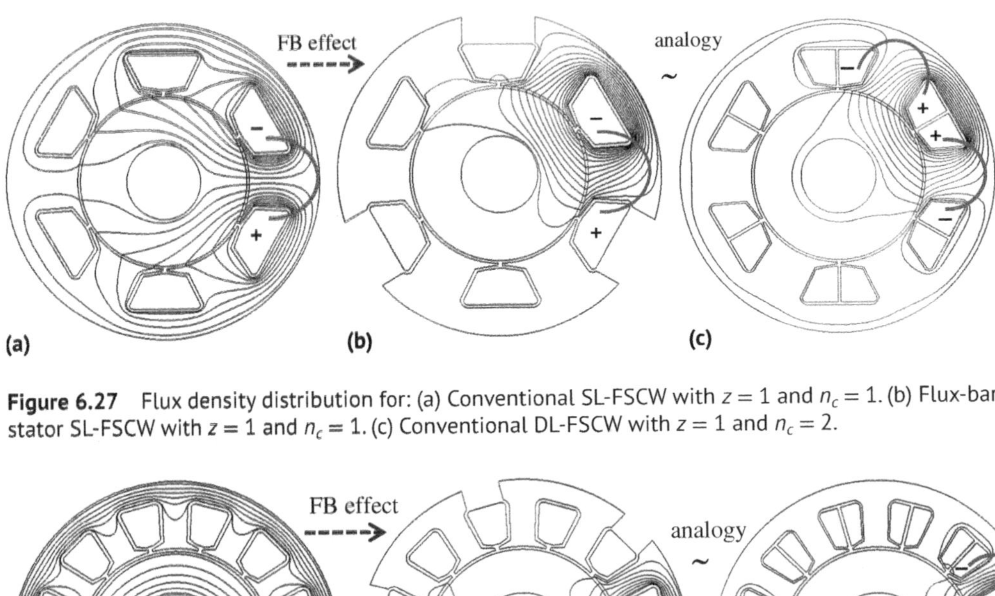

Figure 6.27 Flux density distribution for: (a) Conventional SL-FSCW with $z = 1$ and $n_c = 1$. (b) Flux-barrier stator SL-FSCW with $z = 1$ and $n_c = 1$. (c) Conventional DL-FSCW with $z = 1$ and $n_c = 2$.

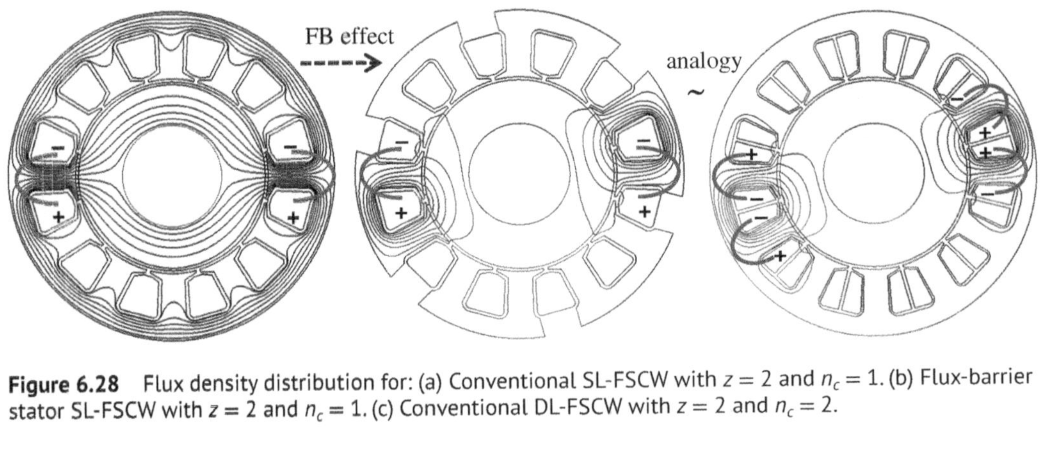

Figure 6.28 Flux density distribution for: (a) Conventional SL-FSCW with $z = 2$ and $n_c = 1$. (b) Flux-barrier stator SL-FSCW with $z = 2$ and $n_c = 1$. (c) Conventional DL-FSCW with $z = 2$ and $n_c = 2$.

Table 6.6 Winding factor (in %) for various FSCWs with $z = 1, n_c = (1\,\text{for SL}, 2\,\text{for DL})$, and $Q_S = 6$.

	$v = 1$	2	3	4	5	6	7	8	9	10	11	12	13	14	15	16
SL, conv.	50.0	**86.6**	—	**86.6**	50.0	—	50.0	**86.6**		**86.6**	50.0	—	50.0	**86.6**	—	**86.6**
SL, FB	25.0	**75.0**	—	**75.0**	25.0	—	25.0	**75.0**		**75.0**	25.0	—	25.0	**75.0**	—	**75.0**
DL, conv.	25.0	**75.0**	—	**75.0**	25.0	—	25.0	**75.0**		**75.0**	25.0	—	25.0	**75.0**	—	**75.0**

Table 6.7 Winding factor (in %) for various FSCWs with $z = 2, n_c = (1\,\text{for SL}, 2\,\text{for DL})$, and $Q_S = 12$.

	$v = 1$	2	3	4	5	6	7	8	9	10	11	12	13	14	15	16
SL, conv.	25.9	—	—	—	**96.6**		**96.6**	—	—	—	25.9	—	25.9	—	—	—
SL, FB	6.70	—	—	—	**93.3**		**93.3**	—	—	—	6.70	—	6.70	—	—	—
DL, conv.	6.70	—	—	—	**93.3**		**93.3**	—	—	—	6.70	—	6.70	—	—	—

redistribute through the adjacent teeth, resulting in a significantly altered flux pattern. For further comparison, the flux density distribution of the corresponding DL-FSCWs are shown in (c). This comparison reveals that the flux-barrier technique effectively transforms the flux density distribution of a single-layer concentrated winding into one that closely matches that of the equivalent double-layer winding with a conventional stator. This transformation is not only visually apparent but also quantitatively confirmed by the calculated corresponding winding factors listed in Tables 6.6 and 6.7. These results validate that the stator-yoke flux-barrier technique successfully converts single-layer concentrated windings into their double-layer equivalents, thereby extending the applicability of SL-FSCWs and enabling their use as a viable alternative to conventional double-layer windings.

6.3 FSCWs with Stator-teeth Flux-barriers

An alternative to modifying the stator yoke is to insert flux-barriers directly within the stator teeth, as illustrated in Figure 6.29. In this arrangement, slot-shaped gaps are inserted alternately within the teeth and their associated yoke sections. To maintain both geometric and magnetic symmetry, each gap is positioned in the center of the tooth. Similar to the yoke-based approach, the main objective of this technique is to introduce localized magnetic barriers that selectively suppress the subharmonic components of the winding MMF while minimally affecting the desired operating harmonic.

In the geometry under consideration, the term fb_w denotes the barrier width and $\alpha_{\mathbf{fb}}$ represent the angle between adjacent flux-barriers. The tooth carrying the coil is labeled as region "1," the adjacent tooth segments separated by flux-barriers and connected to "1" are referred to as region "2," whereas the remaining stator circumference is referred to as stator area "3." As shown in Figure 6.31, the spatial placement of the flux-barriers restricts $\alpha_{\mathbf{fb}}$ parameter to discrete values relative to a given winding phase (e.g. phase A),

$$a_{\text{fb}} = n_{\text{fb}} \cdot a_\tau = n_{\text{fb}} \cdot \frac{2\pi}{Q_S} \tag{6.42}$$

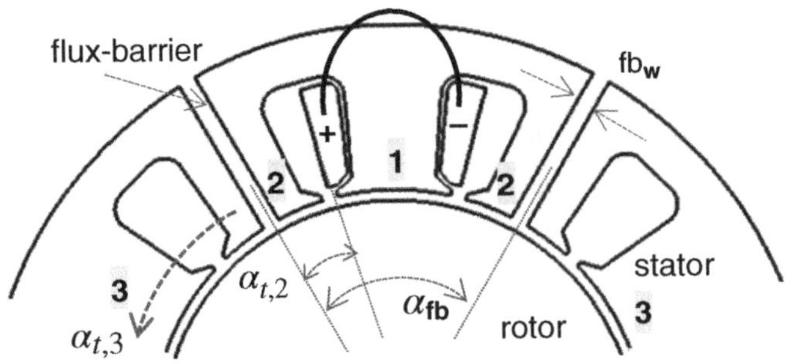

Figure 6.29 Cross section of an electric machine with flux-barriers in stator teeth.

where,

$$n_{\text{fb}} = \begin{cases} 2, & \text{flux-barrier is located in the unwound teeth, or} \\ 4, & \text{flux-barrier is located in the wound teeth} \end{cases}$$

From Figure 6.29, the parameters $\alpha_{t,2}$ and $\alpha_{t,3}$ are defined as:

$$\alpha_{t,2} = \alpha_{\text{fb}} - \alpha_{\tau} = (n_{\text{fb}} - 1)\frac{2\pi}{Q_S} \tag{6.43}$$

$$\alpha_{t,3} = 2\pi - \alpha_{\text{fb}} = 2\pi - n_{\text{fb}}\frac{2\pi}{Q} \tag{6.44}$$

In order to evaluate the impact of the stator-teeth flux-barriers on the MMF distribution, two stator configurations are examined in Figure 6.30. In the conventional stator, the flux density produced by the coil wound around tooth "1" is uniformly distributed over the rest of the stator circumference, as shown in Figure 6.30a. The corresponding MMF waveform for this case is approximated by a rectangular pulse of width $2\pi/Q_S$ (mechanical radians), shown by the solid line in Figure 6.30c), and described using the Fourier series in Equation (6.1). In the second design, however, the flux-barriers in the stator teeth introduce higher magnetic resistances in the magnetic flux-path, forcing the major part of the flux generated by the excitation coil to flow through the adjacent tooth elements "2." Assuming that the gaps behave as ideal magnetic flux-barriers and completely block the magnetic flux, the resulting MMF distribution can be approximated by the dashed line waveform shown in Figure 6.30c. The corresponding MMF magnitudes are denoted by F^*_{11} and F^*_{12}.

Following the modeling approach used in Section 6.2, the total MMF excitation distributed along the stator circumference is constrained to remain equal to that of the conventional design. Consequently, the increase in MMF amplitude observed in tooth segments "2," arising from the flux concentration effect, must be compensated by a proportional reduction in the excitation magnitude F^*_{11} at tooth "1." This condition is implemented via the following formulation

$$F^*_{11} = \begin{cases} \dfrac{1}{2}N_w, & \alpha_{\text{fb}} = 2\alpha_{\tau} \\ F_1 = \dfrac{2\pi - \alpha_{\tau}}{2\pi}N_w, & \alpha_{\text{fb}} = 2\pi \end{cases} \tag{6.45}$$

From Equation (6.45), F^*_{11} is determined as

$$F^*_{11} = \frac{n_{\text{fb}} - 1}{n_{\text{fb}}}N_w \tag{6.46}$$

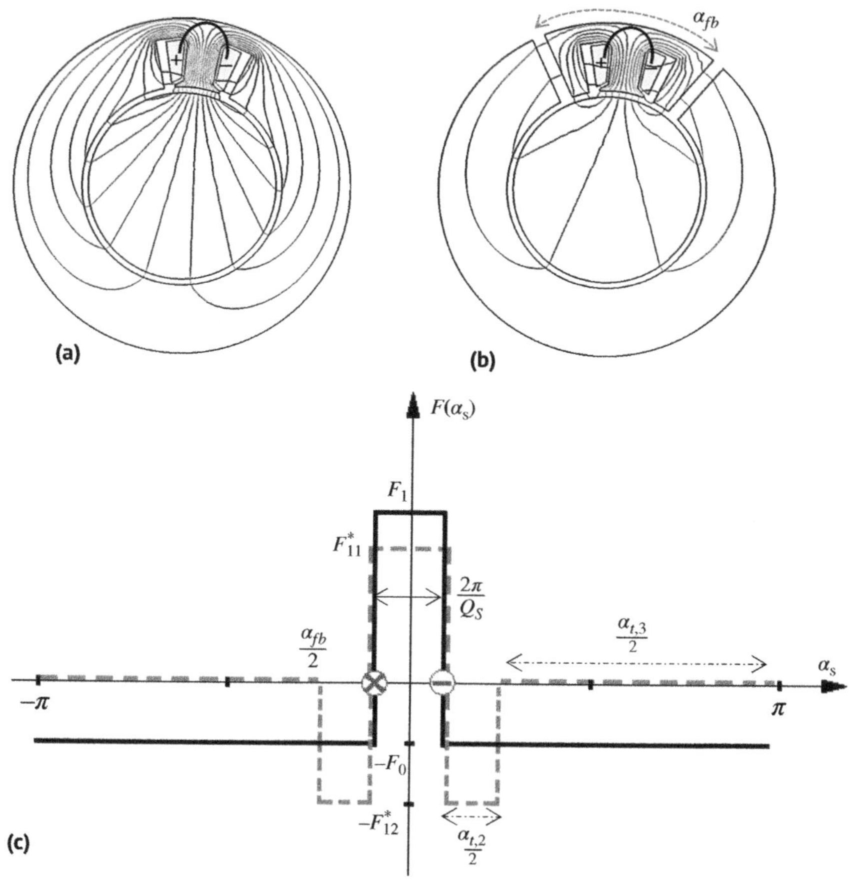

Figure 6.30 (a) Conventional stator with a single coil. (b) Flux-barrier stator with a single coil. (c) Coil MMF distribution for the conventional stator (solid line) and the flux-barrier stator (dashed line).

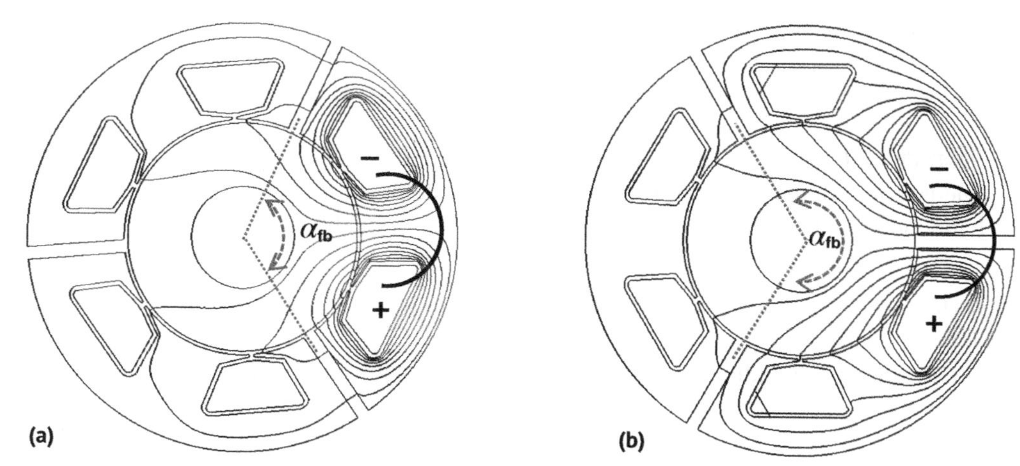

Figure 6.31 (a) Stator configuration with a flux-barrier angle of $\alpha_{fb} = 120°$. (b) Stator configuration with a flux-barrier angle of $\alpha_{fb} = 240°$.

Similarly, F_{12}^{*} is given by

$$F_{12}^{*} = \frac{1}{n_{\text{fb}}} N_w \tag{6.47}$$

Using Equations (6.46) and (6.47), the modified MMF function for the stator with teeth flux-barriers is expressed as

$$F_{\text{fb}}(t, a_S) = i(t) \sum_{\nu} {}^{\nu}\widehat{F}_{\text{FB}} \cdot \cos(\nu a_S) \tag{6.48}$$

According to Fourier series theory, the harmonic coefficient ${}^{\nu}\widehat{F}_{\text{FB}}$, is determined from

$$\begin{aligned} {}^{\nu}\widehat{F}_{\text{FB}} &= \frac{2}{\nu\pi}\left\{ \left(F_{11}^{*} + F_{12}^{*}\right)\sin\left(\nu\frac{a_\tau}{2}\right) - F_{12}^{*}\sin\left(\nu\frac{a_{\text{FB}}}{2}\right)\right\} \\ &= \frac{2N_w}{\nu\pi}\left\{ \sin\left(\nu\frac{a_\tau}{2}\right) - \frac{1}{n_{\text{fb}}}\sin\left(\nu\frac{a_{\text{FB}}}{2}\right)\right\} \\ &= \frac{2N_w}{\nu\pi}\left\{ {}^{\nu}\xi_p - \frac{1}{n_{\text{fb}}}\sin\left(\nu\frac{a_{\text{FB}}}{2}\right)\right\} \end{aligned} \tag{6.49}$$

and the resulting closed-form expression for the modified coil MMF function is

$$F_{\text{fb}}(t, a_S) = \frac{2N_w i(t)}{\pi} \sum_{\nu} \frac{{}^{\nu}\xi_{p,\text{fbt}}}{\nu} \cos(\nu a_S) \tag{6.50}$$

In (6.50), ${}^{\nu}\xi_{p,\text{fbt}}$ represents the equivalent coil-pitch factor due to flux-barrier effect, defined as

$${}^{\nu}\xi_{p,\text{fbt}} = {}^{\nu}\xi_p - \frac{1}{n_{\text{fb}}}\sin\left(\nu\frac{a_{\text{FB}}}{2}\right) \tag{6.51}$$

6.3.1 SL-FSCWs with Stator-teeth Flux-barriers

Before proceeding with a detailed analysis, it should be noted that applying this method to double-layer windings introduces asymmetric effects on the respective coil function within a zone, leading to degraded MMF performance. Accordingly, the following analysis is limited to single-layer windings. Moreover, although this technique is applicable to any single-layer winding topology, the following examples focus exclusively on winding configurations with higher winding factors, particularly SL-FSCWs with the same-polarity coil arrangements. The term "fb-stator" in the following results (legend label) refers to the new stator including flux-barriers in the stator teeth.

6.3.1.1 SL-FSCW with $z = 1$

For a single-layer concentrated winding with $z = 1$ and a stator-teeth flux-barrier configuration, the MMF function is obtained from Equations (2.23) and (6.50), yielding

$$F_{j,\text{fb}}(t, a_S) = \frac{2n_c N_w}{\pi} i_j(t) \sum_{\nu} \frac{{}^{\nu}\xi_{w,\text{fbt}}}{\nu} \cdot \cos\left(\nu\left(a_S - (n_c - 1)\frac{2\pi}{Q_S}\right)\right) \tag{6.52}$$

where the corresponding winding factor is the product of the equivalent coil-pitch and the distribution factor,

$${}^{\nu}\xi_{w,\text{fbt}} = {}^{\nu}\xi_{p,\text{fbt}} \cdot {}^{\nu}\xi_d = {}^{\nu}\xi_d\left\{ {}^{\nu}\xi_p - \frac{1}{n_{\text{fb}}}\sin\left(\nu\frac{a_{\text{FB}}}{2}\right)\right\} \tag{6.53}$$

Example 6.6 SL-FSCW with $z = 1$, $n_c = 1$, and stator-teeth flux-barriers
The first winding example considers an SL-FSCW with $z = 1$, $n_c = 1$, and three stator-teeth flux-barriers placed in alternating teeth. As mentioned earlier, flux-barriers can be positioned either in unwound or wound stator teeth, resulting in distinct MMF characteristics.

- *Flux-barriers in unwound teeth*: As shown in Figure 6.32, this arrangement significantly reduces the 1st and 2nd subharmonics while at the same time notably enhances the 4th MMF harmonic. Due to this influence on the harmonic contents, this configuration is well suited for an 8-pole PMSM.
- *Flux-barriers in wound teeth*: In the configuration of Figure 6.33, the 1st MMF harmonic is reduced again, however, the 2nd MMF harmonic is amplified. Consequently, this arrangement is more suitable for a 4-pole PMSM.

Example 6.7 SL-FSCW with $z = 1$, $n_c = 2$, and stator-teeth flux-barriers
The second example examines an SL-FSCW with $z = 1$ and $n_c = 2$. Depending on the location of the stator-teeth flux-barriers, the resulting winding configurations and their MMF characteristics are presented in Figures 6.34 and 6.35.

- *Flux-barriers in unwound teeth*: All MMF subharmonics from the 1st through the 5th are strongly suppressed, while the 7th harmonic is increased by approximately 26%, making this configuration suitable for a 14-pole PMSM.

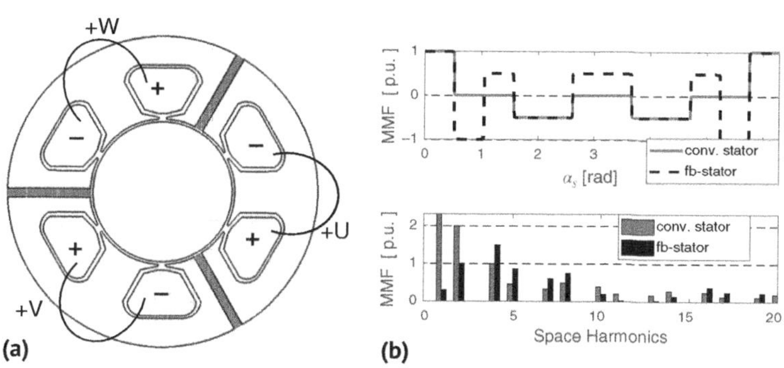

Figure 6.32 (a) Three-phase SL-FSCW with $z = 1$, $n_c = 1$, and with flux-barriers in unwound teeth. (b) Comparison of MMF characteristics between the conventional and flux-barrier stator.

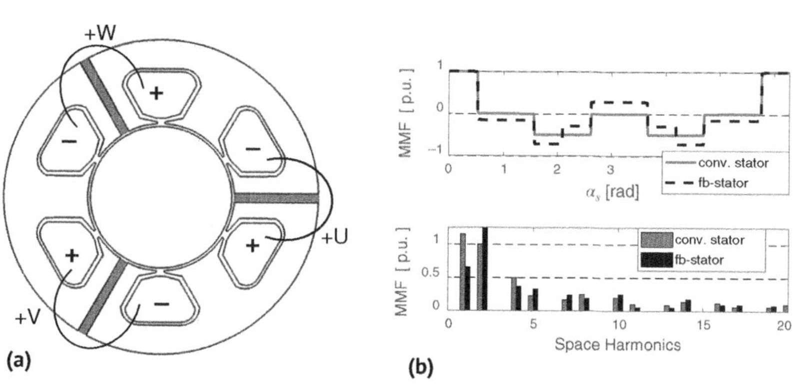

Figure 6.33 (a) Three-phase SL-FSCW with $z = 1$, $n_c = 1$, and with a flux-barrier in wounds teeth. (b) Comparison of MMF characteristics between the conventional and flux-barrier stator.

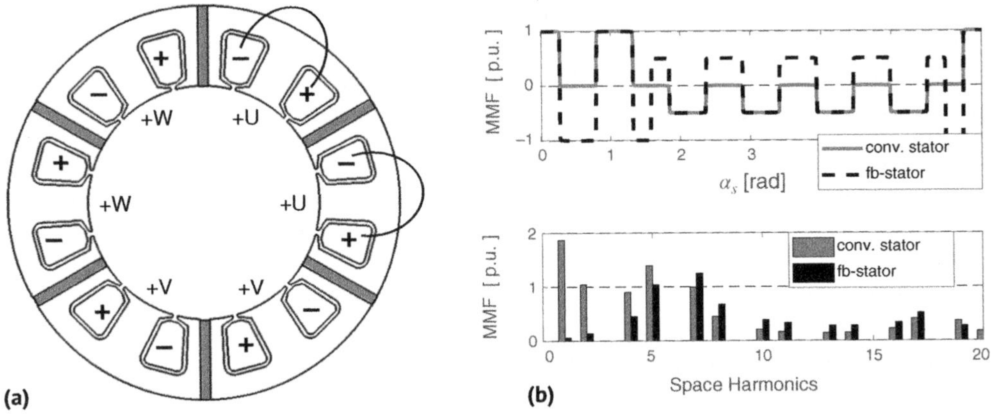

Figure 6.34 (a) Three-phase SL-FSCW with $z = 1, n_c = 2$, and with flux-barriers in unwound teeth. (b) Comparison of MMF characteristics between the conventional and flux-barrier stator.

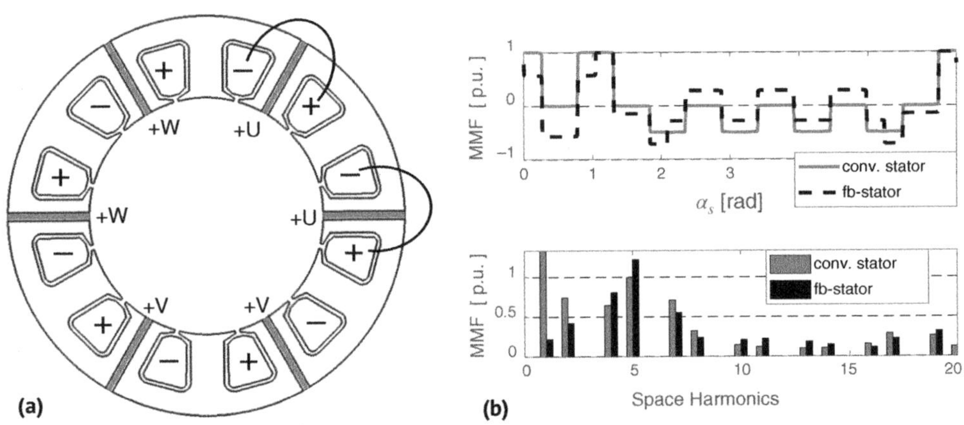

Figure 6.35 (a) Three-phase SL-FSCW with $z = 1, n_c = 2$, and with flux-barriers in wound teeth. (b) Comparison of MMF characteristics between the conventional and flux-barrier stator.

- *Flux-barriers in wound teeth*: Here, most subharmonics and higher MMF harmonics are reduced, except of the 5th harmonic, which is amplified by about 22%. This topology is therefore preferred for a 10-pole PMSM.

6.3.1.1.1 Impact of Flux-barriers on Winding Factors

Table 6.8 shows the impact of this technique on the winding factors, comparing conventional stators (white rows) to configurations with flux-barriers in unwound teeth (first gray row) and wound teeth (second gray row). These results lead to the following conclusions:

- The stator-teeth flux-barrier technique is beneficial for all one-zone SL-FSCWs.
- Proper positioning is crucial: barriers in unwound or wound teeth selectively enhance either the first or second MMF operating harmonic.

Table 6.8 Winding factors (in %) for SL-FSCWs with $z = 1$ and coils with the same polarity. Comparison of the conventional stator and the flux-barrier stator (first gray row for flux-barrier in unwound teeth, second gray row for flux-barrier in wound teeth).

n_c	Q_s	$\nu = 1$	2	3	4	5	6	7	8	9	10	11	12	13	14	15	16
1	6	50.0	**86.6**	—	**86.6**	50.0	—	50.0	**86.6**	—	**86.6**	50.0	—	50.0	**86.6**	—	**86.6**
		6.70	43.3	—	**129.9**	93.3	—	93.3	**129.9**	—	43.3	6.70	—	6.70	43.3	—	**129.9**
		28.3	**108.2**	—	64.9	71.6	—	71.6	64.9	—	**108.2**	28.3	—	28.3	**108.2**	—	64.9
2	12	22.4	25.0	—	43.3	**83.7**	—	**83.7**	43.3	—	25.0	22.4	—	22.4	25.0	—	43.3
		0.76	3.35	—	21.6	62.0	—	**105.3**	64.9	—	46.6	44.0	—	44.0	46.6	—	64.9
		3.66	14.2	—	54.1	**102.4**	—	64.9	32.5	—	35.8	41.2	—	41.2	35.8	—	32.5
3	18	14.7	15.4	—	18.8	22.5	—	42.2	**83.1**	—	**83.1**	42.2	—	22.5	18.8	—	15.4
		0.22	0.92	—	4.40	8.0	—	27.7	68.7	—	**97.5**	56.6	—	36.8	33.3	—	29.8
		1.09	4.30	—	16.3	24.9	—	53.2	**96.7**	—	69.5	31.1	—	19.9	21.3	—	26.4

- This technique provides dual advantages: it amplifies the desired working harmonic and simultaneously reduce the subharmonic components.

In summary, the use of flux-barriers in stator-teeth provides a systematic way of improving the MMF performance of SL-FSCWs and represents an innovative alternative to conventional winding solutions.

6.3.1.2 SL-FSCW with $z = 2$

For the single-layer concentrated winding with $z = 2$ and an flux-barrier stator, the MMF expression derived from Equations (2.23) and (6.50) is,

$$F_{j,\text{fb}}(t, a_S) = \frac{2zn_cN_w}{\pi} i_j(t) \sum_{\nu} \frac{{}^{\nu}\xi_{w,\text{fbt}}}{\nu} \cdot \cos\left(\nu\left(a_S - (n_c - 1)\frac{2\pi}{Q_S}\right)\right) \tag{6.54}$$

Here, the equivalent winding factor is expressed as the product of the equivalent coil-pitch factor (6.51), the distribution factor (2.24), and the zone factor (2.31), resulting in

$${}^{\nu}\xi_{w,\text{fbt}} = {}^{\nu}\xi_{p,\text{fbt}} {}^{\nu}\xi_d {}^{\nu}\xi_z = {}^{\nu}\xi_d {}^{\nu}\xi_z \left\{ {}^{\nu}\xi_p - \frac{1}{n_{\text{fb}}} \sin\left(\nu \frac{a_{\text{FB}}}{2}\right)\right\} \tag{6.55}$$

Example 6.8 SL-FSCW with $z = 2$, $n_c = 1$, and stator-teeth flux-barriers
In its conventional form, this winding use either the 5th or the 7th MMF harmonic as the working harmonic, and also contains a significant subharmonic component.

- *Flux-barriers in unwound teeth*: As shown in Figure 6.36, the 1st harmonic is reduced by more than 90%, the 5th harmonic by approximately 26%, and the 7th harmonic is increased by 26%. This configuration is therefore suitable for a 14-pole PMSMs.
- *Flux-barriers in wound teeth*: When flux-barriers are placed in wound stator teeth, as illustrated in Figure 6.37, the enhancement shifts to the 5th MMF harmonic, which is increased by about 22%, while the 1st subharmonic and 7th harmonic are reduced. As result, this topology is suitable for a 10-pole PMSM.

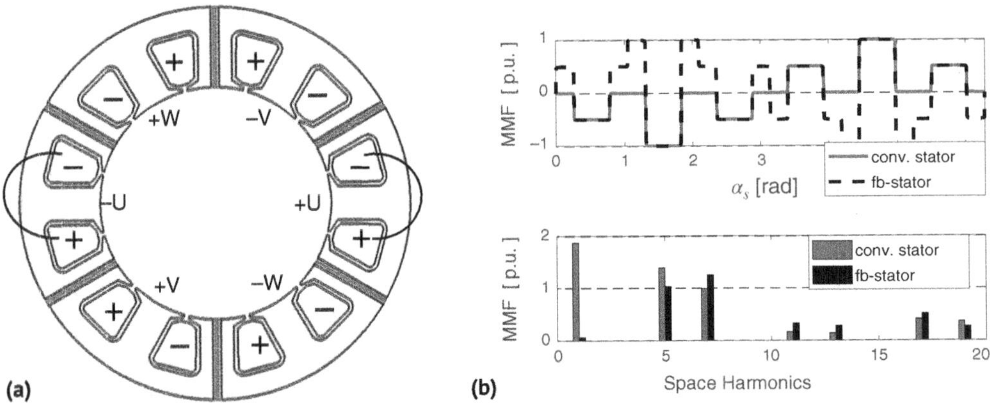

Figure 6.36 (a) Three-phase SL-FSCW with $z = 2, n_c = 1$, and with a flux-barrier in unwound teeth. (b) Comparison of MMF characteristics between the conventional and flux-barrier stator.

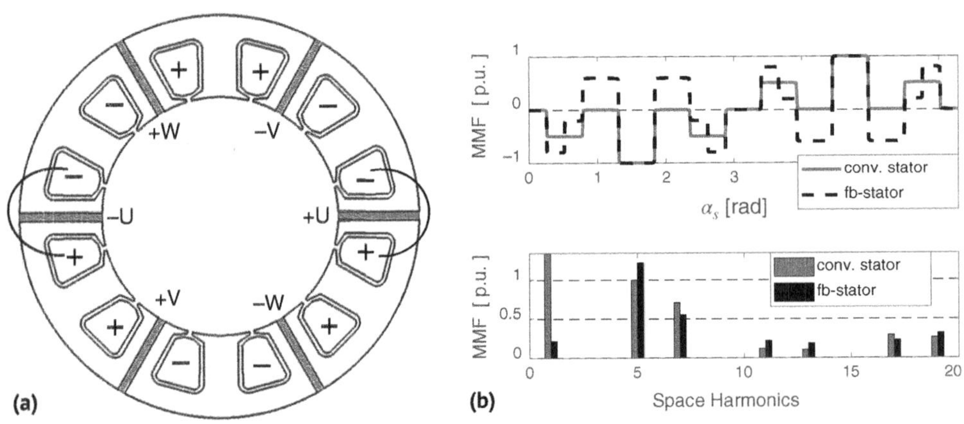

Figure 6.37 (a) Three-phase SL-FSCW with $z = 2, n_c = 1$, and with flux-barrier in wound teeth. (b) Comparison of MMF characteristics between the conventional and flux-barrier stator.

Example 6.9 SL-FSCW with $z = 2$, $n_c = 2$, and stator-teeth flux-barriers
For the conventional stator, the operating wave for this winding type is either the 11th or the 13th harmonic, while its subharmonic component is high.

- *Flux-barriers in unwound teeth*: Figure 6.38a shows that the 1st subharmonic is strongly reduced, while the 13th MMF harmonic increases by about 13.5%, making this configuration suitable for a 26 pole PMSM
- *Flux-barriers in wound teeth*: With this arrangement, subharmonics are again suppressed, while the 11th harmonic increases by approximately 12.5%. This stator design is therefore better suited for a 22-pole PMSM. The comparison of the MMF characteristic is presented in Figure 6.38b.

6.3.1.2.1 Impact of Flux-barriers on Winding Factors

The results summarized in Table 6.9 indicate that this winding group, which is already known for its relatively high winding factors, can be further improved through the implementation of the stator-teeth flux-barrier technique. This approach increases the winding factor of the operating

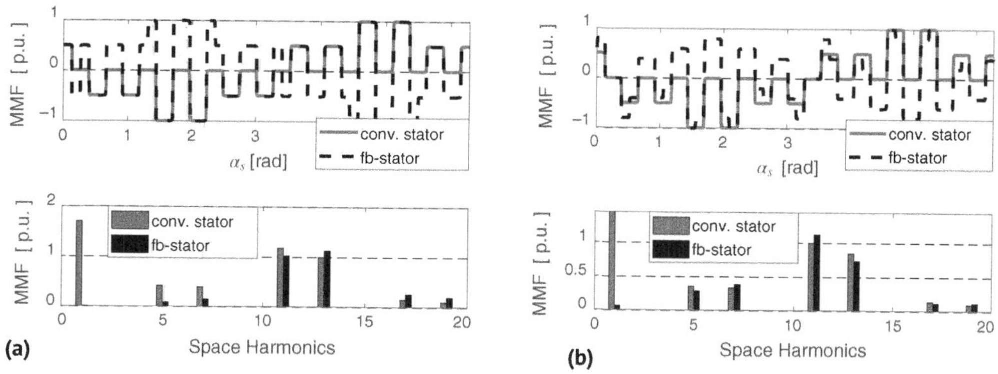

Figure 6.38 Three-phase SL-FSCW with $z = 2$ and $n_c = 2$. (a) Comparison of MMF characteristics between the conventional stator and flux-barrier stator (FB in unwound teeth). (b) Comparison of MMF characteristics between the conventional stator and flux-barrier stator (FB in wound teeth).

Table 6.9 Winding factor (in %) for SL-FSCWs with $z = 2$ and coils with same polarity. Comparison of the conventional and the flux-barrier stator (first gray row for flux-barrier in unwound teeth, second gray row for flux-barrier in wound teeth).

n_c	Q_s	$\nu = 1$	2	3	4	5	6	7	8	9	10	11	12	13	14	15	16
1	12	25.9	—	—	—	**96.6**	—	**96.6**	—	—	—	25.9	—	25.9	—	—	—
		0.88	—	—	—	71.6	—	**121.6**	—	—	—	50.8	—	50.8	—	—	—
		4.23	—	—	—	**118.2**	—	74.9	—	—	—	47.5	—	47.5	—	—	—
2	24	12.6	—	—	—	15.7	—	20.5	—	—	—	**95.7**	—	**95.7**	—	—	—
		0.10	—	—	—	3.25	—	8.03	—	—	—	83.3	—	**108.3**	—	—	—
		0.53	—	—	—	12.5	—	23.8	—	—	—	**107.8**	—	83.7	—	—	—

harmonics while simultaneously suppressing subharmonic components. As observed in the single-zone case, this dual effect makes these windings particularly attractive for applications that require high torque density and high efficiency.

6.3.2 SL-FSCWs with Unequal Zones and Stator-teeth Flux-barriers

Beyond conventional FSCWs, Chapter 3 introduced various winding configurations featuring unequal zones and also asymmetric zone distributions. These specific designs further expand the range of concentrated winding topologies and their potential applications in electrical machines. Similar to standard FSCWs, however, their main limitation lies in the high harmonic content of the MMF, particularly the presence of subharmonics with relatively large amplitudes. To address this issue, the stator-teeth flux-barrier approach, also here proves to be an effective means of improving their MMF characteristics.

For an SL-FSCW with a flux-barrier stator featuring two unequal zones per phase ($z_{1,2}$ and $z_{2,1}$) as well as zones with coils of the same polarity, the MMF function, derived from (3.13) and (6.50), is given by

$$F_{j,\text{fb}}(t, a_S) = \frac{2z(n_{c1} + n_{c2})N_w i(t)}{\pi} \sum_{\nu} \frac{{}^{\nu}\xi_{w,\text{fbt}}}{\nu} \cos\left(\nu a_S - \nu(n_{c1} - 1)\frac{2\pi}{Q_S}\right) \tag{6.56}$$

The corresponding winding factor, including the flux-barrier effect, is defined as the product of the equivalent coil-pitch factor (6.51) and the winding factor of the reference winding defined in (3.14), resulting in

$$^{\nu}\xi_{w,\text{fbt}} = \frac{n_{c1}}{n_{c1}+n_{c2}}\,{}^{\nu}\xi_{w1}\,{}^{\nu}\xi_{p,\text{fbt}} \pm \frac{n_{c2}}{n_{c1}+n_{c2}}\,{}^{\nu}\xi_{w2}\,{}^{\nu}\xi_{p,\text{fbt}}\cos(\nu\pi) \tag{6.57}$$

According to Equations (3.8)–(3.10), the winding factors of the individual subzones are

$$^{\nu}\xi_{w1} = \sin\left(\nu\frac{\pi}{Q_S}\right)\cdot\frac{\sin\left(\nu n_{c1}\frac{2\pi}{Q_S}\right)}{n_{c1}\cdot\sin\left(\nu\frac{2\pi}{Q_S}\right)},\quad {}^{\nu}\xi_{w2} = \sin\left(\nu\frac{\pi}{Q_S}\right)\cdot\frac{\sin\left(\nu n_{c2}\frac{2\pi}{Q_S}\right)}{n_{c2}\cdot\sin\left(\nu\frac{2\pi}{Q_S}\right)} \tag{6.58}$$

Example 6.10 SL-FSCW with unequal zones ($n_{c1}=2,\ n_{c2}=1$) and stator-teeth flux-barriers
Figures 6.39 and 6.40 illustrate an SL-FSCW with two unequal zones and a flux-barrier stator. The first zone contains two coils per zone ($n_{c1}=2$) with same coil polarity, the second zone has only one coil per zone ($n_{c2}=1$), whereas the two zones are arranged with opposite polarity. This winding topology allows either the 8th or the 10th harmonic to be selected as the working harmonic, both achieving a high winding factor of 94.5%.

- *Flux-barriers in unwound teeth* (Figure 6.39): This arrangement strongly suppress the MMF subharmonics while increasing the 10th harmonic by approximately 16%, making it an ideal choice for a 20-pole PMSM.
- *Flux-barriers in wound teeth* (Figure 6.40): Here, a similar reduction in subharmonics can be observed; however, the flux-barriers now enhance the 8th harmonic instead. Consequently, this configuration is well suited for a 16-pole PMSM.

A second unequal-zone FSCW with winding zone parameters $n_{c1}=3$ and $n_{c2}=2$ is examined in Table 6.10. In this case as well, the same positive flux-barrier effect is observed, namely a significant reduction in subharmonics with simultaneous enhancement in the operating harmonic.

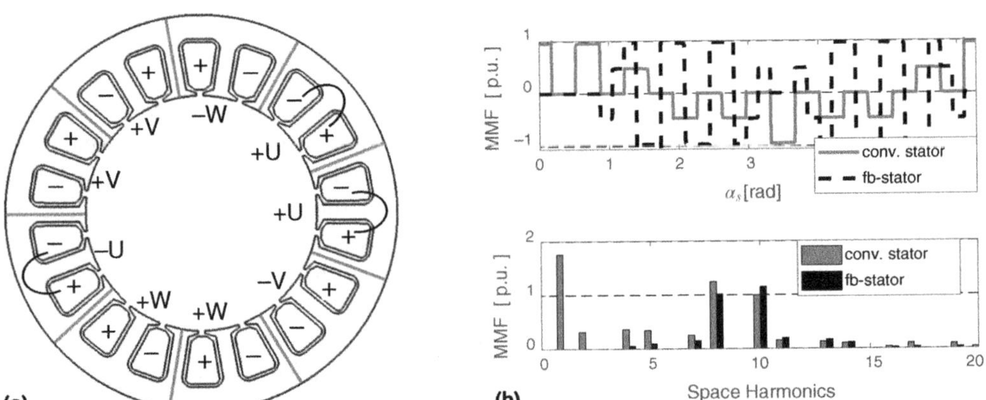

Figure 6.39 Three-phase SL-FSCW with two unequal zones per phase, with $n_{c1}=2, n_{c2}=1$, and with opposite polarity zones. (a) Winding configuration with flux-barriers in unwound teeth. (b) Comparison of MMF characteristics for the conventional stator and the flux-barrier stator.

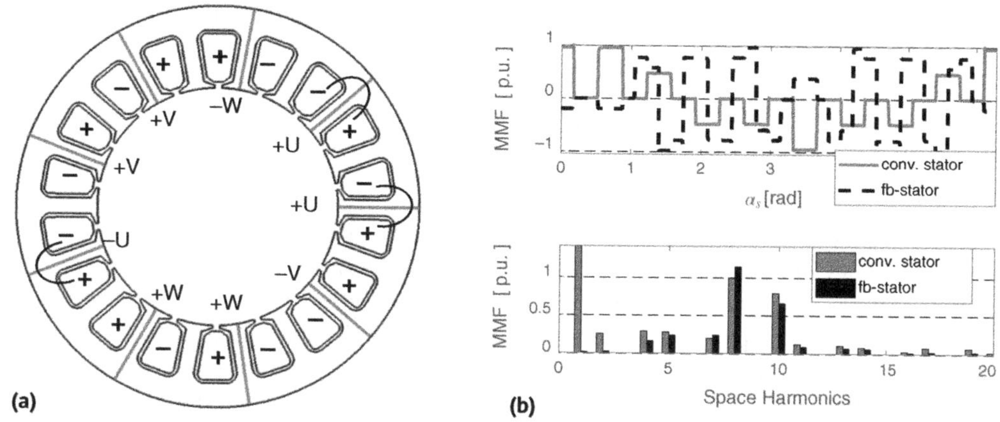

Figure 6.40 Three-phase SL-FSCW with two unequal zones per phase, with $n_{c1} = 2, n_{c2} = 1$, and with opposite polarity zones. (a) Winding configuration with flux-barriers in wound teeth. (b) Comparison of MMF characteristics for the conventional stator and the flux-barrier stator.

Table 6.10 Winding factors (in %) for the SL-FSCWs with unequal zones per phase and with opposite polarity zones. Comparison of the conventional stator and the flux-barrier stator (first gray row for flux-barriers in unwound teeth, second gray row for flux-barriers in wound teeth).

n_c	Q_s	$\nu = 1$	2	3	4	5	6	7	8	9	10	11	12	13	14	15	16
2&1	18	16.6	6.06	—	13.9	16.6	—	16.6	**94.5**	—	**94.5**	16.6	—	16.6	13.9	—	6.06
		0.04	0.12	—	2.10	4.56	—	10.3	76.9	—	**109.2**	21.0	—	20.9	15.8	—	4.02
		0.22	0.58	—	7.79	14.2	—	19.8	**108.3**	—	77.9	11.6	—	11.3	10.2	—	3.56
3&2	30	10.0	2.12	—	4.45	10.0	—	10.0	11.1	—	17.3	10.0	—	10.0	**95.1**	—	**95.1**
		0	0.	—	0.15	0.7	—	1.72	2.73	—	7.50	5.42	—	7.74	84.7	—	**104.5**
		0.02	0.05	—	0.70	2.83	—	6.17	8.83	—	18.75	11.6	—	11.6	**104.3**	—	84.9

6.4 Multiphase SL-FSCWs with Flux-barrier Stator

In addition to conventional multiphase windings, such as 2 × three-phase or 3 × three-phase windings, Chapter 5 introduced a class of multiphase FSCWs characterized by a simple structure and an arbitrary number of phases. These windings offer numerous advantages, including redundancy, enhanced flexibility in MMF characteristics, and improved control capabilities. However, similar to their three-phase counterparts, they also suffer from the high harmonic content in the MMF spectrum. This disadvantage can be effectively mitigated by using the flux-barrier technique, with both stator configurations offering effective solutions.

6.4.1 Multiphase FSCWs with Stator-yoke Flux-barriers

As described in section 6.2, the implementation of flux-barriers at specific locations in the stator yoke modifies the MMF distribution of the coil, and consequently, the overall MMF distribution of the winding. When this technique is applied to a single-layer winding with an arbitrary number of phases, the resulting MMF expression, derived from (5.21) and (6.19), is

$$F_{j,\text{fb}}(t, a_S) = \frac{2N_w}{\pi} i_j(t) \sum_{\nu} \frac{1}{\nu} {}^{\nu}\xi_{w,\text{fby}} \cos\left(\nu a_S + (\nu_0 - \nu)\frac{2\pi}{m}\right) \tag{6.59}$$

In this case, the winding factor is equal to the equivalent coil-pitch factor defined in (6.21) and (6.22), i.e.

$$^{\nu}\xi_{w,\mathrm{fby}} = {^{\nu}\xi_{p,\mathrm{fby}}} = \frac{|^{\nu}\xi_{p}|}{2}\sqrt{\left(2-\frac{h_{\mathrm{fb}}}{h_{\mathrm{yoke}}}\right)^{2}+\left(\frac{h_{\mathrm{fb}}}{h_{\mathrm{yoke}}}\right)^{2}-2\left(2-\frac{h_{\mathrm{fb}}}{h_{\mathrm{yoke}}}\right)\frac{h_{\mathrm{fb}}}{h_{\mathrm{yoke}}}\cos\left(\nu\frac{2\pi}{Q_S}\right)} \quad (6.60)$$

Example 6.11 Various multiphase SL-FSCWs with stator-yoke flux-barriers
Figure 6.41 illustrates the winding configurations for two representative multiphase SL-FSCWs employing stator-yoke flux-barriers. According to Equation (5.16), this winding type generates two MMF waveforms in the five-phase case and three waveforms in the seven-phase case. Let's first examine the five-phase winding topology depicted in Figure 6.41a. A comparison of the MMF characteristics for the case $\nu_0 = 1$ and $\nu_0 = 2$ is presented in Figure 6.42, while the corresponding

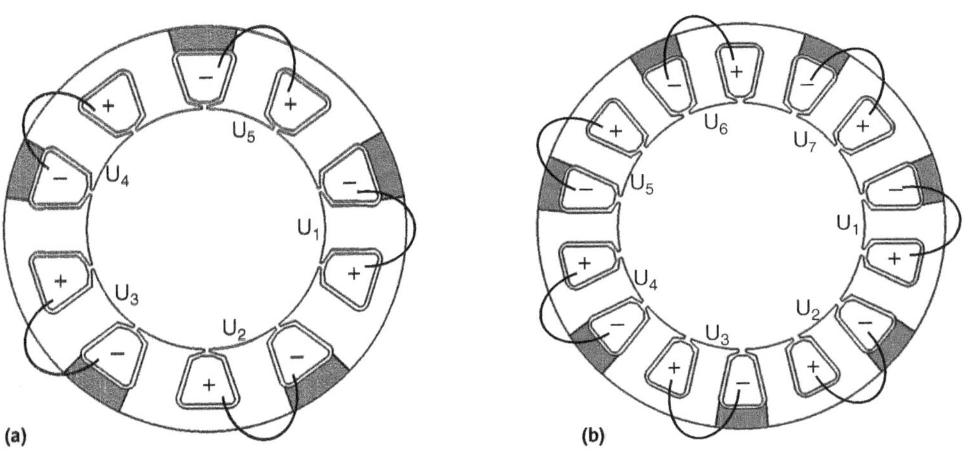

Figure 6.41 Multiphase SL-FSCWs with stator-yoke flux-barriers. (a) Five-phase SL-FSCW. (b) Seven-phase SL-FSCW.

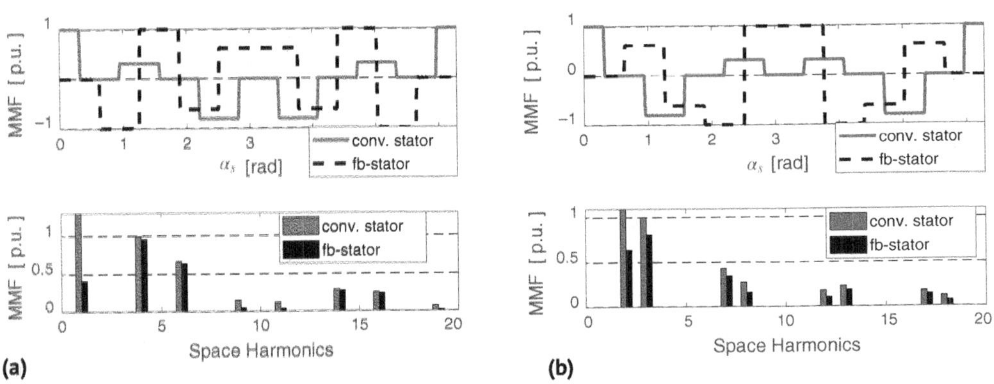

Figure 6.42 Comparison of MMF characteristics for the five-phase SL-FSCW with conventional stator and flux-barrier stator. (a) Excitation mode $\nu_0 = 1$. (b) Excitation mode $\nu_0 = 2$.

winding factors are summarized in Table 6.11. As before, the winding factors of the conventional stator are listed in the white rows, whereas those of the flux-barrier stator are highlighted in gray. The results indicate that the use of this technique effectively mitigates the MMF subharmonics, but it also leads to a reduction in the amplitude of the working harmonic.

Similar trends can also be observed for the seven-phase, eight-phase, and 11-phase configurations, as shown in Figures 6.43–6.45 and Tables 6.12–6.14. Based on these results, the following conclusions can be drawn:

- Parameter ν_0 determines the waveform of the MMF.
- The maximum winding factor is achieved when $\nu_0 = 1$.
- The flux-barrier effect is most effective when $\nu_0 = 1$.
- The flux-barrier effect on the reduction of the operating harmonic increases with increasing ν_0.

Table 6.11 Winding factors (in %) for the five-phase SL-FSCW ($\nu_0 = 1$ and $\nu_0 = 2$). Comparison of the conventional stator (white row) and the flux-barrier stator (gray row) with $h_{fb} = h_{yoke}$.

ν_0	$\nu = 1$	2	3	4	5	6	7	8	9	10	11	12	13	14	15	16
1	30.9	—	—	**95.1**		**95.1**	—	—	30.9		30.9	—	—	95.1	—	95.1
	9.5	—	—	**90.4**		**90.4**	—	—	9.55		90.4	—	—	90.4	—	90.4
2	—	58.7	**80.9**	—	—	—	**80.9**	58.7	—	—	—	58.7	80.9	—	—	—
	—	34.5	**65.4**	—	—	—	**65.4**	34.5	—	—	—	34.5	65.4	—	—	—

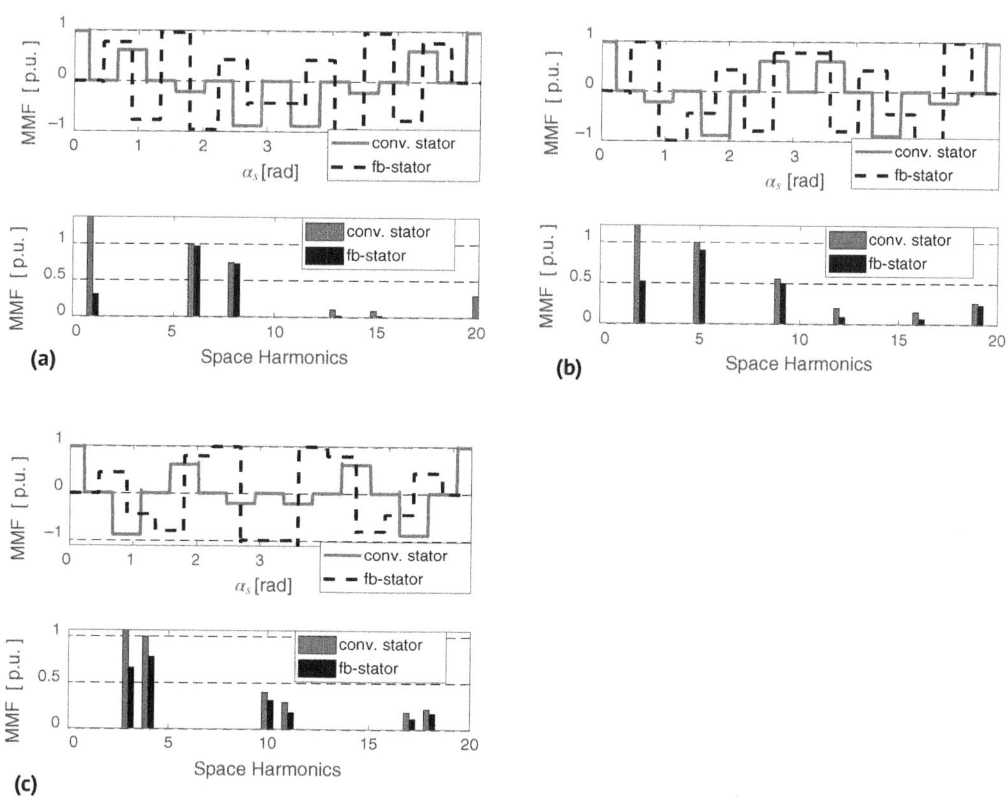

Figure 6.43 Comparison of MMF characteristics for the seven-phase SL-FSCW with conventional stator and flux-barrier stator. (a) Excitation mode $\nu_0 = 1$. (b) Excitation mode $\nu_0 = 2$. (c) Excitation mode $\nu_0 = 3$.

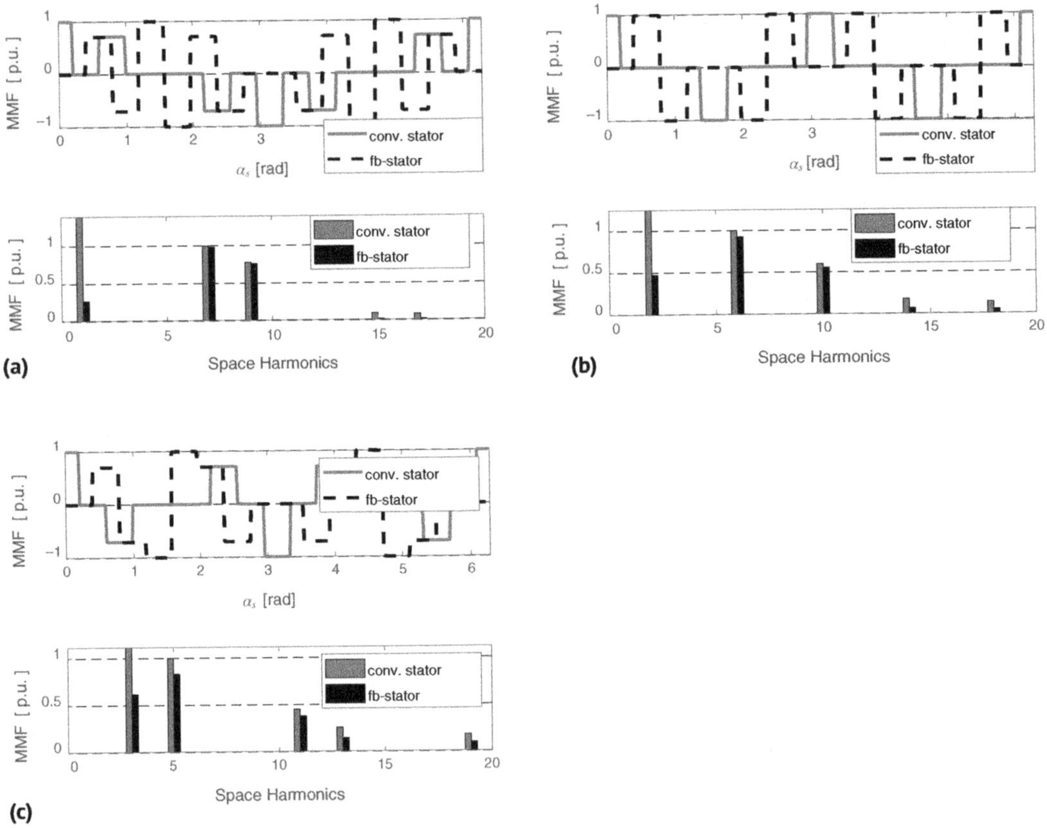

Figure 6.44 Comparison of MMF characteristics for the eight-phase SL-FSCW with conventional stator and flux-barrier stator. (a) Excitation mode $v_0 = 1$. (b) Excitation mode $v_0 = 2$. (c) Excitation mode $v_0 = 3$.

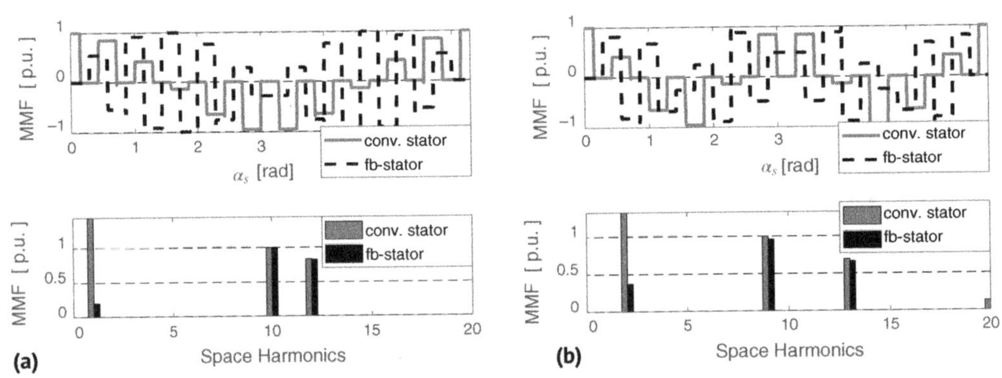

Figure 6.45 Comparison of MMF characteristics for the 11-phase SL-FSCW with conventional stator and flux-barrier stator. (a) Excitation mode $v_0 = 1$. (b) Excitation mode $v_0 = 2$. (c) Excitation mode $v_0 = 3$. (d) Excitation mode $v_0 = 4$.

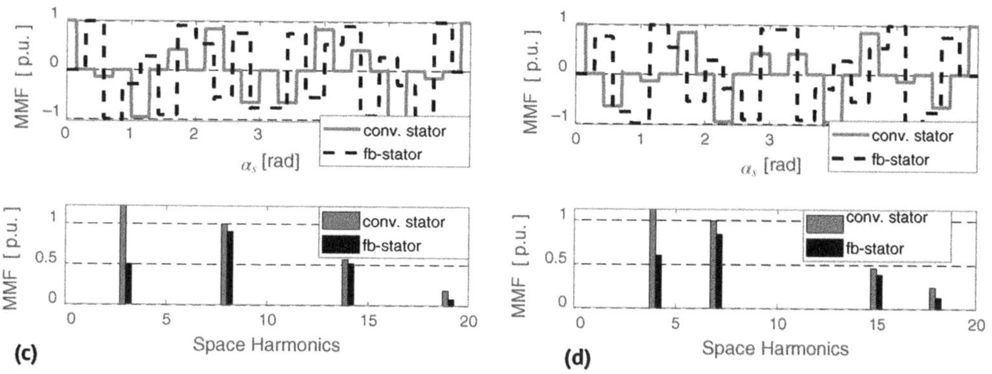

Figure 6.45 (Continued)

Table 6.12 Winding factors (in %) for the seven-phase SL-FSCW ($v_0 = 1 - v_0 = 3$). Comparison of the conventional stator (white row) and the flux-barrier stator (gray row) with $h_{fb} = h_{yoke}$.

v_0	$v = 1$	2	3	4	5	6	7	8		9	10	11	12	13	14	15	16
1	22.2	—	—	—	—	**97.5**	—	**97.5**	—	—	—	—	—	22.2	—	22.2	—
	4.95		—	—	—	**95.1**	—	**95.1**	—	—	—	—	—	4.95	—	4.95	—
2	—	43.4	—	—	**90.1**	—	—	—	—	**90.1**	—	—	43.4	—	—	—	43.4
	—	18.8	—	—	**81.2**	—	—	—	—	**81.2**	—	—	18.8	—	—	—	18.8
3	—	—	62.3	**78.2**		—	—	—	—		**78.2**	62.3	—	—	—	—	—
	—	—	38.9	**61.1**	—	—	—	—	—	—	**61.1**	38.9	—	—	—	—	—

Table 6.13 Winding factors (in %) for the eight-phase SL-FSCW ($v_0 = 1 - v_0 = 3$). Comparison of the conventional stator (white row) and the flux-barrier stator (gray row) with $h_{fb} = h_{yoke}$.

v_0	$v = 1$	2	3	4	5	6	7	8	9	10	11	12	13	14	15	16
1	19.5	—	—	—	—	—	**98.1**	—	**98.1**	—	—	—	—	—	19.5	—
	3.8		—	—	—	—	**96.2**	—	**96.2**	—	—	—	—	—	3.8	—
2	—	38.3	—	—	—	**92.4**	—	—	—	**92.4**	—	—	—	—	38.3	—
	—	14.6	—	—	—	**85.4**	—	—	—	**85.4**	—	—	—	—	14.6	—
3	—	—	55.5	—	**83.3**	—	—	—	—	—	**83.2**	—	55.5	—	—	—
	—	—	30.9	—	**69.1**	—	—	—	—	—	**69.1**	—	30.9	—	—	—

6.4.2 Multiphase FSCWs with Stator-teeth Flux-barriers

For a multiphase single-layer winding with an arbitrary number of phases and a stator-teeth flux-barrier configuration, the resulting MMF function, obtained from Equations (5.21) and (6.50), can be expressed as

$$F_{j,\mathrm{fb}}(t, a_S) = \frac{2N_w}{\pi} i_j(t) \sum_{\nu} \frac{1}{\nu} {}^{\nu}\xi_{w,\mathrm{fbt}} \cos\left(\nu a_S + (\nu_0 - \nu)\frac{2\pi}{m}\right) \tag{6.61}$$

Table 6.14 Winding factors (in %) for the 11-phase SL-FSCW ($v_0 = 1 - v_0 = 4$). Comparison of results for the conventional stator (white row) and the flux-barrier stator (gray row) with $h_{fb} = h_{yoke}$.

v_0	$v = 1$	2	3	4	5	6	7	8	9	10	11	12	13	14	15	16
1	14.2	—	—	—	—	—	—	—	—	**98.9**	—	**98.9**	—	—	—	—
	2.02	—	—	—	—	—	—	—	—	**98.0**	—	**98.0**	—	—	—	—
2	—	28.1	—	—	—	—	—	—	**95.9**	—	—	—	**95.9**	—	—	—
	—	7.9	—	—	—	—	—	—	**92.1**	—	—	—	**92.1**	—	—	—
3	—	—	41.5	—	—	—	—	**91.0**	—	—	—	—	—	**91.0**	—	—
	—	—	17.3	—	—	—	—	**83.0**	—	—	—	—	—	**83.0**	—	—
4	—	—	—	54.1	—	—	**84.1**	—	—	—	—	—	—	—	**84.1**	—
	—	—	—	29.2	—	—	**70.8**	—	—	—	—	—	—	—	**70.8**	—

For this winding group, the winding factor is equal to the equivalent coil-pitch factor defined in Equation (6.51), which includes the flux-barrier effect, namely

$$
\begin{aligned}
{}^{\nu}\xi_{w,\text{fbt}} &= {}^{\nu}\xi_{p,\text{fbt}} \\
&= {}^{\nu}\xi_{p} - \frac{1}{n_{\text{fb}}}\sin\left(\nu\frac{a_{\text{FB}}}{2}\right)
\end{aligned}
\tag{6.62}
$$

Example 6.12 Various multiphase SL-FSCWs with stator-teeth flux-barriers
As already mentioned, the stator-teeth flux-barrier technique may be implemented in two distinct ways, depending on whether the flux gaps are positioned in unwound or in wound teeth. These two implementations are demonstrated in Figures 6.46 and 6.47 for two exemplary multiphase FSCWs. Additionally, when combined with the various possible excitation modes, controlled by ν_0 parameter, this technique offers a wide range of possible solutions for selecting the appropriate rotor pole number to meet specific electrical machine requirements and design specifications. This versatility can be observed in the MMF properties of various multiphase windings, which are examined below.

Let us first look at the five-phase configuration shown in Figure 6.46. For this winding, two excitation modes are available, namely $\nu_0 = 1$ and $\nu_0 = 2$. When combined with a stator employing stator-teeth flux-barriers, this configuration is capable to generate four distinct MMF waveforms. Consequently, the new five-phase winding arrangement can effectively serve four different machine topologies with different number of pole-pairs. The resulting MMF characteristic and corresponding winding factors are compared in Figures 6.48 and 6.49 and summarized in Table 6.15. These results clearly demonstrate that both stator arrangements lead to a significant improvement in the MMF characteristics. Moreover, unlike the stator-yoke modification, the stator-teeth flux-barrier technique provides a consistent enhancement of MMF performance across all excitation modes. Additionally, although the winding factor for the operating wave decreases as ν_0 increases, this reduction is effectively compensated due to increased flux-barrier effect, which becomes more pronounced at higher ν_0 values.

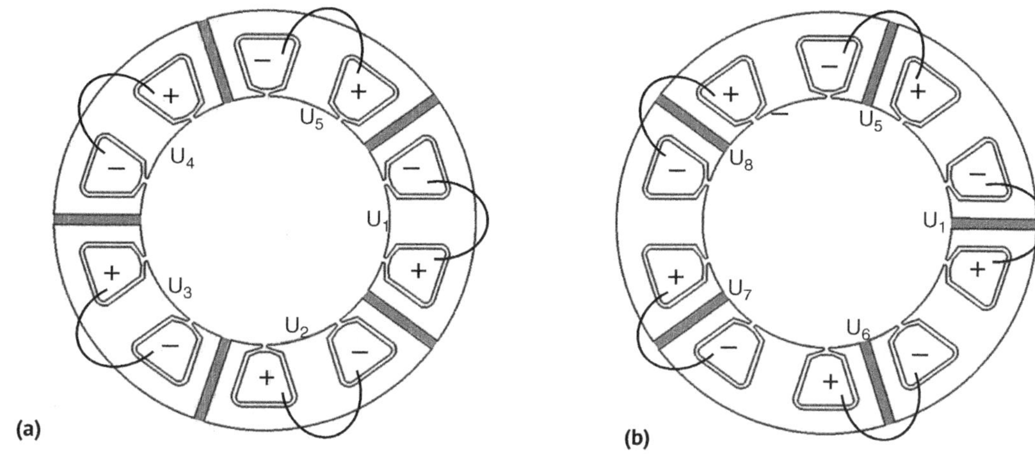

Figure 6.46 Five-phase SL-FSCW with a flux-barrier stator. (a) Flux-barriers in unwound teeth. (b) Flux-barriers in wound teeth.

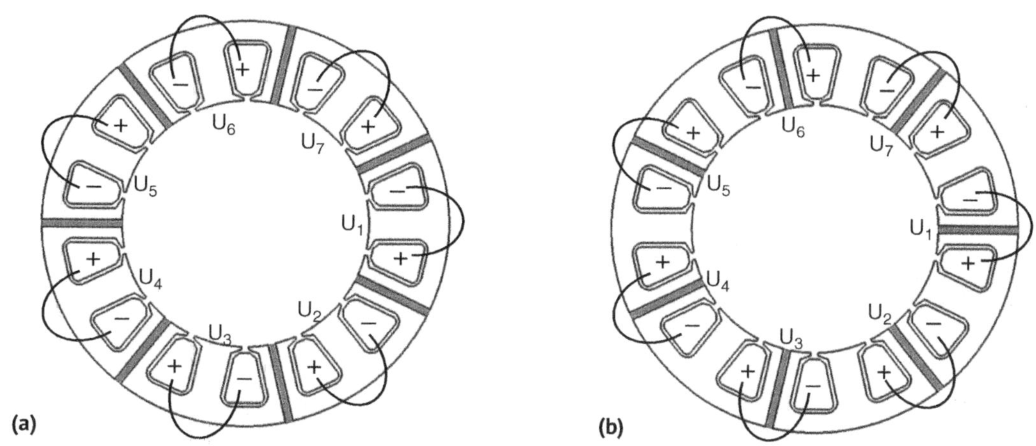

Figure 6.47 Seven-phase SL-FSCW with a flux-barrier stator. (a) Flux-barriers in unwound teeth. (b) Flux-barriers in wound teeth.

Comparable trends are observed also for the seven-, eight-, and the eleven-phase configuration. The corresponding results are presented in Figures 6.50–6.59 and Tables 6.16–6.18. Based on this analysis, the following conclusions can be drawn:

- The parameter $\max(\nu_0)$ determines the maximum number of distinct MMF waveforms available for a given winding configuration.
- Winding factor for the operating harmonic decreases with increasing ν_0.
- The flux-barrier positive effect on the operating harmonic increases with ν_0.
- For each m-phase FSCW, the stator-teeth flux-barrier technique enables up to $2^* \max(\nu_0)$ highly efficient winding solutions.

6.4.2.1 MMF Characteristics for the Five-phase SL-FSCW

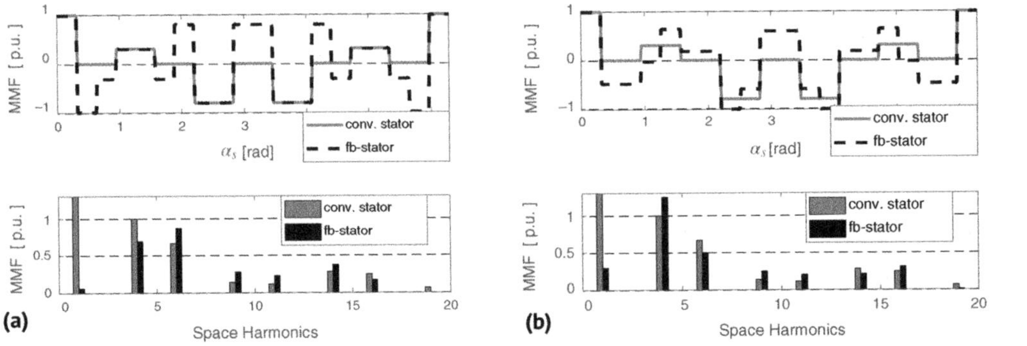

Figure 6.48 Comparison of MMF characteristics for the five-phase SL-FSCW with $\nu_0 = 1$ excitation mode. (a) Flux-barriers in unwound teeth. (b) Flux-barriers in wound teeth.

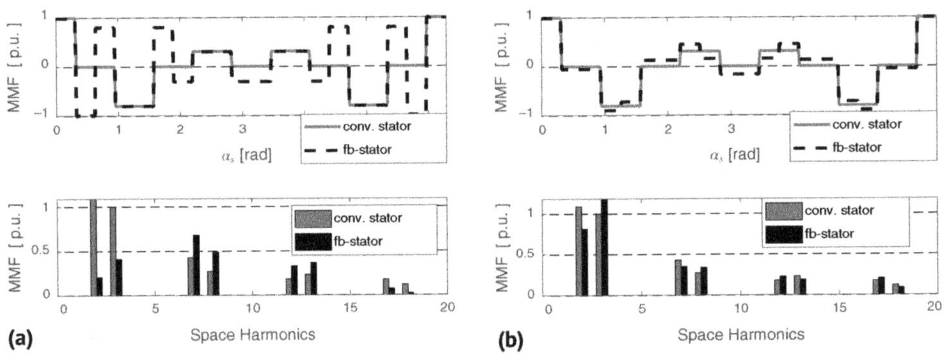

Figure 6.49 Comparison of MMF characteristics for the five-phase SL-FSCW with $\nu_0 = 2$ excitation mode. (a) Flux-barriers in unwound teeth. (b) Flux-barriers in wound teeth.

Table 6.15 Winding factors (in %) for the five-phase SL-FSCW ($\nu_0 = 1$ and $\nu_0 = 2$). Comparison of the conventional stator and the flux-barrier stator (first gray row for flux-barrier in unwound teeth, second gray row for flux-barrier in wound teeth).

ν_0	$\nu = 1$	2	3	4	5	6	7	8	9	10	11	12	13	14	15	16
1	30.9	—	—	**95.1**	—	**95.1**	—	—	30.9	—	30.9	—	—	95.1	—	95.1
	1.51	—	—	65.7	—	**124.5**	—	—	60.3	—	60.3	—	—	124.5	—	65.7
	7.12	—	—	**118.9**	—	71.3	—	—	54.7	—	54.7	—	—	71.3	—	118.8
2	—	58.7	**80.9**	—	—	—	**80.9**	58.7	—	—	—	58.7	80.9	—	—	—
	—	11.2	33.3	—	—	—	**128.5**	106.3	—	—	—	106.3	128.4	—	—	—
	—	44.1	**95.6**	—	—	—	**66.2**	73.4	—	—	—	73.4	66.2	—	—	—

6.4.2.2 MMF Characteristics for the Seven-phase SL-FSCW

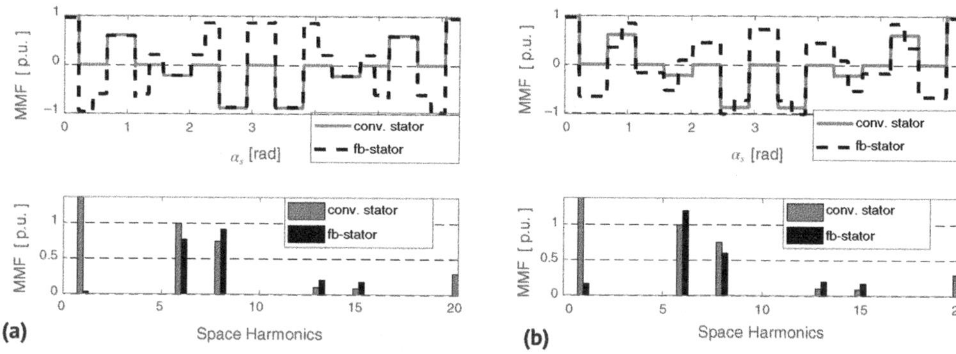

Figure 6.50 Comparison of MMF characteristics for the seven-phase SL-FSCW with $v_0 = 1$ excitation mode. (a) Flux-barriers in unwound teeth. (b) Flux-barriers in wound teeth.

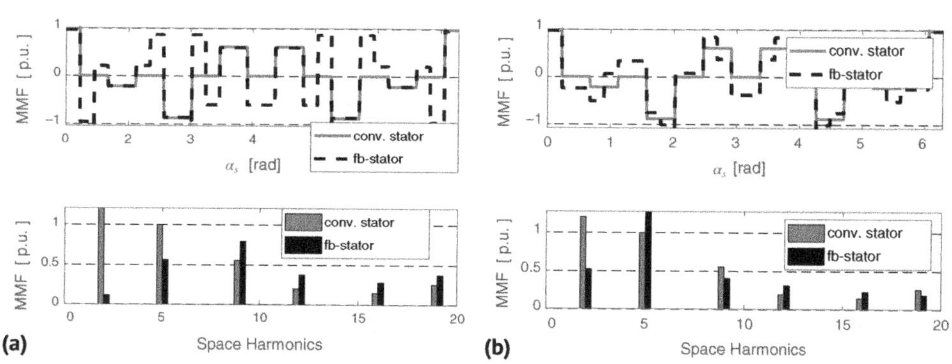

Figure 6.51 Comparison of MMF characteristics for the seven-phase SL-FSCW with $v_0 = 2$ excitation mode. (a) Flux-barriers in unwound teeth. (b) Flux-barriers in wound teeth.

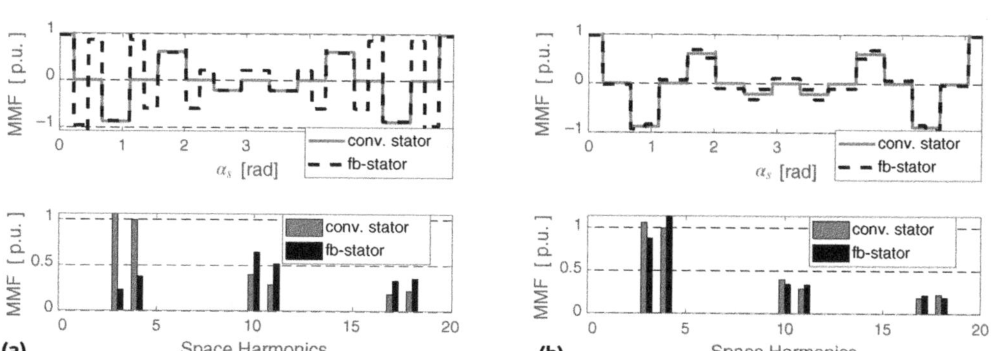

Figure 6.52 Comparison of MMF characteristics for the seven-phase SL-FSCW with $v_0 = 3$ excitation mode. (a) Flux-barriers in unwound teeth. (b) Flux-barriers in wound teeth.

Table 6.16 Winding factors (in %) for the seven-phase SL-FSCW ($v_0 = 1$ to $v_0 = 3$). Comparison of the conventional stator and the flux-barrier stator (first gray row for flux-barrier in unwound teeth, second gray row for flux-barrier in wound teeth).

v_0	$v = 1$	2	3	4	5	6	7	8		9	10	11	12	13	14	15	16
1	22.2	—	—	—	—	**97.5**	—	**97.5**	—	—	—	—	—	22.2	—	22.2	—
	0.55	—	—	—	—	75.8	—	**119.2**	—	—	—	—	—	43.9	—	43.9	—
	2.71	—	—	—	—	**117.0**	—	77.9	—	—	—	—	—	41.8	—	41.8	—
2	—	43.4	—	—	**90.1**	—	—	—	—	**90.1**	—	—	43.4	—	—	—	43.4
	—	4.30	—	—	51.0	—	—	—	—	**129.2**	—	—	82.5	—	—	—	82.5
	—	19.0	—	—	**114.5**	—	—	—	—	65.7	—	—	67.8	—	—	—	67.8
3	—	—	62.3	**78.2**	—	—	—	—	—	—	**78.2**	62.3	—	—	—	—	—
	—	—	13.6	29.4	—	—	—	—	—	—	**126.9**	111.1	—	—	—	—	—
	—	—	51.5	**89.0**	—	—	—	—	—	—	67.3	73.2	—	—	—	—	—

6.4.2.3 MMF Characteristics for the eight-phase SL-FSCW

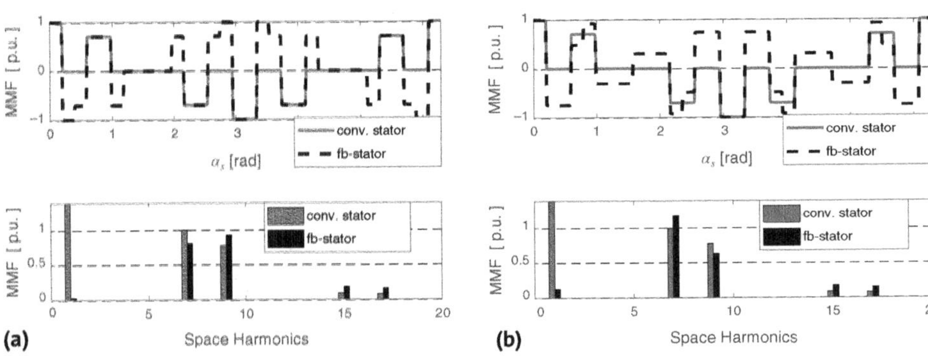

Figure 6.53 Comparison of MMF characteristics for the eight-phase SL-FSCW with $v_0 = 1$ excitation mode. (a) Flux-barriers in unwound teeth. (b) Flux-barriers in wound teeth.

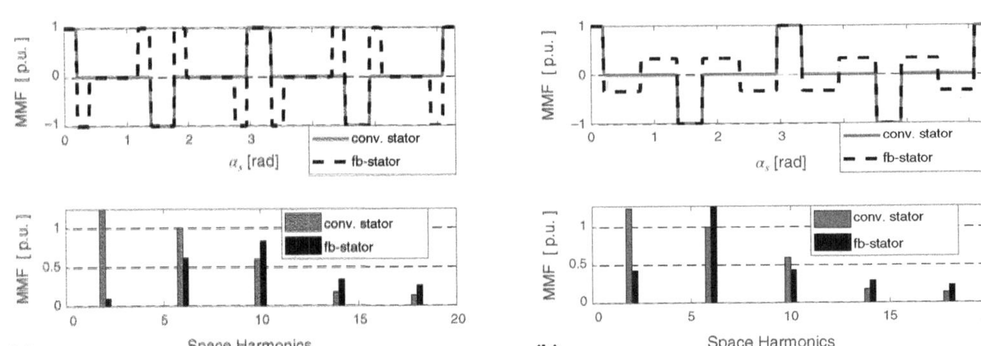

Figure 6.54 Comparison of MMF characteristics for the eight-phase SL-FSCW with $v_0 = 2$ excitation mode. (a) Flux-barriers in unwound teeth. (b) Flux-barriers in wound teeth.

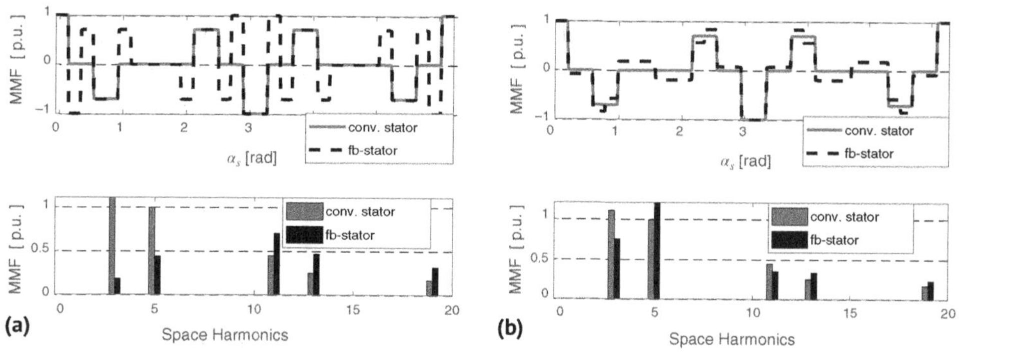

Figure 6.55 Comparison of MMF characteristics for the eight-phase SL-FSCW with $v_0 = 3$ excitation mode. (a) Flux-barriers in unwound teeth. (b) Flux-barriers in wound teeth.

Table 6.17 Winding factors (in %) for the eight-phase SL-FSCW ($v_0 = 1 - v_0 = 3$). Comparison of the conventional stator and the flux-barrier stator (first gray row for flux-barrier in unwound teeth, second gray row for flux-barrier in wound teeth).

v_0	$v = 1$	2	3	4	5	6	7	8	9	10	11	12	13	14	15	16
1	19.5	—	—	—	—	—	**98.1**	—	**98.1**	—	—	—	—	—	19.5	—
	0.37	—	—	—	—	—	78.9	—	**117.2**	—	—	—	—	—	38.6	—
	1.83	—	—	—	—	—	**115.7**	—	80.4	—	—	—	—	—	37.2	—
2	—	38.3	—	—	—	**92.4**	—	—	—	**92.4**	—	—	—	38.3	—	—
	—	2.91	—	—	—	57.0	—	—	—	**127.7**	—	—	—	73.6	—	—
	—	13.3	—	—	—	**117.4**	—	—	—	67.4	—	—	—	63.3	—	—
3	—	—	55.5	—	**83.3**	—	—	—	—	—	**83.2**	—	55.5	—	—	—
	—	—	9.36	—	36.9	—	—	—	—	—	**129.3**	—	101.7	—	—	—
	—	—	37.8	—	**100.8**	—	—	—	—	—	65.5	—	73.2	—	—	—

6.4.2.4 MMF Characteristics for the 11-phase SL-FSCW

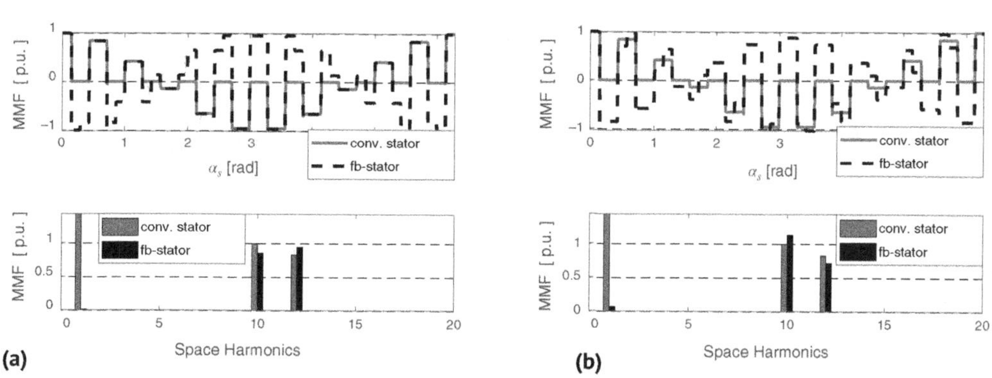

Figure 6.56 Comparison of MMF characteristics for the 11-phase SL-FSCW with $v_0 = 1$ excitation mode. (a) Flux-barriers in unwound teeth. (b) Flux-barriers in wound teeth.

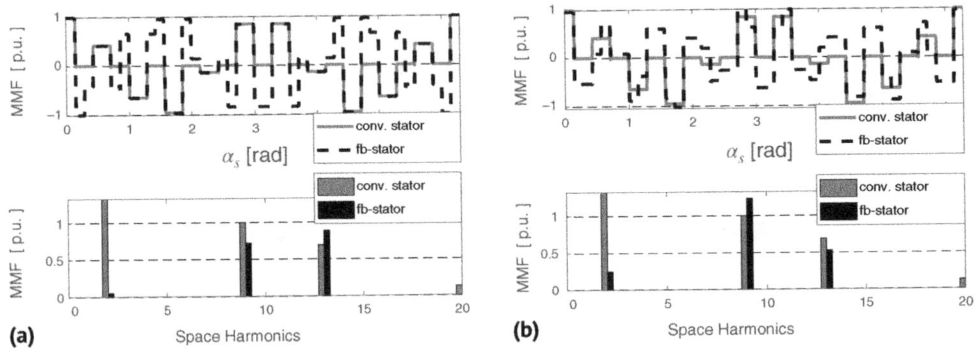

Figure 6.57 Comparison of MMF characteristics for the 11-phase SL-FSCW with $v_0 = 2$ excitation mode. (a) Flux-barriers in unwound teeth. (b) Flux-barriers in wound teeth.

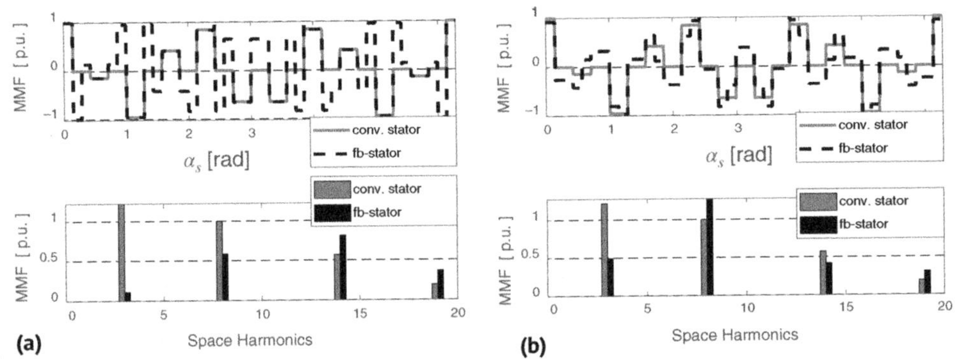

Figure 6.58 Comparison of MMF characteristics for the 11-phase SL-FSCW with $v_0 = 3$ excitation mode. (a) Flux-barriers in unwound teeth. (b) Flux-barriers in wound teeth.

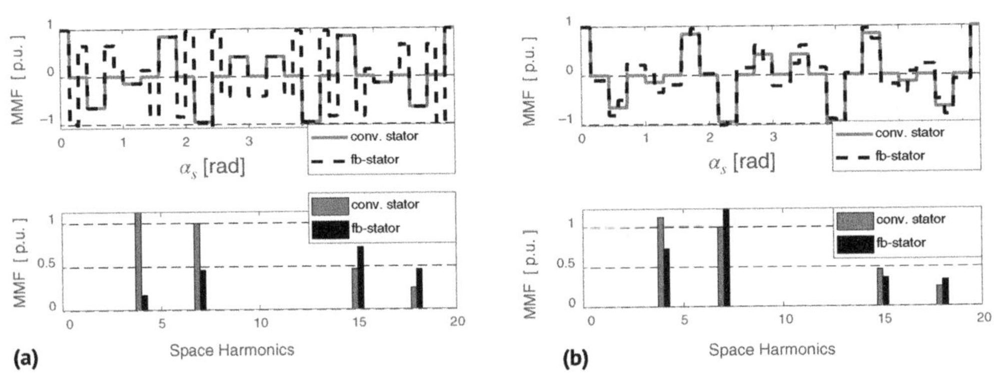

Figure 6.59 Comparison of MMF characteristics for the 11-phase SL-FSCW with $v_0 = 4$ excitation mode. (a) Flux-barriers in unwound teeth. (b) Flux-barriers in wound teeth.

Table 6.18 Winding factors (in %) for the 11-phase SL-FSCW ($\nu_0 = 1 - \nu_0 = 4$). Comparison of the conventional stator and the flux-barrier stator (first gray row for flux-barrier in unwound teeth, second gray row for flux-barrier in wound teeth).

ν_0	$\nu = 1$	2	3	4	5	6	7	8	9	10	11	12	13	14	15	16
1	14.2	—	—	—	—	—	—	—	—	**98.9**	—	**98.9**	—	—	—	—
	0.14	—	—	—	—	—	—	—	—	84.8	—	**113.1**	—	—	—	—
	0.72		—	—	—	—	—	—	—	**112.5**	—	85.5	—	—	—	—
2	—	28.1	—	—	—	—	—	—	**95.9**	—	—	—	**95.9**	—	—	—
	—	1.14	—	—	—	—	—	—	68.9	—	—	—	**123.0**	—	—	—
	—	5.43	—	—	—	—	—	—	**118.7**	—	—	—	73.2	—	—	—
3	—	—	41.5	—	—	—	—	**91.0**	—	—	—	—	—	**91.0**	—	—
	—	—	3.75	—	—	—	—	53.2	—	—	—	—	—	**128.7**	—	—
	—	—	16.8	—	—	—	—	**115.7**	—	—	—	—	—	66.2	—	—
4	—	—	—	54.1	—	—	**84.1**	—	—	—	—	—	—	—	**84.1**	—
	—	—	—	8.58	—	—	38.6	—	—	—	—	—	—	—	**129.6**	—
	—	—	—	35.2	—	—	**103.9**	—	—	—	—	—	—	—	65.3	—

7

Two-tooth Fractional Slot Winding

Since the winding arrangement and its discrete distribution across the stator slots are the main sources of magnetomotive force (MMF) harmonics, an effective way to improve the MMF characteristics of fractional slot concentrated winding (FSCWs), beyond the flux-barrier technique described in the previous chapter, is to directly modify the winding arrangement by changing its layout structure. This approach encompasses several optimization techniques, many of which have already been presented, including

- windings with asymmetrical coils (unequal turns per coil-side),
- zones with different turns per coil,
- windings with unequal coil-pitch,
- multilayer windings (more than two layers), or
- multiphase windings.

It should be noted that these techniques primarily affect certain MMF winding harmonics, with particular effectiveness in mitigating subharmonic components. However, from the standpoint of loss reduction and acoustic noise mitigation, it is essential to reduce both the low- and high-order MMF harmonics simultaneously. One of the most effective approaches for achieving a broad reduction of harmful space harmonics in FSCWs is the stator-shifting concept introduced by Dajaku and Gerling [51, 52]. In this work, the standard 12-slot/10-pole DL-FSCW has been considered as a reference winding for optimization. The fundamental principle involves doubling the number of stator slots and integrating two identical winding systems, with one shifted relative to the other by a defined angular offset. In contrast to tooth-concentrated windings, the coils in this arrangement span two adjacent stator teeth, hence the resulting structure is commonly referred to as a "two-tooth" fractional slot winding (FSW). Further improvements in the winding performances can be obtained by combining the stator-shifting concept with the aforementioned techniques. Moreover, when a specific offset angle between the two winding systems is required, the method can be complemented by a stator core featuring unequal tooth widths.

Since its introduction, the stator-shifting technique for harmonic cancellation in concentrated windings has been the subject of extensive investigations. Numerous studies have demonstrated its effectiveness in optimizing a wide range of FSCW configurations. For example, Reddy et al. [92, 93] and Kum-Kang et al. [94] examined its impact on stator MMF harmonics, winding factors, power density, efficiency, and torque ripples for various slot/pole combinations. Wang et al. [95] and Wang et al. [96] applied the stator-shifting concept to convert a common 9-slot/8-pole FSCW into an 18-slot/8-poles two-tooth FSW exhibiting substantially reduced harmonic content. This design was subsequently realized as a six-phase machine supplied by two independently controlled inverters to improve operational safety and efficiency in traction applications, as reported by Patel et al. [97, 98]. A related multiphase solution based on a 24-slot/10-pole FSW was proposed by Domann and Henke (2014) [99]. Another notable example is the 6-slot/2-pole two-tooth FSW introduced by Spas et al. [100], which demonstrates significant improvements in the MMF characteristics compared with the widely used 3-slot/2-pole conventional FSCW with $q = 0.5$. Moreover, the stator-shifting technique is not limited to double-layer windings; it has also been successfully applied to various single-layer concentrated winding topologies, as reported by Xie et al. [101] and Rudden et al. [102].

7.1 Implementation and Operating Principles

The stator-shifting concept is based on extending a reference stator by doubling the number of slots and adding a second identical stator m-phase winding set that is spatially displaced from the first by a certain angle (α_w). The fundamental objective of the additional winding set is to generate MMF harmonics whose spatial phase can be adjusted with respect to those of the first winding. Thus, when a given harmonic component generated by the two winding sets has equal amplitudes and is displaced by 180 electrical degrees, complete cancellation of that harmonic can be achieved.

An example of a typical DL-FSCW with $Q_S = 3$ and +U+V+W winding layout is illustrated in Figure 7.1a. This winding type is well known for its high harmonic content, as shown by the MMF spectrum in Figure 7.1b. In this case, the fundamental (1st order) MMF harmonic is typically selected as the operating wave, whereas the 2nd, 4th, 5th, and higher-order harmonics are regarded as undesirable. The transformation of this reference FSCW into the corresponding two-tooth FSW is illustrated in Figure 7.2 and proceeds as follows:

- first, the original FSCW is divided into two identical winding sets,
- then, the number of stator slots is doubled,
- and finally, the two winding sets are spatially shifted with respect to each other by a specific angle α_w.

Accordingly, the number of stator slots in the resulting winding is

$$Q'_S = 2Q_S \tag{7.1}$$

where Q_S denotes the number of stator slots in the original FSCW.

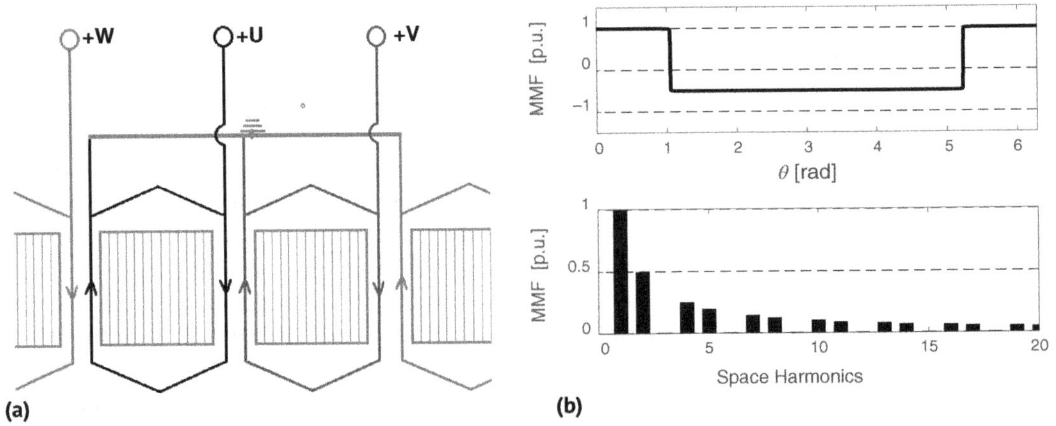

Figure 7.1 DL-FSCW with $z = 1$, $n_c = 1$, and $Q_s = 3$. (a) Winding layout. (b) Three-phase spatial MMF distribution and harmonic contents.

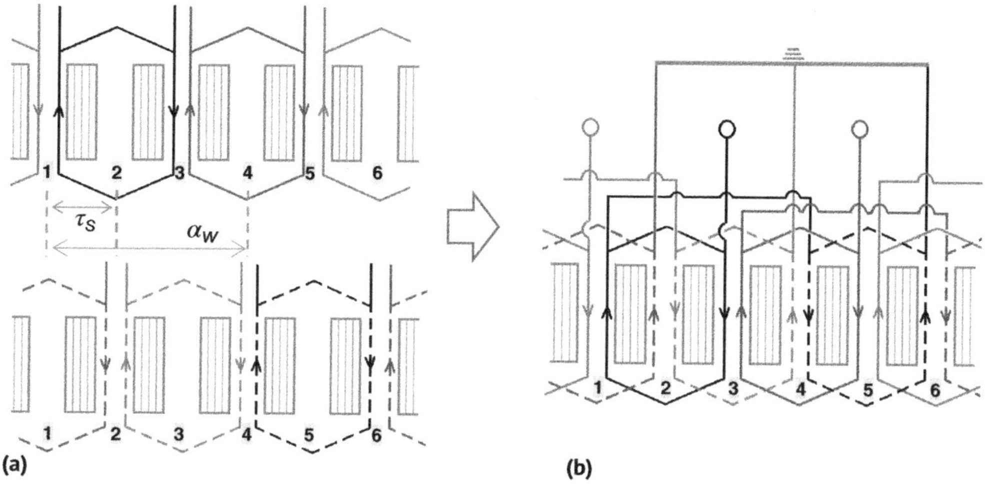

Figure 7.2 Realization of the two-tooth FSW. (a) Two winding sets shown separately; (b) Resulting two-tooth winding with a spatial winding sets displacement of $\alpha_w = 3\tau_{s2}$.

In order to maintain the total number of turns per phase, the number of turns per coil in each winding set must be reduced to half the value in the reference design, i.e.

$$N'_w = 0.5N_w \tag{7.2}$$

Based on the winding layouts shown in Figures 7.1 and 7.2, the coil-span angles for the original winding and the two-tooth winding are defined by Equations (7.3) and (7.4), respectively.

$$\alpha_\tau = \frac{2\pi}{Q_S} \tag{7.3}$$

$$\alpha'_\tau = 2\frac{2\pi}{Q'_S} = \frac{2\pi}{Q_S} = \alpha_\tau \tag{7.4}$$

As a result, the coil-pitch factor of the two-tooth FSW remains identical to that of the original winding, as shown in Equations (7.5) and (7.6).

$${}^{\nu}\xi_p = \sin\left(\nu\frac{\alpha_\tau}{2}\right) \tag{7.5}$$

$${}^{\nu}\xi'_p = \sin\left(\nu\frac{\alpha'_\tau}{2}\right) = \sin\left(\nu\frac{\alpha_\tau}{2}\right) = {}^{\nu}\xi_p \tag{7.6}$$

Under these conditions, the MMF functions of individual winding sets can be expressed as follows

$$F_{j,I}(t,\alpha_S) = \frac{2N'_w}{\pi} i_j(t) \sum_\nu \frac{{}^{\nu}\xi_p}{\nu} \cdot \sin(\nu\alpha_S) \tag{7.7}$$

$$F_{j,II}(t,\alpha_S) = \frac{2N'_w}{\pi} i_j(t) \sum_\nu \frac{{}^{\nu}\xi_p}{\nu} \cdot \sin(\nu(\alpha_S - \alpha_w)) \tag{7.8}$$

Depending on whether the two windings are connected with the same- or opposite coil polarities, as illustrated in Figure 7.3, the resulting MMF function is given by Equation (7.9). In this expression, ${}^{\nu}\xi_{z12}$ represents the additional distribution factor resulting from the spatial displacement of

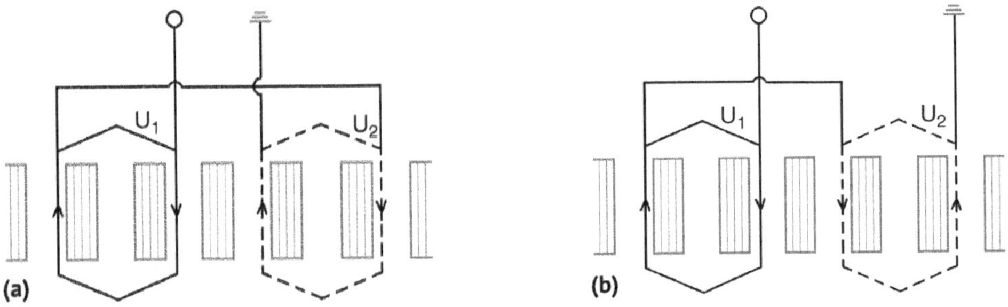

Figure 7.3 (a) Same polarity winding sets. (b) Opposite polarity winding sets.

winding sets. This factor expresses the effect of shift angle α_w on the individual MMF harmonics, as defined in Equation (7.10).

$$F_j(t, a_S) = F_{j,I}(t, a_S) \pm F_{j,II}(t, a_S) = \frac{2N_w}{\pi} i_j(t) \sum_\nu \frac{{}^\nu\xi_p\,{}^\nu\xi_{z12}}{\nu} \cdot \sin\left(\nu\left(a_S - \frac{a_w}{2}\right)\right) \tag{7.9}$$

$${}^\nu\xi_{z_{12}} = \begin{cases} \cos\left(\nu\frac{a_w}{2}\right), & \text{for same polarity} \\ \sin\left(\nu\frac{a_w}{2}\right), & \text{for opposite polarity} \end{cases} \tag{7.10}$$

As can be observed from Figure 7.2, the shift angle α_w is always an integer multiple of the slot-pitch angle τ_s, i.e.

$$a_w = k_q \cdot \tau_s = k_q \frac{2\pi}{Q_s'} \tag{7.11}$$

The stator-shifting technique is applicable to all FSCW configurations and provides an efficient means of improving their MMF characteristics. In its general form, the MMF function of the resulting two-tooth FSW is expressed by Equation (7.12), with the corresponding winding factor defined in Equation (7.13). The respective coil-pitch, distribution, and zone factors can be derived from the analytical relationships provided in the preceding chapters.

$$F_j(t, a_S) = \frac{2zn_cN_w}{\pi} i_j(t) \sum_\nu \frac{{}^\nu\xi_w}{\nu} \cdot \sin\left(\nu\left(a_S - \frac{a_w}{2}\right)\right) \tag{7.12}$$

$${}^\nu\xi_w = {}^\nu\xi_p\,{}^\nu\xi_d\,{}^\nu\xi_z\,{}^\nu\xi_{z_{12}} \tag{7.13}$$

Figure 7.4 examines the influence of the shift-angle α_w on the dominant MMF harmonics of the winding under investigation. For the topology employing the same polarity winding sets and with a shift angle of $3\tau_s$, the 1st, and 5th harmonics are completely suppressed, while the 2nd and 4th harmonics remain unaffected. When, on the other hand, winding sets of opposite polarities are connected using the same shift angle, the 1st and 5th MMF harmonics remain unaffected, whereas the 2nd and 4th harmonics are effectively eliminated.

The cross-section of an exemplary two-tooth FSW is illustrated in Figure 7.5a. For clarity, the end-windings of the first and second winding sets are depicted using solid-black and dashed-gray lines, respectively. From the winding arrangement and the foregoing analysis, the main features of a two-tooth FSW compared to conventional FSCWs are:

- double the number of the stator slots,
- double the number of coils per phase,

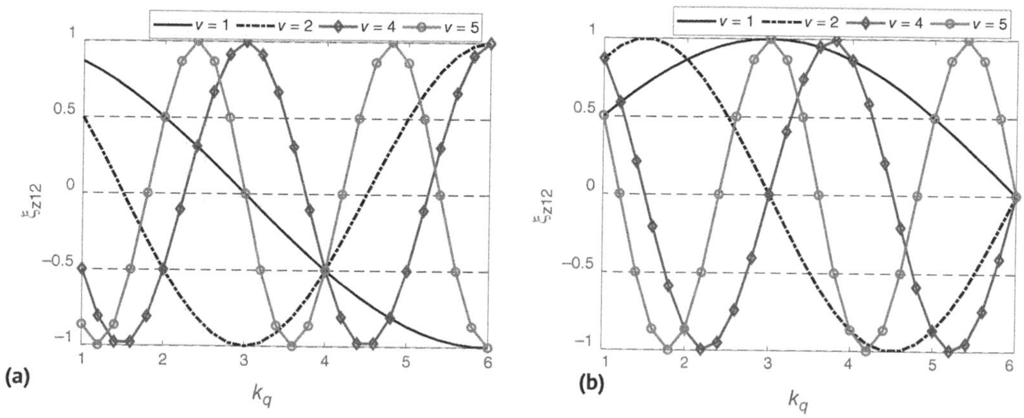

Figure 7.4 Distribution factors vs. shift angle $\alpha_w = k_q\tau_s$ for different MMF harmonics. (a) Same polarity winding sets. (b) Opposite polarity winding sets.

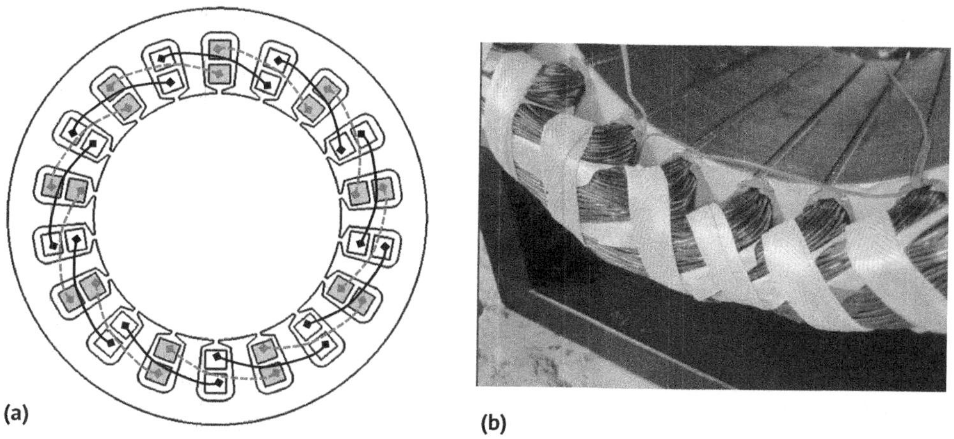

Figure 7.5 (a) Cross section of an example two-tooth FSW. (b) Prototype stator with two-tooth FSW.

- two-tooth coil-pitch structure,
- identical coil-pitch relative to the original FSCW,
- nonoverlapping end-windings, and
- simultaneous reduction of multiple space MMF harmonics.

On the other hand, when compared with fully distributed windings, two-tooth FSWs exhibit MMF waveforms that are less sinusoidal but still substantially improved relative to conventional concentrated windings. From a manufacturing standpoint, two-tooth FSWs combine the structural simplicity of concentrated windings with a reduced number of coils per pole and phase compared to distributed windings. In addition, as shown in Figure 7.5, the absence of end-winding overlap leads to shorter end-winding lengths and a more compact machine structure. Consequently, it can be concluded that two-tooth FSWs may be regarded as an alternative solution between purely concentrated and fully distributed windings.

7.2 Single-layer Two-tooth/Three-phase FSWs

7.2.1 Single-layer FSWs with One Zone/Phase

Example 7.1 Optimization of an SL-FSCW with $Q_S = 6$, $z = 1$, and $n_c = 1$

Figure 7.6 illustrates the winding layouts for both the reference concentrated winding and the corresponding two-tooth winding obtained through application of the stator-shifting technique. In the reference configuration, the operating MMF harmonic is either the 2nd or the 4th, both exhibiting a winding factor of 86.6%. Despite this fundamental behavior, this winding produces a high overall harmonic content, including significant subharmonics. The influence of the shift angle on the dominant MMF harmonics for cases of the same and opposite polarity is presented in Figure 7.7. The results indicate that a shift angle of $\alpha_w = 3\tau_s$ is optimal for suppressing one of the critical harmonics, namely the 2nd or the 4th. The resulting two-tooth winding with the same polarity winding sets and $\alpha_w = 3\tau_s$ is illustrated in Figure 7.6b. A comparison of the MMF characteristics and winding factors is provided in Figure 7.8 and summarized in Table 7.1, with the reference winding results listed in white rows,

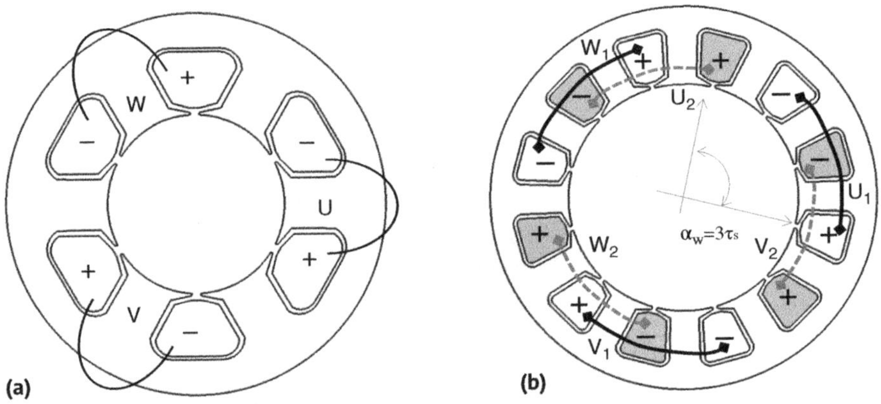

Figure 7.6 (a) SL-FSCW with $Q_S = 6, z = 1$, and $n_c = 1$. (b) Two-tooth FSW with $Q_S = 12$ and with the same polarity winding sets.

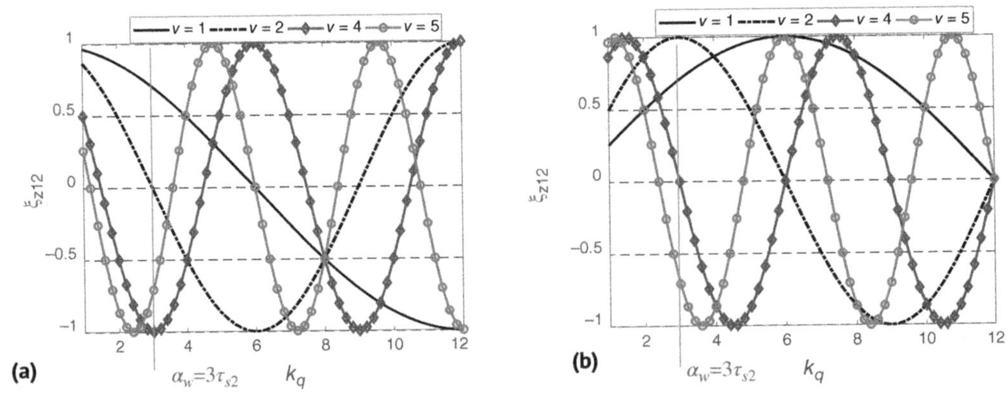

Figure 7.7 Distribution factors vs. shift angle $\alpha_w = k_q\tau_s$ for different MMF harmonics. (a) Same polarity winding sets. (b) Opposite polarity winding sets.

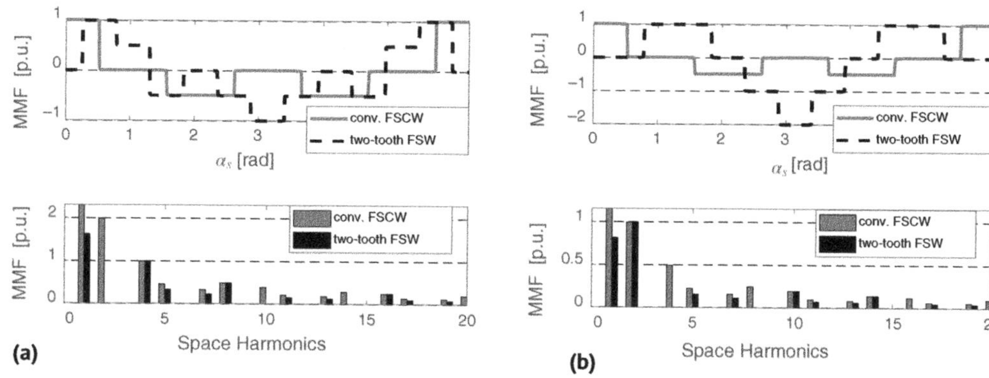

Figure 7.8 Comparison of MMF characteristics for the reference ($Q_S = 6, z = 1, n_c = 1$) and the two-tooth ($Q_S = 12$) winding with $3\tau_s$ shift angle. (a) Same polarity winding sets. (b) Opposite polarity winding sets.

Table 7.1 Winding factors (in %) of one-zone single-layer windings. Comparison of results for conventional FSCW and two-tooth FSW (first gray row: same polarity, second gray row: opposite polarity).

n_c	α_w	$v = 1$	2	3	4	5	6	7	8	9	10	11	12	13	14	15
1	Ref.	50.0	**86.6**	—	**86.6**	50.0	—	50.0	86.6	—	86.6	50.0	—	50.0	86.6	—
	$3\tau_s$	35.3	0	—	**86.6**	35.3	—	35.3	86.6	—	0	35.3	—	35.3	0	—
		35.3	**86.6**	—	0	35.3	—	35.3	0	—	86.6	35.3	—	35.3	86.6	—
2	Ref.	22.4	25.5	—	43.3	**83.6**	—	**83.6**	43.3	—	25.0	22.4	—	22.4	25.0	—
	$5\tau_s$	17.8	6.47	—	37.5	**82.9**	—	10.9	21.6	—	24.1	13.6	—	13.6	24.1	—
		13.6	24.1	—	21.6	10.9	—	**82.9**	37.5	—	6.47	17.8	—	17.8	6.47	—

followed by the same and opposite polarity with two-tooth configurations in gray rows. Values shown in bold correspond to the maximum winding factors for the respective winding configurations.

For winding sets of the same polarity, the 1st, 5th, and 7th MMF harmonics are reduced by approximately 30%, the 2nd, 10th, and 14th harmonics are completely eliminated, while the 4th harmonic (operating harmonic) remains unaffected. Conversely, when the winding sets are connected with opposite polarity, the 4th and the 8th harmonics are completely canceled, whereas the 1st, 5th, and 7th MMF harmonics are again reduced by about 30%. In this case, the 2nd harmonic remains unchanged and serves as the operating wave.

Example 7.2 Optimization of an SL-FSCW with $Q_S = 12$, $z = 1$, and $n_c = 2$

The second example considers an SL-FSCW with the following winding parameters: $Q_S = 12$, $z = 1$, and $n_c = 2$, as shown in Figure 7.9a. This winding produces two MMF harmonics with relatively high winding factors, specifically the 5th and the 7th, which may be selected as the operating harmonics. As is typical for concentrated windings, however, significant subharmonics and higher-order components are also present.

Applying the stator-shifting technique yields a 24-slot two-tooth winding, illustrated in Figure 7.9b. Based on Equation (7.10), the optimal shift angle for this winding type is $\alpha_w = 5\tau_s$. The resulting MMF characteristics for both winding-set polarities are presented in Figure 7.10 and summarized in Table 7.1. These results clearly demonstrate the ability of the stator-shifting concept to selectively reduce multiple undesirable MMF harmonics without affecting the desired operating harmonic component.

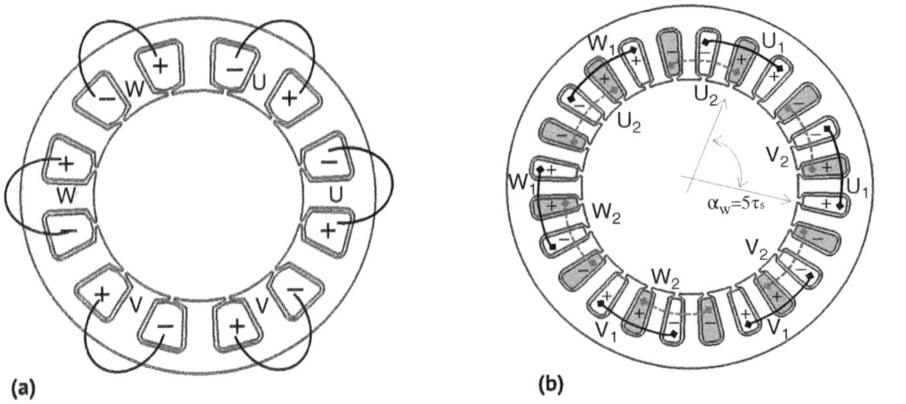

Figure 7.9 (a) SL-FSCW with $Q_S = 12, z = 1$, and $n_c = 2$. (b) Two-tooth FSW with $Q_S = 24$ and with the same polarity winding sets.

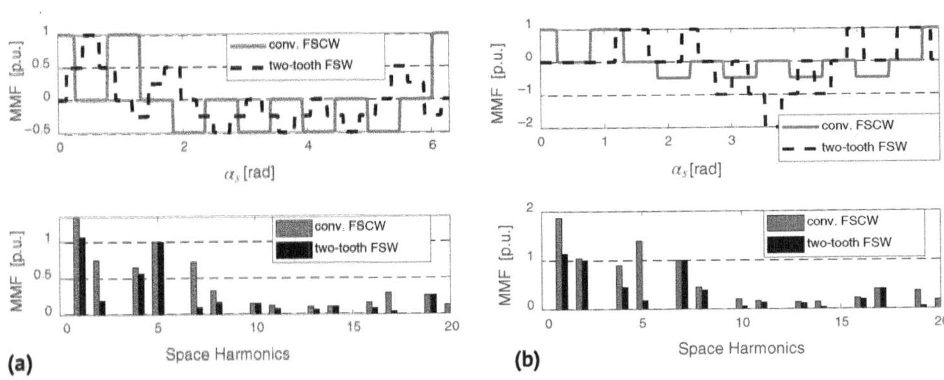

Figure 7.10 Comparison of MMF characteristics for the reference ($Q_S = 12, z = 1, n_c = 2$) and the two-tooth ($Q_S = 24$) winding with $5\tau_s$ shift angle. (a) Same polarity winding sets. (b) Opposite polarity winding sets.

7.2.2 Single-layer FSWs with Two Zones/Phase

Example 7.3 Optimization of an SL-FSCW with $Q_S = 12$, $z = 2$, and $n_c = 1$

Similar to the previously investigated one-zone configuration, the stator-shifting technique is equally effective for SL-FSCWs employing two zones per phase. A commonly used winding with parameters $Q_S = 12$, $z = 1$, and $n_c = 1$ is illustrated in Figure 7.11a. By applying the stator-shifting technique, this winding type is transformed into a 24-slot two-tooth FSW, illustrated in Figure 7.11b. In the reference winding, either the 5th or 7th harmonic may be selected as the operating wave. Consequently, to suppress one of the critical harmonics, the optimal shift angle according to Equation (7.10), should be $\alpha_w = 5\tau_s$. The resulting MMF characteristics and harmonic winding factors are shown in Figure 7.12 and summarized in Table 7.2. With the same polarity winding sets, the 1st harmonic is reduced by 20%, while the 7th and 17th harmonics are suppressed by more than 87%, leaving the 5th harmonic essentially unaffected. This configuration is therefore well suited for a 10-pole electric machine. Alternatively, opposite polarity winding sets cancel or significantly reduce the 1st, 5th, and 19th harmonics, making this configuration ideal for a 14-pole electric machine.

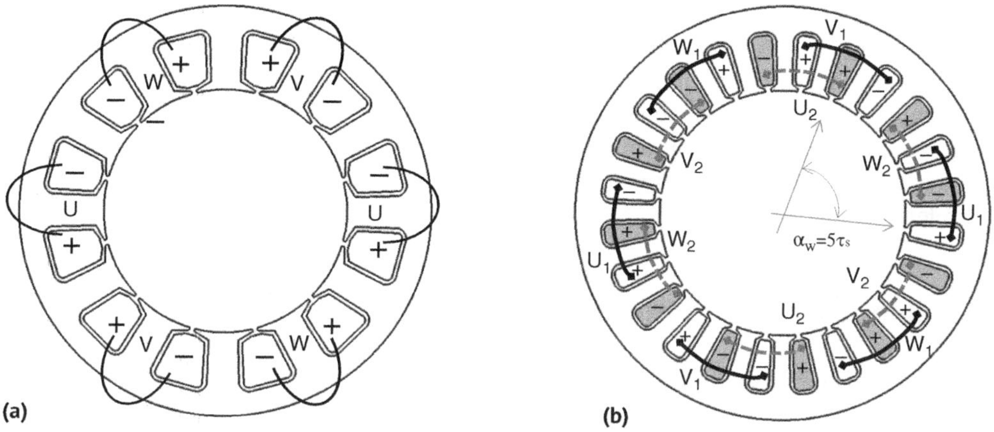

Figure 7.11 (a) SL-FSCW with $Q_S = 12, z = 1$, and $n_c = 1$. (b) Two-tooth FSW with $Q_S = 24$ and with the same polarity winding sets.

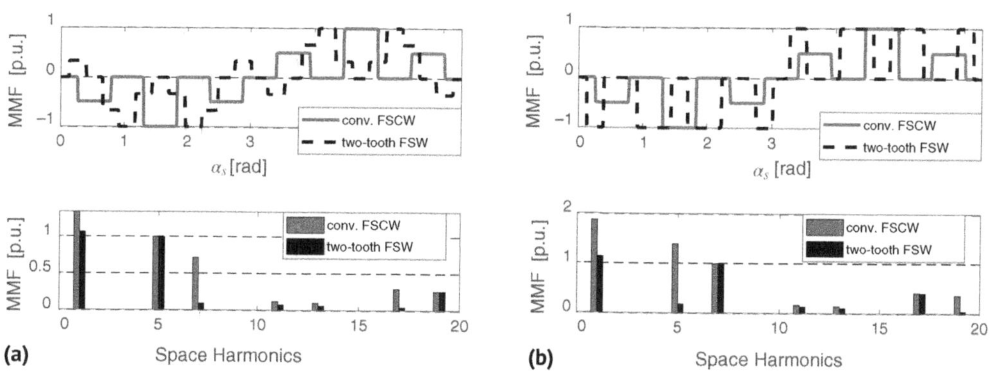

Figure 7.12 Comparison of MMF characteristics for the reference ($Q_S = 12, z = 2, n_c = 1$) and the two-tooth ($Q_S = 24$) winding with $5\tau_s$ shift angle. (a) Same polarity winding sets. (b) Opposite polarity winding sets.

Table 7.2 Winding factors (in %) of two-zone single-layer windings. Comparison of results for conventional FSCW and two-tooth FSW (first gray row: same polarity, second gray row: opposite polarity).

n_c	α_w	$v = 1$	2	3	4	5	6	7	8	9	10	11	12	13	14	15
1	Ref.	25.9	—	—	—	**96.6**	—	**96.6**	—	—	—	25.9	—	25.9	—	—
	$5\tau_s$	20.5	—	—	—	**95.7**	—	12.6	—	—	—	15.7	—	15.7	—	—
		15.7	—	—	—	12.6	—	**95.7**	—	—	—	20.5	—	20.5	—	—
2	Ref.	12.6	—	—	—	15.7	—	20.5	—	—	—	**95.7**	—	**95.7**	—	—
	$9\tau_s$	10.5	—	—	—	15.4	—	11.4	—	—	—	**93.9**	—	18.7	—	—
		7.0	—	—	—	3.1	—	17.1	—	—	—	18.7	—	**93.9**	—	—

Example 7.4 Optimization of an SL-FSCW with $Q_S = 24$, $z = 2$, and $n_c = 2$
A further two-zone FSCW example with parameters $Q_S = 12$, $z = 2$, and $n_c = 2$ is presented in Figure 2.24. In this configuration, the 11th or the 13th MMF harmonics exhibit the highest winding factors and therefore can be used as operating waves. By selecting a suitable shift angle, i.e. for $\alpha_w = 9\tau_s$, one of these dominant harmonics can be selectively suppressed. The resulting MMF characteristics, shown in Figure 7.13 and Table 7.2, demonstrate a significant improvement over the reference winding.

7.3 Double-layer Two-tooth/Three-phase FSWs

7.3.1 Double-layer FSWs with One Zone/Phase

Example 7.5 Optimization of an DL-FSCW with $Q_S = 3$, $z = 1$, and $n_c = 1$
Figure 7.14 illustrates a standard 3-slot/2-pole FSCW and its corresponding optimized two-tooth configuration. This winding type typically operates with either the 1st or 2nd MMF harmonic. To suppress the associated critical harmonics, a shift angle of $\alpha_w = 3\tau_s$ is required. The resulting MMF characteristics for both winding-set polarities are plotted in Figure 7.15 and summarized in Table 7.3. When winding sets of the same polarity are employed, the 1st,

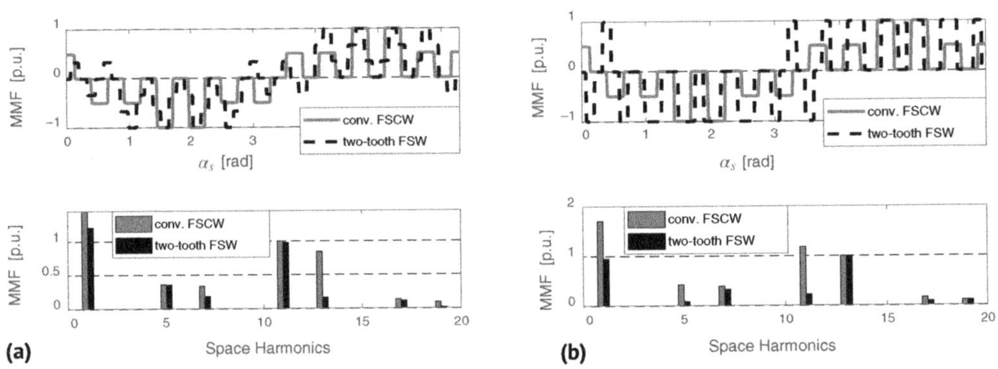

Figure 7.13 Comparison of MMF characteristics for the reference ($Q_S = 24, z = 2, n_c = 2$) and the two-tooth ($Q_S = 48$) winding with $9\tau_s$ shift angle. (a) Same polarity winding sets. (b) Opposite polarity winding sets.

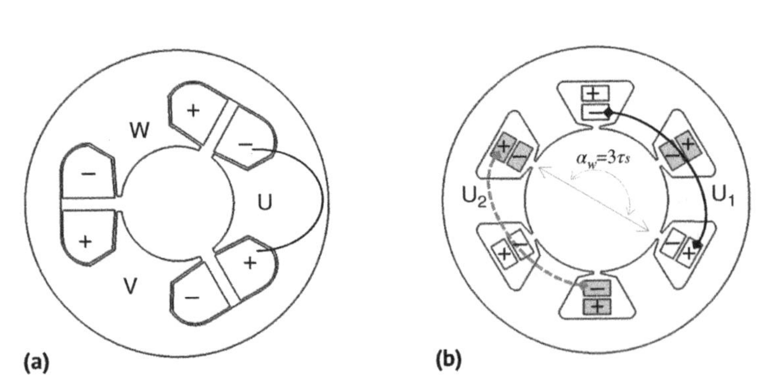

Figure 7.14 (a) DL-FSCW with $Q_S = 3, z = 1$, and $n_c = 1$. (b) Two-tooth FSW with $Q_S = 6$ and with opposite polarity winding sets.

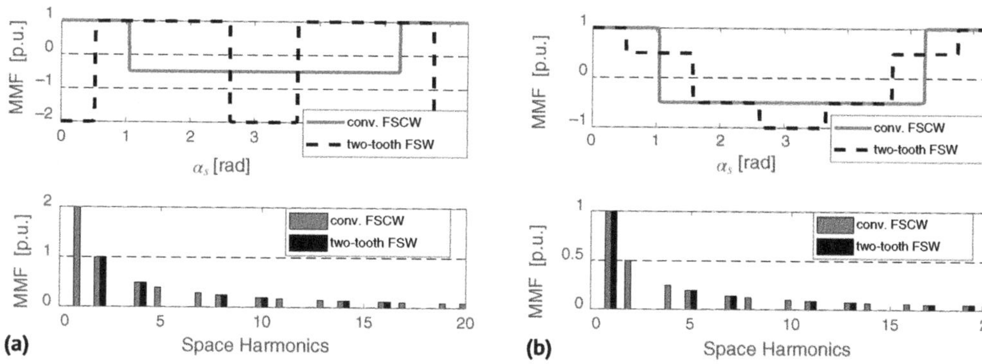

Figure 7.15 Comparison of MMF characteristics for the reference ($Q_S = 3, z = 1, n_c = 1$) and a new two-tooth ($Q_S = 6$) winding with $3\tau_s$ shift angle. (a) Same polarity winding sets. (b) Opposite polarity winding sets.

Table 7.3 Winding factors (in %) of one-zone double-layer windings. Comparison of results for conventional FSCW and two-tooth FSW (first gray row: same polarity, second gray row: opposite polarity).

n_c	α_{sh}	$v = 1$	2	3	4	5	6	7	8	9	10	11	12	13	14	15
1	Ref.	**86.6**	**86.6**	—	86.6	86.6	—	86.6	86.6	—	86.6	86.6	—	86.6	86.6	—
	$3\tau_s$	0	**86.6**	—	86.6	0	—	0	86.6	—	86.6	0	—	0	86.6	—
		86.6	0	—	0	86.6	—	86.6	0	—	0	86.6	—	86.6	0	—
2	Ref.	25.0	**75.0**	—	**75.0**	25.0	—	25.0	75.0	—	75.0	25.0	—	25.0	75.0	—
	$3\tau_s$	16.7	0	—	**75.0**	17.7	—	17.7	75.0	—	0	17.7	—	17.7	0	—
		16.7	**75.0**	—	0	17.7	—	17.7	0	—	75.0	17.7	—	17.7	75.0	—
3	Ref.	6.06	14.0	—	**94.5**	**94.5**	—	14.0	6.06	—	6.06	14.0	—	94.5	94.5	—
	$9\tau_s$	0	14.0	—	**94.5**	0	—	0	6.06	—	6.06	0	—	0	94.5	—
		6.06	0	—	0	**94.5**	—	14.0	0	—	0	14.0	—	94.5	0	—

5th, 7th, and other higher-order harmonics are effectively canceled, while the 2nd harmonic remains unaffected. This makes the configuration well suited for 4-pole machines. On the other hand, when winding sets of opposite polarity are used, the 2nd, 4th, 8th, and 10th harmonics are suppressed, whereas the 1st harmonic remains unaffected, leading to ani an effective FSW solution for 2-pole machines.

Example 7.6 Optimization of an DL-FSCW with $Q_S = 6, z = 1$, and $n_c = 2$
In the second example shown in Figure 7.16, the 2nd and 4th MMF harmonics have the highest winding factor (75%) and may serve as operating harmonics. By selecting $\alpha_w = 3\tau_S$, both critical and higher-order harmonics can be effectively mitigated, as demonstrated in Figure 7.17 and Table 7.3. The same polarity configuration is particularly well suited for 8-pole machine, while the opposite polarity arrangement offers an ideal solution for 4-pole machines.

Example 7.7 Optimization of DL-FSCW with $Q_S = 9, z = 1$, and $n_c = 3$
A further example with $Q_S = 9, z = 1$, and $n_c = 3$ is shown in Figure 7.18a. Here, the suppression of the 4th or the 5th harmonic is achieved using a shift angle of $\alpha_w = 9\tau_s$, as illustrated in

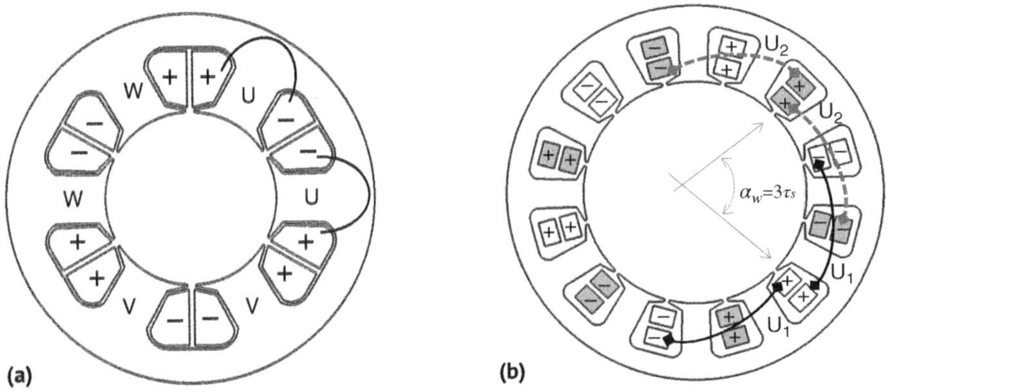

Figure 7.16 (a) DL-FSCW with $Q_S = 6, z = 1$, and $n_c = 2$. (b) Two-tooth FSW with $Q_S = 12$ and with the same polarity winding sets.

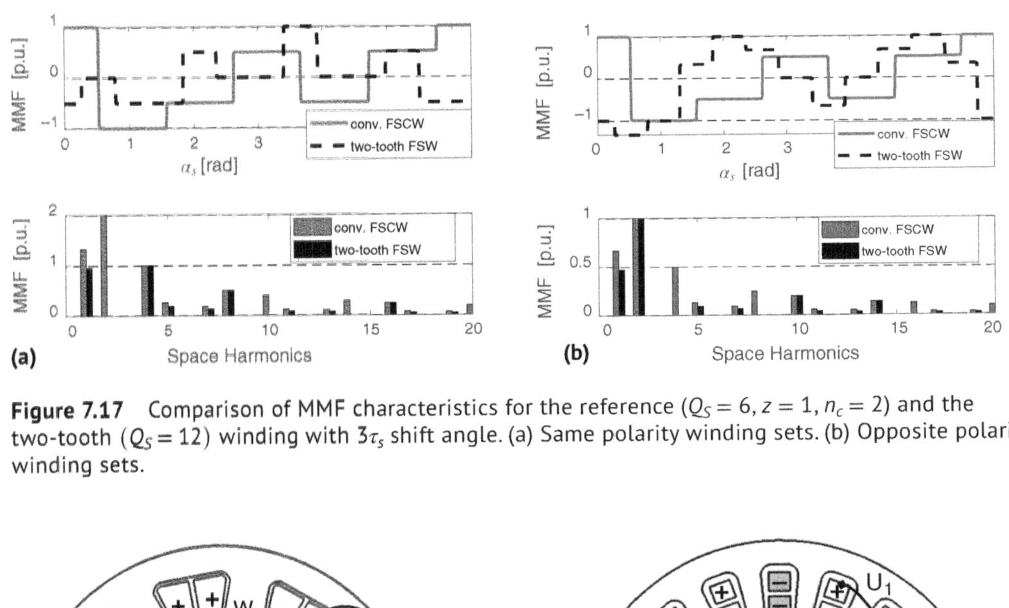

Figure 7.17 Comparison of MMF characteristics for the reference ($Q_S = 6, z = 1, n_c = 2$) and the two-tooth ($Q_S = 12$) winding with $3\tau_s$ shift angle. (a) Same polarity winding sets. (b) Opposite polarity winding sets.

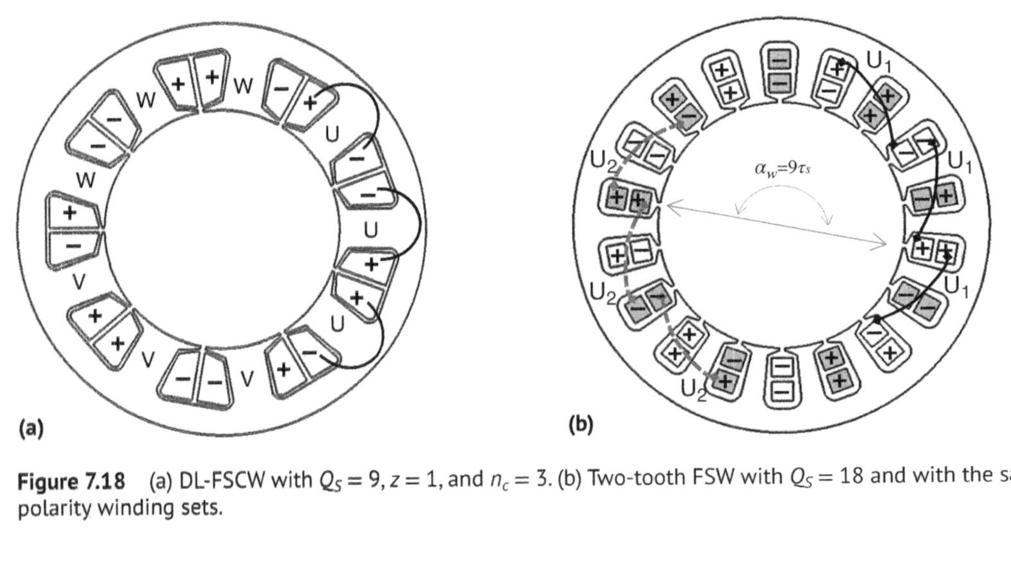

Figure 7.18 (a) DL-FSCW with $Q_S = 9, z = 1$, and $n_c = 3$. (b) Two-tooth FSW with $Q_S = 18$ and with the same polarity winding sets.

Figure 7.18b. The MMF results in Figure 7.19 and Table 7.3 indicate that the configuration with the same polarity winding sets combination is well suited for 8-pole machines, whereas the other case with opposite polarity is suitable for 10-pole machines.

7.3.2 Double-layer FSWs with Two Zone/Phase

Example 7.8 Optimization of an DL-FSCW with $Q_S = 12$, $z = 2$, and $n_c = 2$
The first example of a two-zone FSCW with parameters $Q_S = 12$, $z = 2$, and $n_c = 2$ is depicted in Figure 7.20a. This winding type exhibits pronounced MMF harmonics of the 5th and 7th order, along with a 1st-order subharmonic component. By applying a shift angle of $\alpha_w = 5\tau_s$, the corresponding two-tooth winding illustrated in Figure 7.20b is obtained. The resulting MMF characteristics, presented in Figure 7.21 and summarized in Table 7.4, show that the configuration with winding sets of the same polarity provides an effective solution for 10-pole machines, whereas the opposite polarity configuration is better suited for 14-pole machine applications.

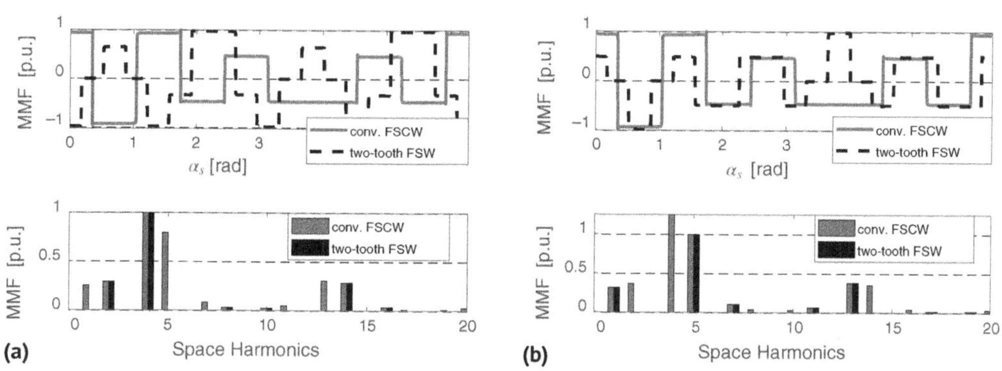

Figure 7.19 Comparison of MMF characteristics for the reference ($Q_S = 9, z = 1, n_c = 3$) and the two-tooth ($Q_S = 18$) winding with $9\tau_s$ shift angle. (a) Same polarity winding sets. (b) Opposite polarity winding sets.

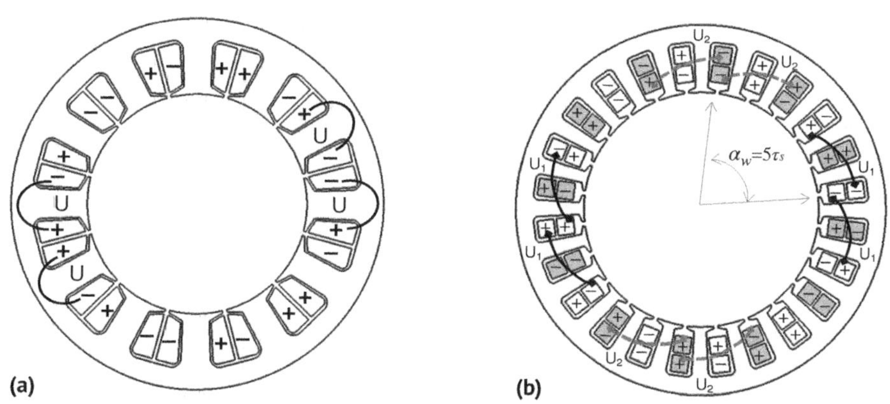

Figure 7.20 (a) DL-FSCW with $Q_S = 12, z = 2$, and $n_c = 2$. (b) Two-tooth FSW with $Q_S = 24$ and with the same polarity winding sets.

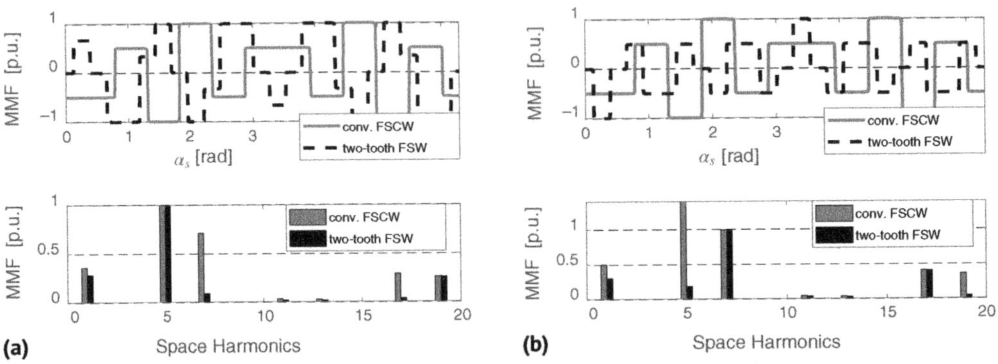

Figure 7.21 Comparison of MMF characteristics for the reference ($Q_S = 12, z = 2, n_c = 2$) and the two-tooth ($Q_S = 24$) winding with $5\tau_s$ shift angle. (a) Same polarity winding sets. (b) Opposite polarity winding sets.

Table 7.4 Winding factors (in %) of two-zone double-layer windings. Comparison of results for conventional FSCW and two-tooth FSW (first gray row: same polarity, second gray row: opposite polarity).

n_c	α_{sh}	$\nu = 1$	2	3	4	5	6	7	8	9	10	11	12	13	14	15
2	Ref.	6.7	—	—	—	**93.3**		**93.3**	—	—	—	6.7	—	6.7	—	—
	$5\tau_s$	5.3	—	—	—	**92.5**		12.2	—	—	—	4.1	—	4.1	—	—
		4.1	—	—	—	12.2		**92.5**	—	—	—	5.3	—	5.3	—	—
4	Ref.	1.6	—	—	—	9.6		16.3	—	—	—	**94.9**	—	**94.9**	—	—
	$11\tau_s$	1.2	—	—	—	8.6		5.2	—	—	—	6.2	—	**94.7**	—	—
		1.1	—	—	—	4.2		15.4	—	—	—	**94.7**	—	6.2	—	—

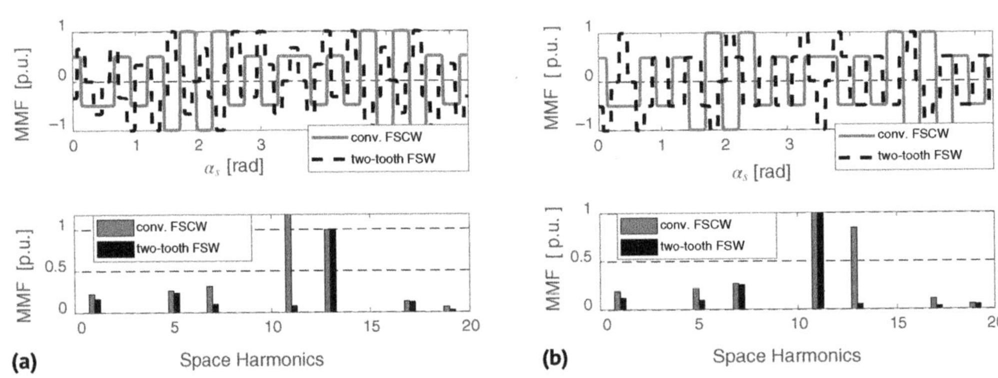

Figure 7.22 Comparison of MMF characteristics for the reference ($Q_S = 24, z = 2, n_c = 4$) and the two-tooth ($Q_S = 48$) winding with $11\tau_s$ shift angle. (a) Same polarity winding sets. (b) Opposite polarity winding sets.

Example 7.9 Optimization of an DL-FSCW with $Q_S = 24$, $z = 2$, and $n_c = 4$

For the second two-zone FSCW examined in Figure 7.22, the suppression of the 11th or the 13th MMF harmonics is achieved by applying a shift angle of $\alpha_w = 11\tau_s$. Depending on the selected winding-set polarity, the resulting two-tooth configuration is suitable for either a 22- or 26-pole machine.

7.3.3 Double-layer FSWs with Unequal-zones

In addition to the typical concentrated windings, FSCWs may also be realized using unequal-zone and asymmetrical zone distributions, which are described in Chapter 3. As with standard concentrated windings, these configurations are characterized by high harmonic content in the MMF spectrum. Therefore, the stator-shifting concept remains an effective approach for improving their MMF performance.

Example 7.10 Optimization of various DL-FSCWs with unequal-zones
Two typical examples of FSCWs with unequal zones are illustrated in Figure 7.23. In the first case, two unequal zones with $n_{c1} = 3$ and $n_{c2} = 1$ coils generate dominant 5th- and 7th-order MMF harmonics with winding factors of approximately 90%, making them suitable as operating harmonics. Applying the stator-shifting technique with $\alpha_w = 5\tau_s$ selectively suppresses one of the critical harmonics, as shown in Figure 7.24 and Table 7.5. The resulting two-tooth FSW configuration with the same polarity winding sets is suitable for 10-pole machines, whereas the opposite polarity configuration 14-pole machines.

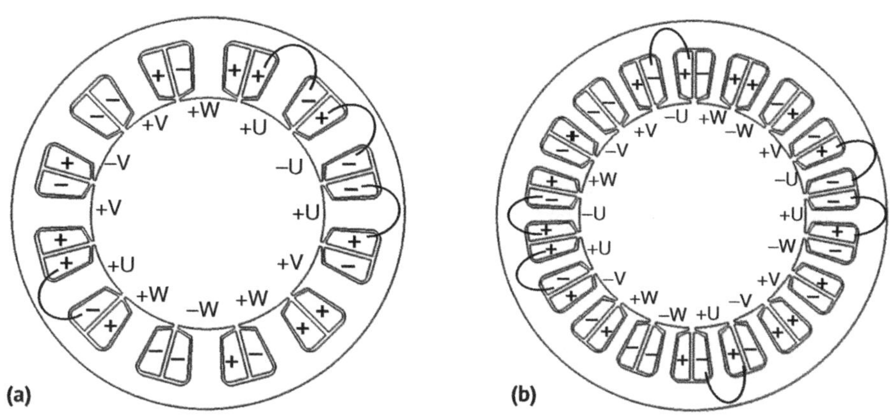

Figure 7.23 (a) DL-FSCW with two unequal zones per phase with $n_{c1} = 3$ and $n_{c2} = 1$. (b) DL-FSCW with two unequal zone-groups per phase with $n_{c1} = 2$ and $n_{c2} = 1$.

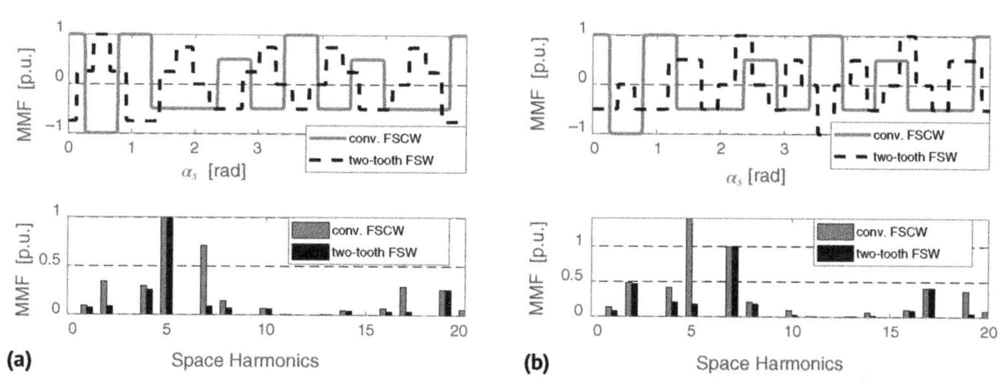

Figure 7.24 Comparison of MMF characteristics for the reference ($n_{c1} = 3, n_{c2} = 1$, and $Q_S = 12$) and the two-tooth ($Q_S = 24$) winding with $5\tau_s$ shift angle. (a) Same polarity winding sets. (b) Opposite polarity winding sets.

Table 7.5 Winding factors (in %) for the double-layer winding with two unequal zones with $n_{c1} = 3$ and $n_{c2} = 1$. Comparison of results for conventional FSCW and two-tooth FSW (first gray row: same polarity, second gray row: opposite polarity).

α_{sh}	$\nu = 1$	2	3	4	5	6	7	8	9	10	11	12	13	14	15
Ref.	1.7	12.5		21.6	**90.1**		**90.1**	21.6		12.5	1.7		1.7	12.5	
$5\tau_s$	1.4	3.2		18.7	**89.4**		11.8	10.8		12.1	1.1		1.1	12.1	
	1.1	12.1		10.8	11.8		**89.4**	18.8		3.2	1.4		1.4	3.2	

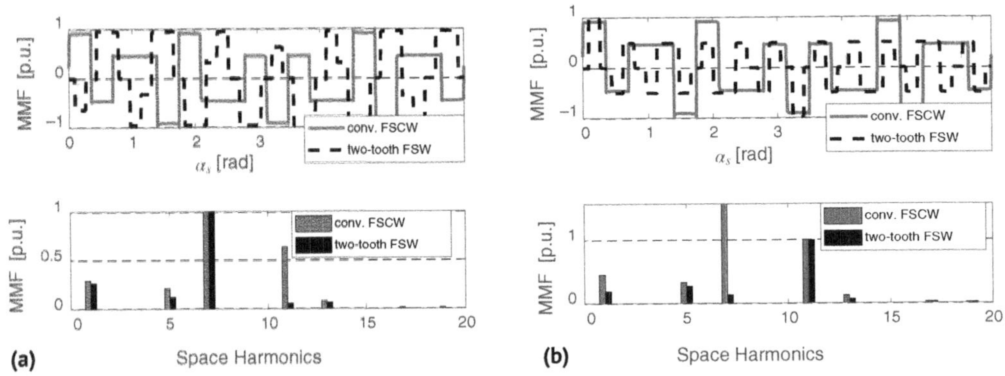

Figure 7.25 Comparison of MMF characteristics for the reference ($n_{c1} = 2$, $n_{c2} = 1$, and $Q_S = 18$) and the two-tooth ($Q_S = 36$) winding with $5\tau_s$ shift angle. (a) Same polarity winding sets. (b) Opposite polarity winding sets.

Table 7.6 Winding factors (in %) for the double-layer winding with two unequal zone-groups with $n_{c1} = 2$ and $n_{c2} = 1$. Comparison of results for conventional FSCW and two-tooth FSW (first gray row: same polarity, second gray row: opposite polarity).

α_{sh}	$\nu = 1$	2	3	4	5	6	7	8	9	10	11	12	13	14	15
Ref.	3.7	—	—	—	13.6	—	**90.2**	—	—	—	**90.2**	—	13.6	—	—
$5\tau_s$	3.4	—	—	—	7.8	—	**89.8**	—	—	—	7.8	—	11.1	—	—
	1.6	—	—	—	11.1	—	7.8	—	—	—	**90.2**	—	7.8	—	—

The second example, shown in Figure 7.23b, presents an FSCW with two unequal zone-groups, and with $n_{c1} = 2$ and $n_{c2} = 1$ coils per respective zones. In this configuration, the 7th and the 11th harmonics have the highest winding factor (approximately 90%), which may be selected as operating harmonics. Again, the application of a shift angle of $\alpha_w = 5\tau_s$ significantly improves the MMF results, as demonstrated in Figure 7.25 and Table 7.6. Depending on the used winding set polarity, the resulting two-tooth FSWs represent an effective solution for a 14- or 22-pole machine design.

7.4 Multiphase Two-tooth FSWs

In direct analogy to the three-phase case, the stator-shifting technique can be extended to any multiphase FSCW configuration. Consider a general winding with an arbitrary number of phases, where each phase consists of a single coil. Under this condition, the resulting MMF function for the multiphase two-tooth FSWs, as described by Equations (5.21) and (7.9), can be expressed as follows:

$$F_j(t, a_S) = \frac{2N_w}{\pi} i_j(t) \sum_\nu {}^\nu\xi_w \cos\left(\nu\left(a_S - \frac{a_w}{2}\right) - (\nu_0 - \nu)\frac{2\pi}{m}\right) \tag{7.14}$$

The winding factor for this winding configuration is obtained as the product of the coil pitch and an additional distribution factor, given by

$$^\nu\xi_w = {}^\nu\xi_p \, {}^\nu\xi_{z12} \tag{7.15}$$

In (7.14), the parameter ν_0 specifies the spatial MMF waveform. Depending on the winding layer arrangement, i.e. single-layer or double-layer configuration, this parameter defines either the working harmonic or a subharmonic component. As indicated in Equation (5.16), the number of distinct MMF waveforms that can be generated with a multiphase winding increases with the number of phases.

This type of winding offers several inherent advantages, including flexibility in MMF distribution, variable pole number, and the possibility of dual-pole operation. Moreover, the application of the stator-shifting technique provides an additional degree of freedom, enabling further improvement of the MMF performance. Consequently, multiphase two-tooth FSWs represent an attractive solution for a wide range of electrical machine applications.

7.4.1 Single-layer Multiphase Two-tooth FSWs

A concentrated single-layer winding with an arbitrary number of phase m is characterized by a total number of stator slots $Q_s = 2m$ and one coil per phase $(n_c = 1)$. The corresponding two-tooth FSW is obtained by doubling the number of stator slots $Q_S' = 4m$ and implementing two identical winding sets shifted in space for α_w. The phase windings of the first set are denoted by U_1 to U_n, whereas of the second set with X_1 to X_n.

Example 7.11 Five-phase SL-FSW
For a five-phase configuration, the stator therefore comprises ten slots, as illustrated in Figure 7.26a. According to Equation (5.16), two distinct excitation modes are possible, corresponding to $\nu_0 = 1$ and $\nu_0 = 2$.

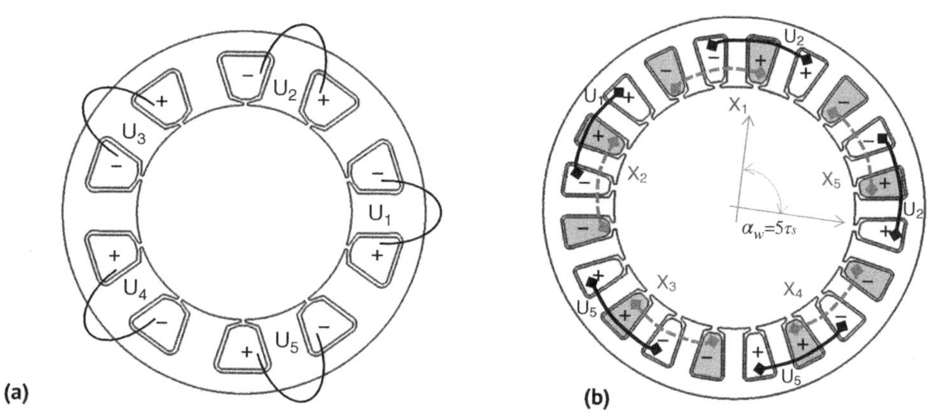

Figure 7.26 (a) Five-phase SL-FSCW with $Q_S = 10$. (b) Two-tooth five-phase SL-FSW with $Q_S = 20$ and with the same polarity winding sets.

Case-1: Excitation with $\nu_0 = 1$

In this excitation mode, the MMF spectrum contains the 1st, 5th, 7th, and higher-order harmonics. Among these components, the 5th and 7th harmonics have the highest winding factor (approximately 95%) and are therefore suitable for use as working harmonics. To suppress or completely eliminate one of the critical harmonics, a stator shift angle of $\alpha_w = 5\tau_s$ between the winding sets is required, as shown in Figure 7.26b. The resulting MMF characteristics are presented in Figure 7.27 and summarized in Table 7.6.

Case-2: Excitation with $\nu_0 = 2$

Under the $\nu_0 = 2$ excitation mode, the resulting MMF contains the 2nd, 3th, 7th, 8th, and higher-order harmonics. Due to its highest winding factor, the 3rd harmonic is suitable as the working harmonic. As in the previous case, stator-shifting technique is applied to suppress the critical harmonics; however, the optimal shift angle for this case is different, namely $\alpha_w = 13\tau_s$. The respective MMF characteristics for both same and opposite polarity winding sets are shown in Figure 7.28 and summarized in Table 7.7.

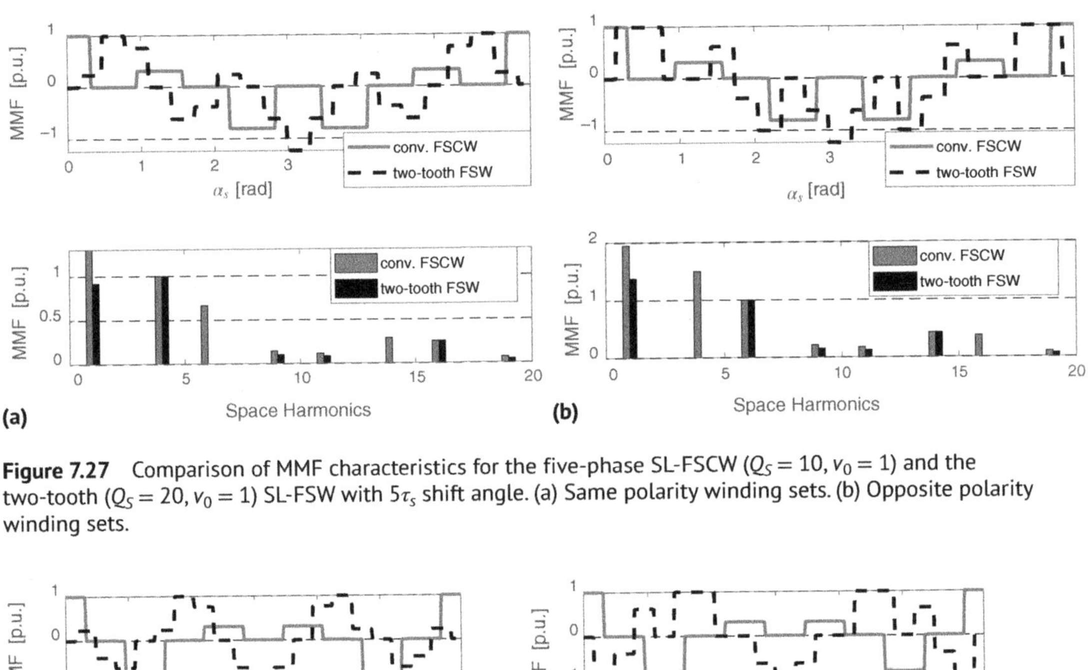

Figure 7.27 Comparison of MMF characteristics for the five-phase SL-FSCW ($Q_S = 10, \nu_0 = 1$) and the two-tooth ($Q_S = 20, \nu_0 = 1$) SL-FSW with $5\tau_s$ shift angle. (a) Same polarity winding sets. (b) Opposite polarity winding sets.

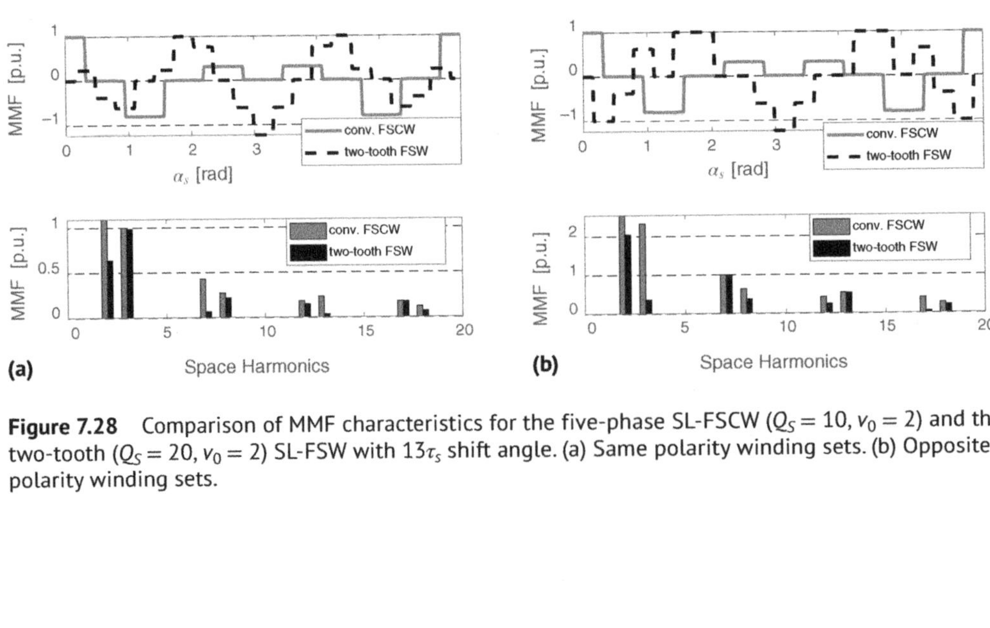

Figure 7.28 Comparison of MMF characteristics for the five-phase SL-FSCW ($Q_S = 10, \nu_0 = 2$) and the two-tooth ($Q_S = 20, \nu_0 = 2$) SL-FSW with $13\tau_s$ shift angle. (a) Same polarity winding sets. (b) Opposite polarity winding sets.

Table 7.7 Winding factors (in %) for the five-phase single-layer winding. Comparison of results for conv. FSCW and two-tooth FSW (first gray row: same polarity, second gray row: opposite polarity).

ν_0	α_{sh}	$\nu = 1$	2	3	4	5	6	7	8	9	10	11	12	13	14	15
1	Ref.	30.9	—	—	**95.1**		**95.1**	—	—	30.9		30.9	—	—	95.1	—
	$5\tau_s$	21.8	—	—	**95.1**		0	—	—	21.8		21.8	—	—	0	—
		21.8	—	—	0		**95.1**	—	—	21.8		21.8	—	—	95.1	—
2	Ref.	—	58.7	**80.9**	—	—	—	**80.9**	58.7	—	—	—	58.7	80.9	—	—
	$13\tau_s$	—	34.5	**79.9**	—	—	—	12.6	47.5	—	—	—	47.5	12.6	—	—
		—	47.5	12.6	—	—	—	**79.9**	34.5	—	—	—	34.5	79.9	—	—

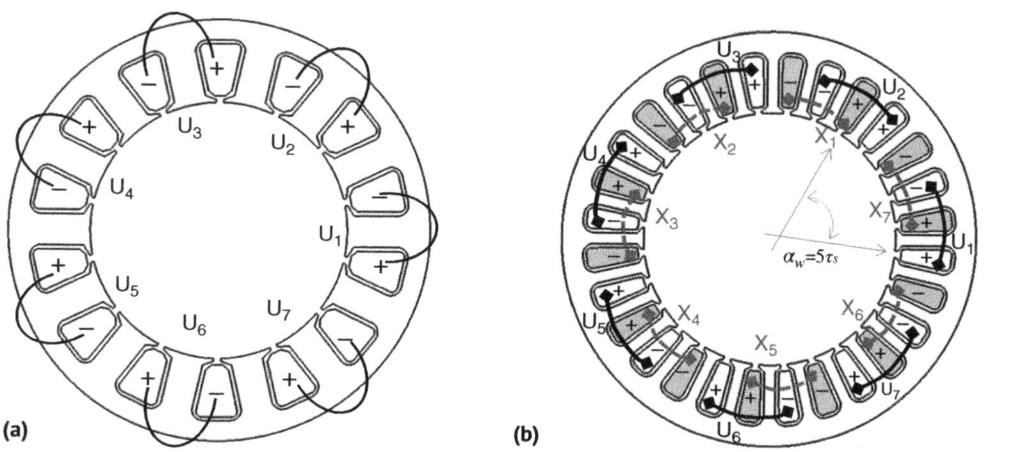

Figure 7.29 (a) Seven-phase SL-FSCW with $Q_S = 7$. (b) Two-tooth seven-phase SL-FSW with $Q_S = 14$ and with the same polarity winding sets.

Example 7.12 Seven-phase SL-FSW
A seven-phase SL-FSCW is defined by $m = 7$ and $Q_S = 14$, as illustrated in Figure 7.29a. Due to the increased number of phases, three distinct MMF waveforms can be generated, corresponding to excitation modes $\nu_0 = 1$, $\nu_0 = 2$, and $\nu_0 = 3$.

Case-1: Excitation with $\nu_0 = 1$
This excitation mode produces the 1st, 6th, 8th, and higher-order MMF harmonics. Because of their relatively high winding factors, either the 6th or the 8th harmonics may be selected as the working harmonics. The corresponding two-tooth winding layout is shown in Figure 7.29b, with an optimal shift angle of $\alpha_w = 5\tau_S$. Depending on the polarity of winding sets, the resulting MMF characteristics are presented in Figure 7.30 and summarized in Table 7.8.

Case-2: Excitation with $\nu_0 = 2$
Under this excitation mode, the generated MMF contains the 2nd, 5th, 9th, and higher order harmonics. The 2nd and 9th harmonics, having the highest winding factors, are suitable for operation. To mitigate critical harmonics, a shift angle of $\alpha_w = 11\tau_s$ is required. The corresponding MMF characteristics for the configuration with the same and opposite polarity winding sets are compared in Figure 7.31 and Table 7.7.

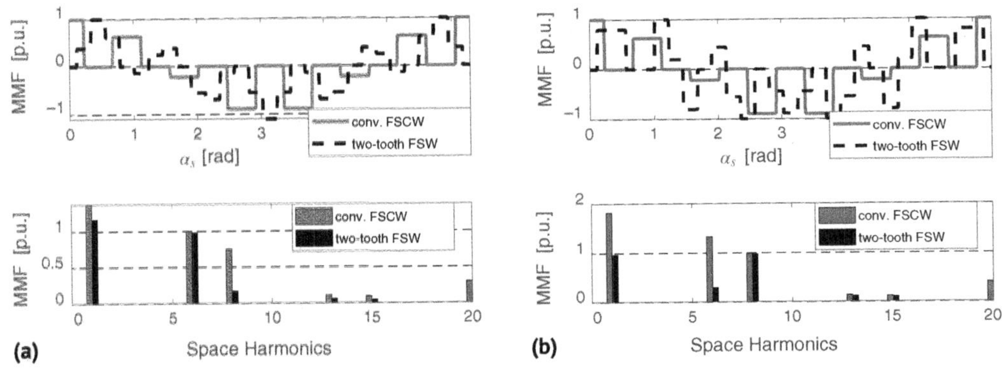

Figure 7.30 Comparison of MMF characteristics for the seven-phase SL-FSCW ($Q_S = 14$, $v_0 = 1$) and the two-tooth ($Q_S = 28$, $v_0 = 1$) SL-FSW with $5\tau_s$ shift angle. (a) Same polarity winding sets. (b) Opposite polarity winding sets.

Table 7.8 Winding factors (in %) for the seven-phase single-layer winding. Comparison of results for conv. FSCW and two-tooth FSW (first gray row: same polarity, second gray row: opposite polarity).

v_0	α_{sh}	$v = 1$	2	3	4	5	6	7	8	9	10	11	12	13	14	15
1	Ref.	22.2	—	—	—	—	**97.5**	—	**97.5**	—	—	—	—	22.3	—	22.2
	$5\tau_s$	18.8	—	—	—	—	**95.0**	—	21.7	—	—	—	—	11.8	—	11.8
		11.8	—	—	—	—	21.7	—	**95.0**	—	—	—	—	18.8	—	18.8
2	Ref.	—	43.4	—	—	**90.1**	—	—	—	**90.1**	—	—	43.4	—	—	—
	$11\tau_s$	—	33.9	—	—	**89.5**	—	—	—	10.1	—	—	27.1	—	—	—
		—	27.1	—	—	10.1	—	—	—	**89.5**	—	—	33.9	—	—	—
3	Ref.	—	—	62.3	**78.2**	—	—	—	—	—	**78.2**	62.3	—	—	—	—
	$7\tau_s$	—	—	44.1	**78.2**	—	—	—	—	—	0	44.1	—	—	—	—
		—	—	44.1	0	—	—	—	—	—	**78.2**	44.1	—	—	—	—

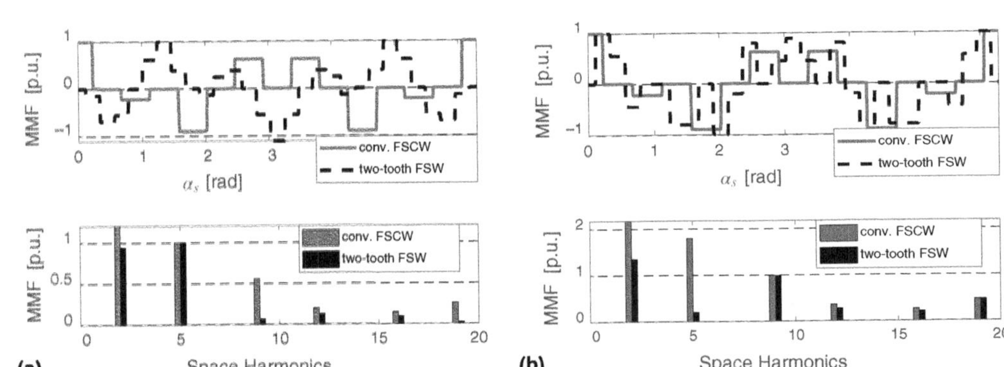

Figure 7.31 Comparison of MMF characteristics for the seven-phase SL-FSCW ($Q_S = 14$, $v_0 = 2$) and the two-tooth ($Q_S = 28$, $v_0 = 2$) SL-FSW with $11\tau_s$ shift angle. (a) Same polarity winding sets. (b) Opposite polarity winding sets.

Case-3: Excitation with $\nu_0 = 3$

When excited with $\nu_0 = 3$, the winding produces the 3rd, 4th, 10th, and higher order harmonics. Here, the 4th and 10th harmonics are potential components for use as working harmonics. In this case, the critical harmonic reduction is achieved by applying a shift angle of $\alpha_w = 7\tau_s$. The resulting MMF characteristics are presented in Figure 7.32 and Table 7.8.

7.4.2 Double-layer Multiphase Two-tooth FSWs

A multiphase double-layer concentrated winding with an arbitrary number of phases is characterized by $Q_S = m$ and $n_c = 1$, whereas the corresponding two-tooth FSW is obtained by doubling the number of stator slots to $Q'_S = 2m$ and implementing two identical winding sets shifted in space for α_w.

Example 7.13 Five-phase DL-FSW

The winding layouts of the five-phase double-layer configuration, realized either as a conventional concentrated winding or as a two-tooth FSW, are illustrated in Figure 7.33. As in the single-layer

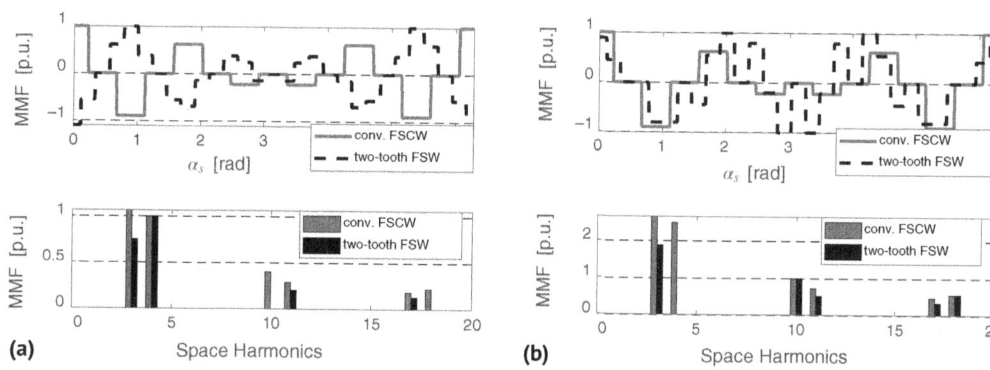

Figure 7.32 Comparison of MMF characteristics for the seven-phase SL-FSCW ($Q_S = 14, \nu_0 = 3$) and the two-tooth ($Q_S = 28, \nu_0 = 3$) SL-FSW with $7\tau_s$ shift angle. (a) Same polarity winding sets. (b) Opposite polarity winding sets.

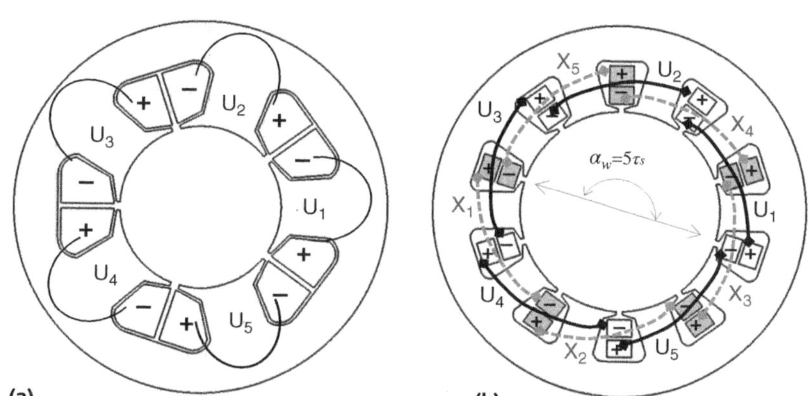

Figure 7.33 (a) Five-phase DL-FSCW with $Q_S = 5$. (b) Two-tooth five-phase DL-FSW with $Q_S = 10$ and with the same polarity winding sets.

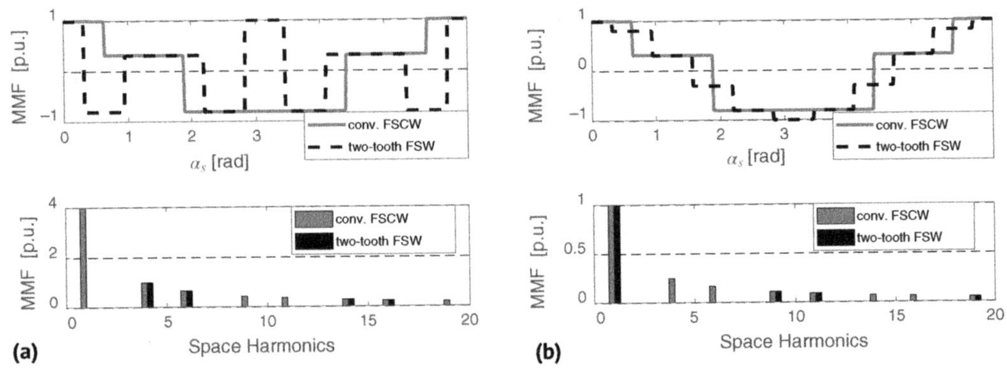

Figure 7.34 Comparison of MMF characteristics for the five-phase DL-FSCW ($Q_S = 5, \nu_0 = 1$) and the two-tooth ($Q_S = 10, \nu_0 = 1$) DL-FSW with $5\tau_s$ shift angle. (a) Same polarity winding sets. (b) Opposite polarity winding sets.

Table 7.9 Winding factors (in %) for the five-phase double-layer winding. Comparison of results for conv. FSCW and two-tooth FSW (first gray row: same polarity, second gray row: opposite polarity).

ν_0	α_{sh}	$\nu = 1$	2	3	4	5	6	7	8	9	10	11	12	13	14	15
1	Ref.	**58.8**	—	—	58.8	—	58.8	—	—	58.8	—	58.8	—	—	58.8	—
	$5\tau_s$	0	—	—	**58.8**	—	58.8	—	—	0	—	0	—	—	58.8	—
		58.8	—	—	0	—	0	—	—	58.8	—	58.8	—	—	0	—
2	Ref.	—	**95.1**	**95.1**	—	—	—	95.1	95.1	—	—	—	95.1	95.1	—	—
	$5\tau_s$	—	**95.1**	0	—	—	—	0	95.1	—	—	—	95.1	0	—	—
		—	0	**95.1**	—	—	—	95.1	0	—	—	—	0	95.1	—	—

case, two distinct excitation modes are available when $m = 5$, corresponding to $\nu_0 = 1$ and $\nu_0 = 2$. These modes lead to different MMF waveforms and pole-number combinations, which are examined in the following analysis.

Case-1: Excitation with $\nu_0 = 1$
The MMF spectrum contains the 1st, 4th, 6th, and subsequent harmonics. To effectively suppress the most unwanted harmonics, an optimal shift angle of $\alpha_w = 5\tau_s$ is applied. A comparison of MMF characteristics for same and opposite polarity winding set is provided in Figure 7.34 and Table 7.9.

Case-2: Excitation with $\nu_0 = 2$
Under this excitation, the MMF contains harmonics of the 2nd, 3rd, 7th, and higher order. In contrast to the single-layer configuration, the optimum shift angle between the winding sets required for harmonic suppression remains the same for all excitation modes, i.e. $\alpha_w = 5\tau_s$. The corresponding MMF results are shown and compared in Figure 7.35 and Table 7.9.

Example 7.14 Optimization of a seven-phase DL-FSCW
A seven-phase DL-FSCW and its corresponding two-tooth winding arrangement are shown in Figure 7.36. For $m = 7$, three excitation modes are possible: $\nu_0 = 1$, $\nu_0 = 2$, and $\nu_0 = 3$, as examined in Figures 7.37–7.39. Although the different harmonic spectra, the results clearly demonstrate that

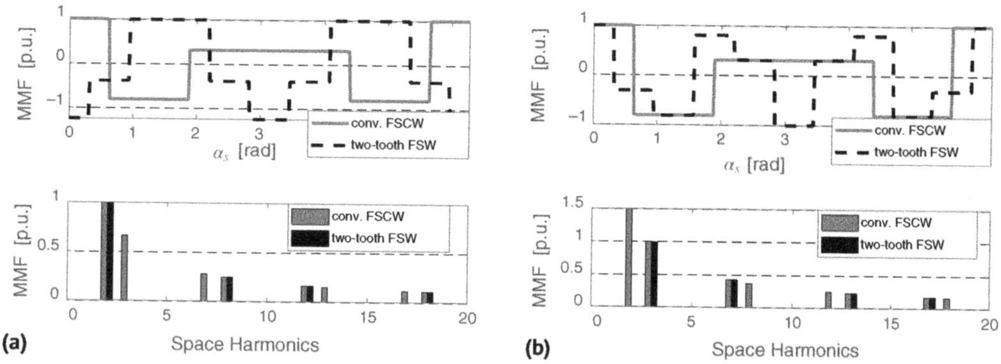

Figure 7.35 Comparison of MMF characteristics for the five-phase DL-FSCW ($Q_S = 5$, $v_0 = 2$) and the two-tooth ($Q_S = 10$, $v_0 = 2$) DL-FSW with $5\tau_s$ shift angle. (a) Same polarity winding sets. (b) Opposite polarity winding sets.

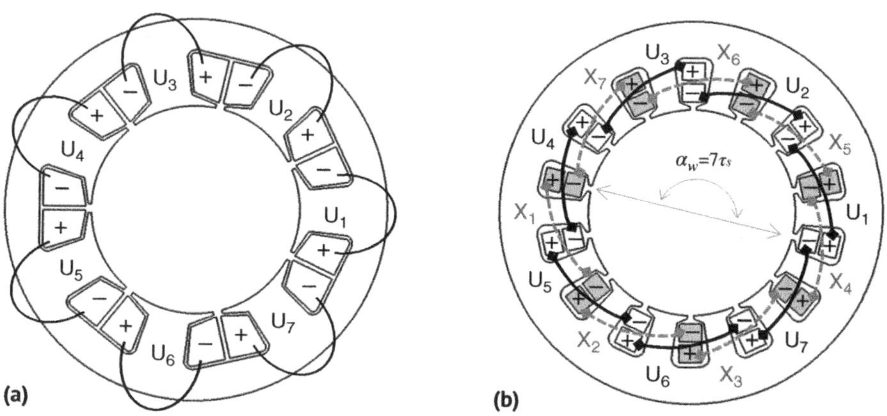

Figure 7.36 (a) Seven-phase DL-FSCW with $Q_S = 7$. (b) Two-tooth seven-phase DL-FSW with $Q_S = 14$ and with the same polarity winding sets.

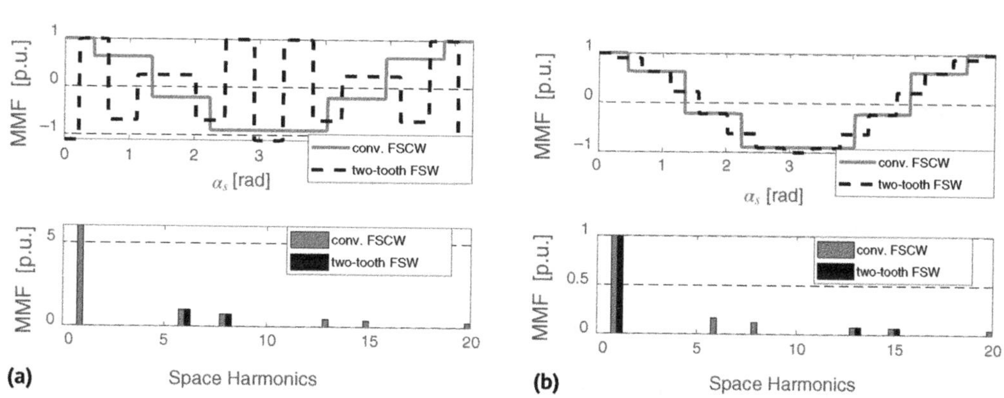

Figure 7.37 Comparison of MMF characteristics for the seven-phase DL-FSCW ($Q_S = 7$, $v_0 = 1$) and the two-tooth ($Q_S = 14$, $v_0 = 1$) DL-FSW with $7\tau_s$ shift angle. (a) Same polarity winding sets. (b) Opposite polarity winding sets.

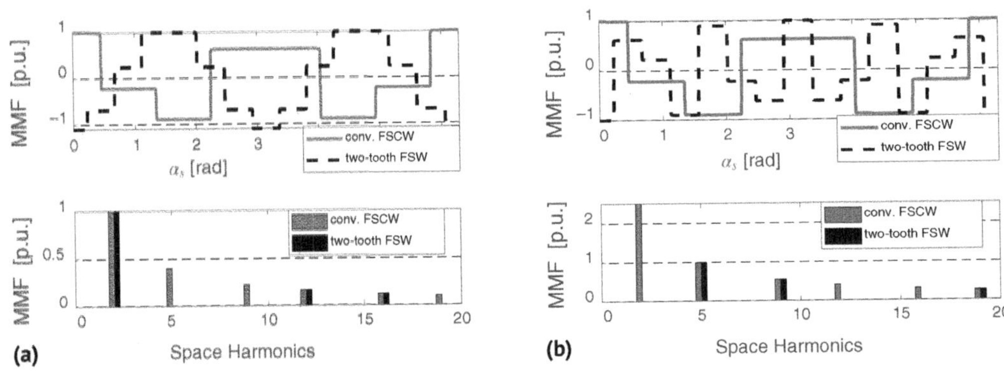

Figure 7.38 Comparison of MMF characteristics for the seven-phase DL-FSCW ($Q_S = 7, v_0 = 2$) and the two-tooth ($Q_S = 14, v_0 = 2$) DL-FCW with $7\tau_s$ shift angle. (a) Same polarity winding sets. (b) Opposite polarity winding sets.

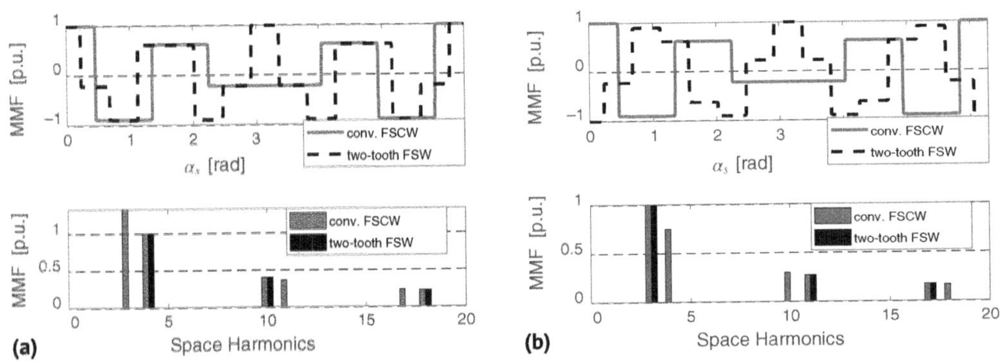

Figure 7.39 Comparison of MMF characteristics for the seven-phase DL-FSCW ($Q_S = 7, v_0 = 3$) and the two-tooth ($Q_S = 14, v_0 = 3$) DL-FSW with $7\tau_s$ shift angle. (a) Same polarity winding sets. (b) Opposite polarity winding sets.

the stator-shifting technique remains effective for all excitation modes. In each case, a common shift angle of $\alpha_w = 7\tau_s$ yields substantial harmonic reduction (Table 7.10).

7.4.3 Concluding Remarks

The above analysis confirms that stator-shifting provides an effective method for optimizing multiphase concentrated windings with arbitrary phase numbers. The principal conclusions may be summarized as follows:

Multiphase SL-FSWs

- The excitation parameter v_0 determine the MMF waveform of the winding.
- Each MMF waveform has unique space harmonics.
- A specific shift angle α_w is required to minimize the dominant critical harmonics associated with v_0 excitation.
- Except the harmonic content reduction, the two-tooth solution does not affect the operating harmonic.

Table 7.10 Winding factors (in %) for the seven-phase double-layer winding. Comparison of results for conv. FSCW and two-tooth FSW (first gray row: same polarity, second gray row: opposite polarity).

ν_0	α_{sh}	$\nu = 1$	2	3	4	5	6	7	8	9	10	11	12	13	14	15
1	Ref.	**43.4**	—	—	—	—	43.4		43.4	—	—	—	—	43.4	—	43.4
	$5\tau_s$	0	—	—	—	—	43.4		43.4	—	—	—	—	0	—	
		43.4	—	—	—	—	0		0	—	—	—	—	43.4	—	43.4
2	Ref.	—	**78.2**	—	—	78.2	—	—	—	78.2	—	—	78.2	—	—	—
	$11\tau_s$	—	**78.2**	—	—	0	—	—	—	0	—	—	78.2	—	—	—
		—	0	—	—	78.2	—	—	—	78.2	—	—	0	—	—	—
3	Ref.	—	—	**97.5**	**97.5**	—	—	—	—	—	97.5	97.5	—	—	—	—
	$7\tau_s$	—	—	0	**97.5**	—	—	—	—	—	97.5	0	—	—	—	—
		—	—	**97.5**	0	—	—	—	—	—	0	97.5	—	—	—	—

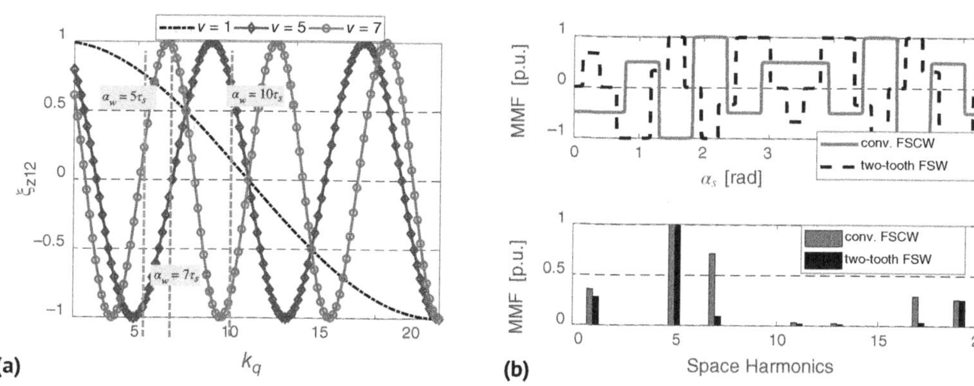

Figure 7.40 Two-tooth FSW with $Q_S = 24$. (a) Distribution factors vs. shift angle $(k_q\tau_s)$ for different space harmonics. (b) Comparison of MMF characteristics for the reference ($Q_S = 12, z = 2, n_c = 2$, double-layer) and the two-tooth ($Q_S = 24$) winding with a shift angle of $5\tau_S$.

Multiphase DL-FSWs

- The excitation parameter ν_0 determine the MMF waveform of the winding.
- Each MMF waveform has unique space harmonics.
- The optimal shift angle α_w remains the same across different excitation modes.
- Except the harmonic content reduction, the two-tooth solution does not affect the operating harmonic.

7.5 Two-tooth FSWs with Specific Coil Configurations

The fundamental objective of the stator-shifting concept is to adjust the spatial phase relationship of selected MMF harmonics such that undesirable components cancel each other through the superposition. This principle is demonstrated in Figure 7.40, which show the influence of the shift angle on the resulting winding factor of specific MMF harmonics. For example, when a shift angle of $5\tau_{S2}$ is applied, the 5th harmonic remains essentially unaffected, while the 1st and

7th harmonics are reduced by 20% and 87%, respectively. Conversely, with a shift angle of $7\tau_{s2}$, the 7th harmonic remains unchanged, whereas the 1st and 5th harmonics can be reduced by around 30% and 87%. A shift angle of $10\tau_{s2}$, on the other hand, has only a minimal influence on the 5th and 7th harmonics but reduce the 1st harmonic by nearly 75%.

Let us consider a winding in which the 5th and 7th space harmonics have the highest winding factors, with one of these being selected as the working harmonic. In such a situation, the adjacent space harmonics, which also have high winding factors, must be carefully examined. These components, commonly referred to as critical harmonics, are known to contribute to machine losses and acoustic noise. Therefore, effective suppression of these harmonics is essential for achieving high-performance machine operation.

Figure 7.40b illustrates a representative case in which the 5th harmonic is selected as the operating harmonic and stator-shifting is applied to suppress the 7th harmonic (i.e. the critical harmonic). Although this approach effectively reduces several undesirable harmonics, a considerable subharmonic component remains in the MMF spectrum, which can further degrade the overall machine efficiency. Therefore, to address this issue, additional design measures are required. For example, the use of the special coil configurations introduced in Chapter 4, or a stator core design with unequal tooth widths, in combination with a two-tooth winding structure, represents an effective solution for the simultaneous suppression of both subharmonic and higher-order MMF harmonics.

7.5.1 Two-tooth FSWs with Asymmetric Coils

As described in Chapter 4, an asymmetric coil is characterized by unequal numbers of turns on its two sides. Specifically, one coil-side contains $N_{w11} = N_w$ turns, while the opposite side contains $N_{w12} = N_w - 1$ turns. Despite the small absolute difference, this arrangement permits a wide range of turn ratios between the two sides, as defined by

$$0 \leq N_{w12}/N_{w11} < 1 \tag{7.16}$$

Previous studies of DL-FSCWs have shown that the use of asymmetric coils can considerably affect their MMF subharmonics. In particular, for two-zone DL windings, complete suppression of the 1st harmonic can be achieved through a suitable selection of turn ratio N_{w11}/N_{w12}.

An illustrative example is shown in Figure 7.41 for a two-tooth winding with $Q_S = 24$ and asymmetric coils. This configuration represents an optimized variant of the conventional FSCW with $Q_S = 12$, $z = 2$, and $n_c = 2$. Using Equations (4.21) and (7.9), the MMF function for the corresponding two-tooth FSW is

$$F_j(t, \alpha_S) = \frac{2zn_cN_{w,\mathrm{eq}}i_j(t)}{\pi} \sum_{\nu} \frac{{}^{\nu}\xi_w}{\nu} \cos\left(\nu\left(\alpha_S - (n_c - 1)\frac{\pi}{Q_S}\right)\right) \tag{7.17}$$

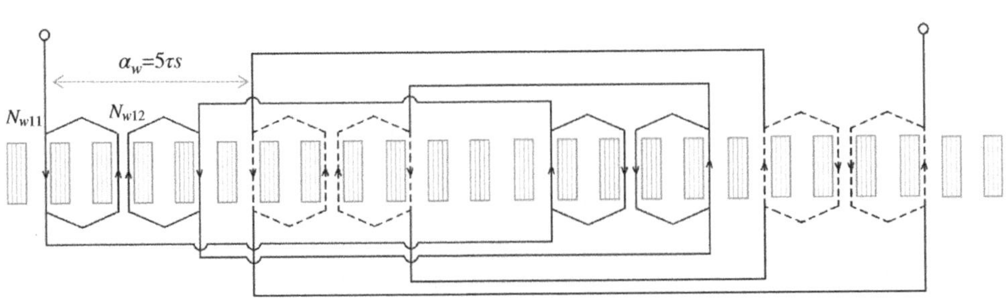

Figure 7.41 The winding layout for the two-tooth FSW with asymmetric coils. Winding parameters are: $Q_S = 24$, $\alpha_w = 5\tau_s$, and with $z = 2$, $n_c = 2$ per winding set.

The resulting winding factor follows from Equations (4.22) and (7.10) as

$$^{\nu}\xi_w = \frac{^{\nu}\xi_{z12}}{N_{w,\text{eq}}}\left(N_w {}^{\nu}\xi_{w,1} \cos\left((n_c/2 - 1)\pi\right) + \Delta n_w {}^{\nu}\xi_{w,2}\right) \tag{7.18}$$

and the equivalent number of turns for the asymmetric coil is given by

$$N_{w,\text{eq}} = \frac{2N_w - \Delta n_w}{2} = N_w - 1/2, \quad \Delta n_w = 1 \tag{7.19}$$

Figure 7.42 compares the MMF characteristic of the reference FSCW with standard coils and the optimized two-tooth winding employing asymmetric coils. A complete elimination of the subharmonic component is achieved when the turn ratio is selected as $N_{w11}/N_{w12} = 7/6$.

7.5.2 Two-tooth FSWs with Unequal Coils per Zone

An alternative approach to mitigate the MMF subharmonics in FSCWs is based on using coil with a different number of turns within the same zone, as described in Section 4.2. An optimized two-tooth winding employing this concept is illustrated in Figure 7.43, where the reference winding is a conventional FSCW with $Q_S = 9$, $z = 1$, and $n_c = 3$. If n_{c1} and n_{c2} represent the number of coils with N_{w1} and N_{w2} turns, the total number of turns per zone is

$$N_{w,z} = n_{c1}N_{w1} + n_{c2}N_{w2} \tag{7.20}$$

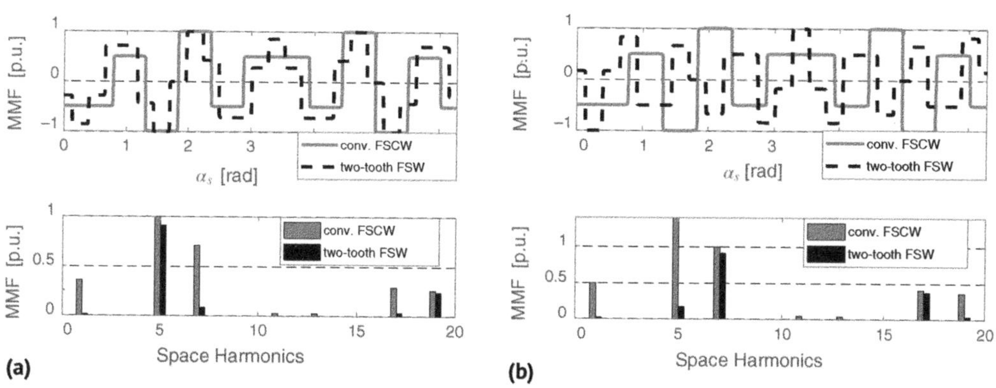

Figure 7.42 Comparison of MMF characteristics for the reference ($Q_S = 12, z = 2, n_c = 2$) and the two-tooth ($Q_S = 24$) winding with unsymmetrical coils. (a) For $5\tau_s$ shift angle. (b) For $7\tau_{s2}$ shift angle.

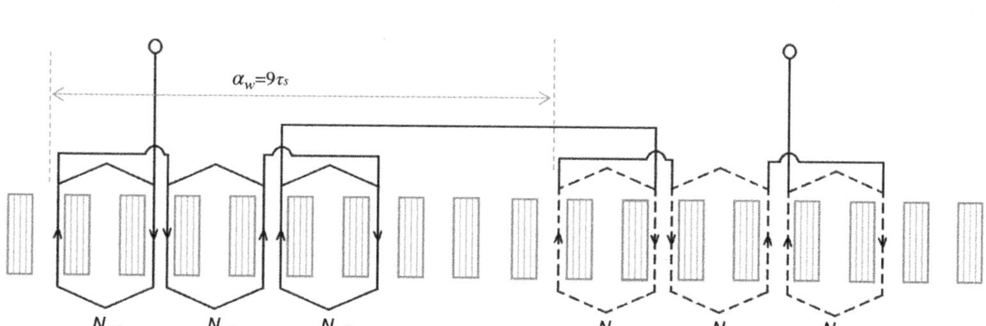

Figure 7.43 The winding layout for the two-tooth FSW with zones with unequal coils. Winding parameters are: $Q_S = 18, z = 1, n_c = 3$, and $\alpha_w = 9\tau_s$.

The resulting MMF function, derived from Equations (4.32) and (7.9), is given by

$$F_j(t, a_S) = \frac{2N_{w,z}i_j(t)}{\pi} \sum_{\nu} \frac{{}^{\nu}\xi_w}{\nu} \cos\left(\nu a_S - \varphi_{1,2}\right) \tag{7.21}$$

and the associated winding factor follows from Equations (4.33) and (7.10) is

$${}^{\nu}\xi_w = \frac{{}^{\nu}\xi_{z12}}{N_{w,z}} \sqrt{\left(n_{c1}N_{w1}{}^{\nu}\xi_{w,1}\right)^2 + \left(n_{c2}N_{w2}{}^{\nu}\xi_{w,2}\right)^2 - 2n_{c1}n_{c2}N_{w1}N_{w2}{}^{\nu}\xi_{w,1}{}^{\nu}\xi_{w,2}\cos\left(\nu(n_{c2} - n_{c1} + 1)\frac{2\pi}{Q_S}\right)} \tag{7.22}$$

where ${}^{\nu}\xi_{w,1}$ and ${}^{\nu}\xi_{w,2}$ denote the winding factor of the respective subzones characterized with n_{c1} and n_{c2} coils, respectively.

A comparison of MMF characteristics for the reference and the optimized two-tooth winding is shown in Figure 7.44. The results indicate that the first subharmonics component is entirely eliminated for a turns ratio of $N_{w1} = 0.65N_{w2}$. In addition, the two-tooth configuration also effectively suppresses most of the other critical harmonics.

A second example is presented in Figure 7.45 for a 24-slot two-tooth winding, representing an optimized version of the widely used FSCW with $Q_S = 12$, $z = 2$, and $n_c = 2$. As previously determined, the optimal shift angle for reducing the critical harmonic is $\alpha_w = 5\tau_s$. Notably, in the case

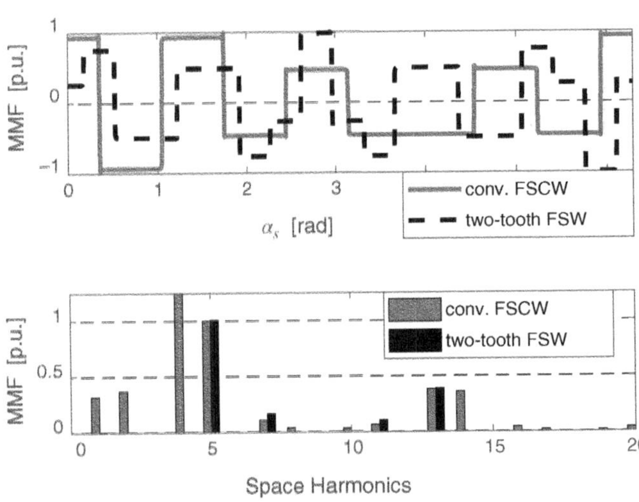

Figure 7.44 Comparison of MMF characteristics for the reference ($Q_S = 9, z = 1, n_c = 3$) and the two-tooth ($Q_S = 18$) winding with $\alpha_w = 9\tau$, $N_{w1} = 0.65N_{w2}$, and opposite polarity winding sets.

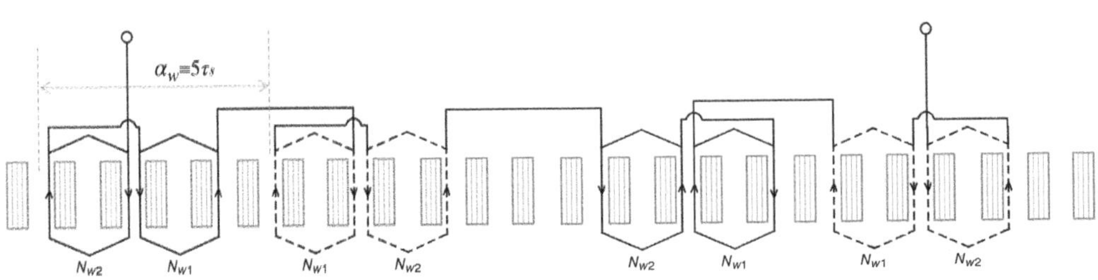

Figure 7.45 The winding layout for the two-tooth FSW with zones with unequal coils. Winding parameters are: $Q_S = 24, z = 2, n_c = 2$, and $\alpha_w = 5\tau_s$.

of a two-zone winding with unequal coils, special attention must be given to the coil sequence (denoted by $N_{w,i}$) of each winding set. Specifically, if the coil sequence in the first winding set is N_{w2} and N_{w1}, the second set must employ N_{w1} and N_{w2}. Based on (4.33), the corresponding winding factors for each winding set are given in Equations (7.23) and (7.24),

$$^{\nu}\xi_{w,I} = \frac{1}{N_{w,z}}\sqrt{\left(n_{c1}N_{w1}{}^{\nu}\xi_{w,1}\right)^2 + \left(n_{c2}N_{w2}{}^{\nu}\xi_{w,2}\right)^2 - 2n_{c1}n_{c2}N_{w1}N_{w2}{}^{\nu}\xi_{w,1}{}^{\nu}\xi_{w,2}\cos\left(\nu\left(n_{c2}-n_{c1}+1\right)\frac{2\pi}{Q_S}\right)} \tag{7.23}$$

$$^{\nu}\xi_{w,II} = \frac{1}{N_{w,z}}\sqrt{\left(n_{c1}N_{w2}{}^{\nu}\xi_{w,1}\right)^2 + \left(n_{c2}N_{w1}{}^{\nu}\xi_{w,2}\right)^2 - 2n_{c1}n_{c2}N_{w1}N_{w2}{}^{\nu}\xi_{w,1}{}^{\nu}\xi_{w,2}\cos\left(\nu\left(n_{c2}-n_{c1}+1\right)\frac{2\pi}{Q_S}\right)} \tag{7.24}$$

leading to the total winding factor expressed in Equation (7.25). The indices "I" and "II" refer to the first and second winding sets, respectively.

$$^{\nu}\xi_{w} = \frac{^{\nu}\xi_{z12}}{N_{w,z}}\sqrt{\left(^{\nu}\xi_{w,I}\right)^2 + \left(^{\nu}\xi_{w,II}\right)^2 + 2^{\nu}\xi_{w,I}{}^{\nu}\xi_{w,II}\cos\left(\nu\alpha_w + \left(\varphi_{z,2}-\varphi_{z,1}\right)\right)} \tag{7.25}$$

Figure 7.46 summarizes two optimization scenarios. In the first case, a shift angle of $\alpha_w = 5\tau_s$ suppress the 7th harmonic, while an appropriate turns ratio of $N_{w2} = 0.5N_{w1}$ completely eliminate the 1st harmonic. In the second scenario, the 7th harmonic is selected as the working harmonic. By applying a shift angle of $\alpha_w = 7\tau_s$ and a turns ratio of $N_{w2} = 0.625N_{w1}$, the 5th harmonic is significantly reduced while the 1st harmonic is fully cancelled. In both cases, the working harmonic remains essentially unaffected.

7.5.3 Two-tooth FSWs Employing an Unequal Tooth-width Stator Core

Although the stator-shifting is highly effective in reducing critical harmonics, its effectiveness is also constrained by the discrete nature of the slot-pitch in conventional stator designs. As shown in Figure 7.46 for the 24-slot two-tooth winding, a complete harmonic cancellation cannot always be achieved when the required shift angle does not correspond to an integer multiple of the slot-pitch angle.

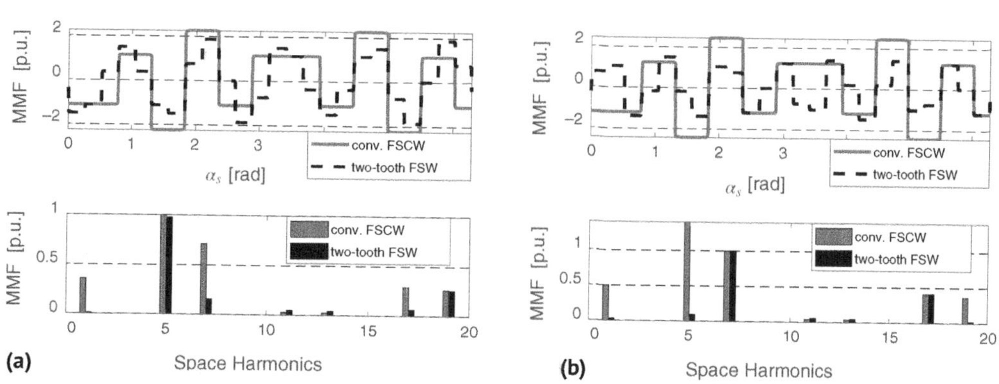

Figure 7.46 Comparison of MMF characteristics for the reference ($Q_S = 12, z = 2, n_c = 2$) and the two-tooth ($Q_S = 24$) winding with unequal coils. (a) The case with $\alpha_w = 5\tau_s$ and $N_{w2} = 0.5N_{w1}$. (b) The second case with $\alpha_w = 7\tau_s$ and $N_{w2} = 0.625N_{w1}$.

According to Equation (7.10), the shift angle required to reduce the distribution factor of a specific harmonic to zero is given by

$$
\begin{aligned}
&{}^{\nu}\xi_{z_{12}} = \cos\left(\nu \cdot \frac{a_w}{2}\right) = 0 \\
&\Rightarrow a_w = \frac{\pi}{\nu}, \frac{3\pi}{\nu}, \frac{5\pi}{\nu}, \ldots
\end{aligned}
\tag{7.26}
$$

Figure 7.47 shows the variation of the distribution factor as a function of the shift angle coefficient k_q, where $\alpha_w = k_q\ \alpha_\tau$. This plot evaluates the behavior of the two dominant MMF harmonics. For instance, when $\alpha_w = 3\pi/\nu$, the corresponding shift angle coefficient can be determined from Equations (7.11) and (7.26) as

$$
\begin{aligned}
&k_q a_\tau = a_w \\
&\Rightarrow k_q \frac{2\pi}{Q_{S2}} = \frac{3\pi}{\nu} \\
&\Rightarrow k_q = \frac{3}{2}\frac{Q_{S2}}{\nu}
\end{aligned}
\tag{7.27}
$$

Consequently, we obtain

$$
\begin{aligned}
&k_q\big|_{\nu=5,\,Q_{S2}=24} = 7.2 \\
&k_q\big|_{\nu=7,\,Q_{S2}=24} = 5.1428
\end{aligned}
$$

The corresponding distribution factors at these specific shift angles are:

Case 1: $a_w = 5.1428 \cdot a_\tau \Rightarrow {}^{1}\xi_{z_{12}} = 0.78, {}^{5}\xi_{z_{12}} = -0.975, {}^{7}\xi_{z_{12}} = 0$

Case 2: $a_w = 7.2 \cdot a_\tau \Rightarrow {}^{1}\xi_{z_{12}} = 0.58, {}^{5}\xi_{z_{12}} = 0, {}^{7}\xi_{z_{12}} = 0.95$

These results indicate that, for certain harmonics, complete cancellation requires a shift angle corresponding to a noninteger value of k_q, which cannot be realized using a conventional stator.

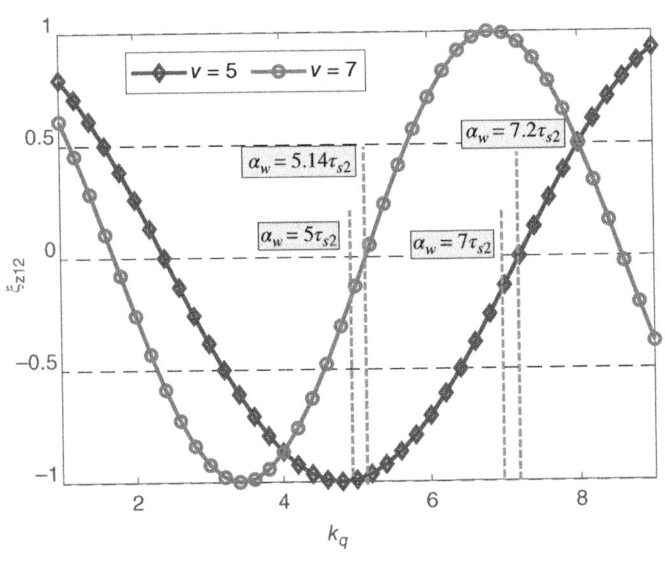

Figure 7.47 Distribution factors vs. shift angle $k_q\tau_s$ for a 24-slot stator FSW.

To overcome this limitation, a stator core with unequal tooth widths is introduced, as illustrated in Figure 7.48. The corresponding slot-pitch angles associated with the wider and narrower stator teeth, denoted by τ_{s1} and τ_{s2}, respectively, are defined as

$$\begin{aligned} \tau_{s1} &= \tau_s + \Delta\tau_s \\ \tau_{s2} &= \tau_s - \Delta\tau_s \end{aligned} \tag{7.28}$$

while ensuring that the average slot-pitch over two adjacent slots remains equal to that of the conventional stator,

$$\tau_{s1} + \tau_{s2} = 2\tau_s \tag{7.29}$$

Let α_{w1} and α_{w2} denote the shift angles associated with the standard and the modified stator structures, respectively. Then, from Equation (7.29) and the illustration in Figure 7.48, the following expression for the modified shift angle is obtained,

$$\alpha_{w1} = 5\tau_s \tag{7.30}$$

$$\alpha_{w2} = 3\tau_{s1} + 2\tau_{s2} = 5\tau_s + \Delta\tau_s \tag{7.31}$$

Equations (7.30) and (7.31) demonstrate that the modified stator structure allows precise adjustment of the relative displacement between winding sets, beyond what is possible with a conventional stator. The MMF characteristics corresponding to shift angles of $5.14\tau_s$ and $7.2\tau_s$, are presented in Figure 7.49, confirming the complete elimination of the critical MMF harmonics. Moreover, when this stator design is combined with the specific coil configurations described earlier, both subharmonics and higher-order critical harmonics can be entirely suppressed. This integrated approach, therefore, provides a powerful and systematic framework for MMF harmonic optimization in two-tooth FSWs.

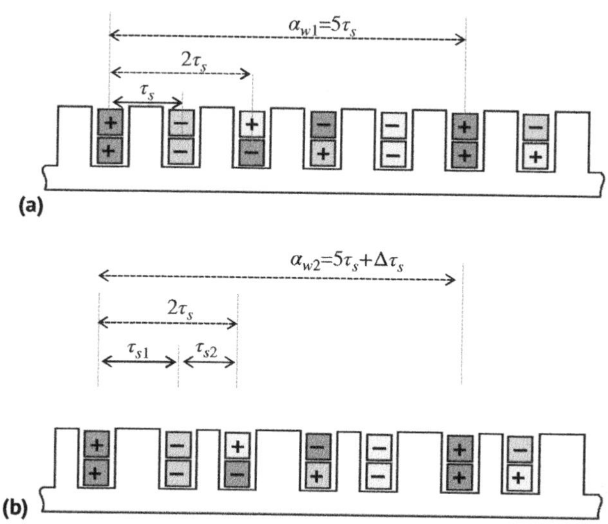

Figure 7.48 (a) Standard core structure. (b) Stator core with unequal tooth width.

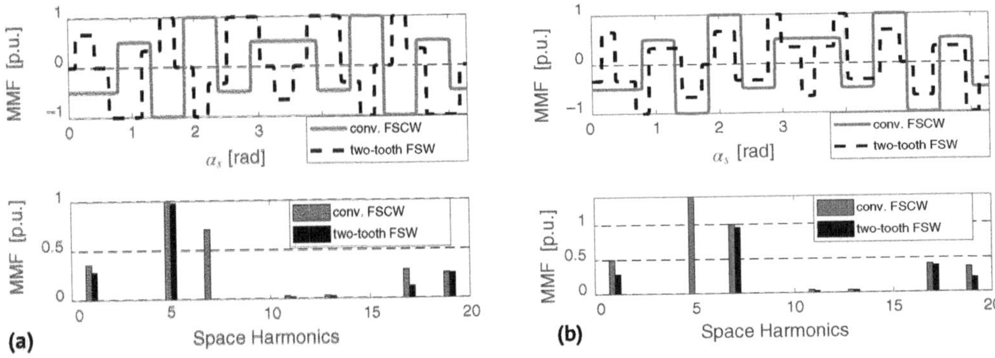

Figure 7.49 Comparison of MMF characteristics for the reference ($Q_S = 12, z = 2, n_c = 2$) and the two-tooth ($Q_S = 24$) winding with unequal tooth width stator. (a) The case with $\alpha_w = 5.14\tau_s$. (b) The second case with $\alpha_w = 7.2\tau_s$.

8

Fractional Slot Concentrated Windings with Star-delta Connection

In the previous chapters, several optimization techniques for fractional slot concentrated windings (FSCWs) have been discussed, including asymmetric and unequal coil arrangements, flux-barrier stators, stator-shifting concept, and multiphase winding configurations. In addition to these methods, the combined star-delta winding represents another effective and well-established approach for improving the magnetomotive force (MMF) properties of concentrated windings. The concept of star-delta winding combination is not new. As early as 1904, Arnold [2] described this connection scheme as a means of adapting three-phase slip-ring DC windings to different voltage levels. A few years later, the star-delta combination was recognized as a powerful technique for improving the MMF characteristics of three-phase AC windings. One of the earlier notable applications can be found in Korthals' 1918 patent [103], where the star-delta combination was employed to achieve a more nearly sinusoidal air-gap MMF distribution in a three-phase distributed winding. In practical implementation, its approach involves splitting the original winding into two equal sets: one set with N_w turns per coil connected in a star configuration, and a second set with $\sqrt{3}N_w$ turns per coil connected in a delta configuration. Korthals' patent not only described this specific arrangement but also introduced the two possible connection variants, namely the outer-star/inner-delta and outer-delta/inner-star [103]. Several decades later, Hughes [104] proposed an alternative star-delta connection for three-phase windings with a limited number of slots per phase. He demonstrated that this configuration yields a higher fundamental winding factor than conventional three-phase windings and produces a smoother MMF waveform. An additional advantage of this approach is the increased flexibility in selecting the number of turns per coil, which is particularly attractive for low-voltage machine designs. The concept was further advanced in the 1987 patent by Gjota [105] who introduced a winding arrangement in which the star and delta winding sets are spatially shifted by an angle of $(\pi/2p)$, where p represents the number of pole pairs. This displacement may be adjusted by $\pm(\pi/6p)$, allowing the proposed arrangement to be applied to machines with different slot numbers. Owing to the significant potential and benefits of this winding concept, interest in its further development and in-depth analysis increased substantially in the following decades. Contributions by Chen and Chen [106], Fogarty [107], Cistelecan et al. [108, 109], Lei et al. [110], Misir and Ponick [111], Ferreira [112], Dotz and Gerling [53, 54], Dotz [113], Raziee et al. [114], and others have considerably expanded the understanding of star-delta winding behavior and design tradeoffs.

Originally, the combined star-delta winding was developed to enhance the fundamental winding factor, and hence the output torque, while simultaneously reducing MMF harmonic content in AC machines with distributed windings. Over the past two decades, however, increasing attention has been paid to the application of this concept to permanent magnet synchronous motors (PMSMs) with concentrated windings, particularly for automotive applications. During this period, numerous studies have explored the optimization of FSCWs through star-delta combinations. Vansompel et al. [115], for example, showed that the combined star-delta connection enhances the fundamental winding factor for most FSCWs topologies when compared with traditional star or delta connections. Subsequently, Dajaku and Gerling [116] introduced a specific star-delta FSCW that effectively reduces both subharmonic and higher-order MMF harmonics in a 12-slot/10-pole winding. In 2015, Abdel-Khalik et al. [70] proposed another star-delta configuration for PMSMs with a 12-slots FSCW stator (or its multiples), achieving complete cancellation of the subharmonic component while slightly increasing the working harmonic. In a further work on this concept, Zhao et al. [117] derived a general relationship between the number of slots and phases suitable for star-delta combinations and investigated the principles of MMF harmonic elimination in such windings. Overall, these studies consistently identify the star-delta combination as a highly effective method for improving the MMF spectrum of FSCWs. The principal benefits of this approach can be summarized as follows:

- complete cancellation of MMF harmonics below the operating (working) harmonic,
- a slight increase in operating harmonic, leading to improved torque density and higher machine efficiency, and
- a reduction in the required number of phases, since the star-delta combination replaces dual three-phase (or multiphase) winding.

It should be noted, however, that the star-delta combination does not offer only benefits. In electric machines where the air-gap flux density contains a considerable 3rd harmonic component (resulting from rotor saliency, magnetic saturation, or rotor flux) circulating currents may be induced in the delta branch, as observed by Hughes (1970) [104]. These circulating currents increase copper losses and consequently reduce the overall efficiency of the machine. When using star-delta windings, careful rotor optimization is therefore essential to minimize these effects and ensure efficient machine operation.

8.1 Star-delta Windings

As previously noted, the star-delta winding consists of two winding sets, one connected in star (Y) and the other in a delta (Δ). As early as 1918, Korthals [103] introduced two basic arrangements of this combined winding, schematically illustrated in Figure 8.1:

a. outer-star and inner-delta connections,
b. inner-star and outer-delta connections.

In the literature, the first combination is commonly referred to as a *star-delta series connection*, whereas the second is known as a *star-delta parallel connection*. From the standpoint of circulating currents, the series-connected star-delta winding has a more limited current due to its series arrangement [110]. As a result, this configuration is generally preferred over the parallel arrangement. Consequently, the following analysis is limited exclusively to the series-connected winding. Therefore, in the following the term "star-delta winding" refers specifically to the outer-star and inner-delta (series connection).

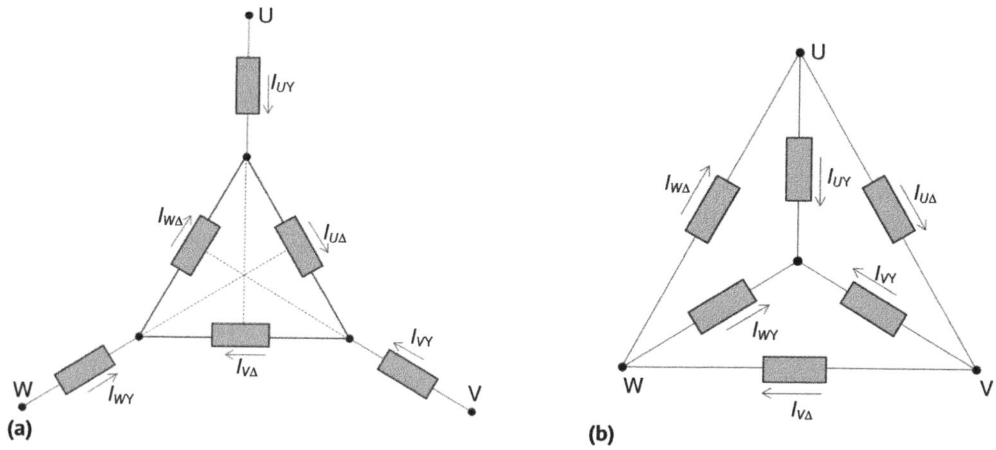

Figure 8.1 (a) Outer-star inner-delta winding (star-delta series connection). (b) Inner-star outer-delta winding (star-delta parallel connection).

8.1.1 Star-delta Phase Currents

Figure 8.1a illustrates the current relationships between the star- and delta-connected winding parts. Let the phase currents in the delta-connected windings be defined as:

$$\begin{aligned} i_{u,\Delta}(t) &= \hat{i} \cdot \cos(\omega t) \\ i_{v,\Delta}(t) &= \hat{i} \cdot \cos\left(\omega t - \frac{2\pi}{3}\right) \\ i_{w,\Delta}(t) &= \hat{i} \cdot \cos\left(\omega t + \frac{2\pi}{3}\right) \end{aligned} \tag{8.1}$$

By applying the Kirchhoff's current law at the interconnection points between the star and delta circuit diagrams, the phase current in the star-connected windings can be expressed as:

$$\begin{aligned} i_{u,Y}(t) &= i_{u,\Delta}(t) - i_{w,\Delta}(t) = \sqrt{3}\hat{i} \cdot \cos\left(\omega t - \frac{\pi}{6}\right) \\ i_{v,Y}(t) &= i_{v,\Delta}(t) - i_{u,\Delta}(t) = \sqrt{3}\hat{i} \cdot \cos\left(\omega t - \frac{\pi}{6} - \frac{2\pi}{3}\right) \\ i_{w,Y}(t) &= i_{w,\Delta}(t) - i_{v,\Delta}(t) = \sqrt{3}\hat{i} \cdot \cos\left(\omega t - \frac{\pi}{6} + \frac{2\pi}{3}\right) \end{aligned} \tag{8.2}$$

Equations (8.1) and (8.2) show that the star-delta combination inherently produces a dual three-phase current system with a relative phase shift of $\pi/6$ (30 degrees) between winding sets, as depicted in Figure 8.2a. An important advantage of this arrangement, when compared to multiphase machines with equivalent properties, is the reduction in the number of phases by a factor of two. In practical terms, instead of using a dual three-phase stator winding fed by two separate three-phase inverters, the star-delta combination provides an alternative solution that requires only three-phase stator terminals as in conventional three-phase machines.

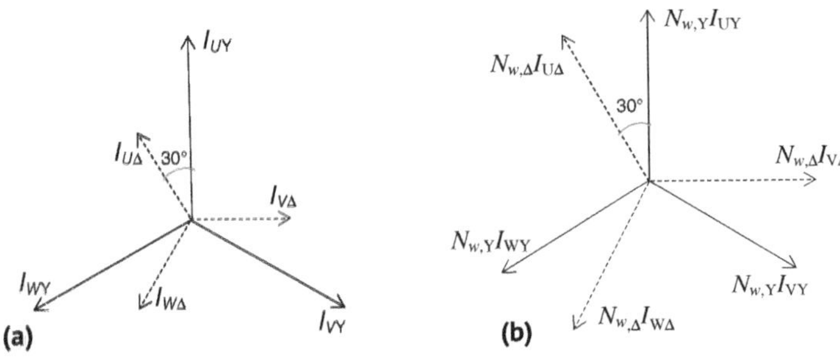

Figure 8.2 (a) Current phasors for the three-phase star-delta connected winding. (b) Balanced stator MMF magnitudes using the turns ratio $N_{w,\Delta} = \sqrt{3}N_{w,Y}$.

8.1.2 Winding MMF Function

With the exception of windings having a single coil per zone ($n_c = 1$), most multiphase windings can be converted into a star-delta combination. This process involves dividing the original winding into two subsets: one connected in delta and the other in star. Moreover, when the reference winding contains more than two coils per zone ($n_c > 2$), the resulting star-delta winding sets may have either identical or different numbers of coils per zone, as described by Equations (8.3) and (8.4) [115, 117]. However, regardless of the specific splitting, the total number of coils in the converted winding remains equal to that of the original winding, as given in Equation (8.5). In these expressions, n_{cY} and $n_{c\Delta}$ denote the number of coils per zone for the star and delta winding sets, respectively.

$$n_{c\Delta} = n_{cY} = \frac{n_c}{2} \tag{8.3}$$

$$n_{c\Delta} \neq n_{cY} \tag{8.4}$$

$$n_{c\Delta} + n_{cY} = n_c \tag{8.5}$$

Since the current flowing in the delta-connected coils is $1/\sqrt{3}$ of the current in the star-connected coils, the number of turns must be adjusted to ensure equal MMF contribution from both winding sets, as illustrated in Figure 8.2b. Accordingly, the number of turns in the delta coils must be increased by a factor of $\sqrt{3}$, as given by:

$$N_{w\Delta} = \sqrt{3}N_{wY} \tag{8.6}$$

To evaluate the MMF characteristics of the star-delta winding, the analytical framework developed in Chapter 2 is employed, accounting for the spatial placement of the coils and the relative angular displacement between the star and delta sets. Let the star-connected winding be spatially shifted by an angle $\varphi_{Y\Delta}$ with respect to the delta-connected winding, then according to Equation (2.41), the MMF contributions of the two winding sets can be expressed as follows:

$$F_\Delta(\alpha_S, t) = \frac{mzn_{c\Delta}N_{w\Delta}\hat{i}_\Delta}{\pi}\sum_\nu \frac{{}^\nu\xi_{w\Delta}}{\nu}\sin\left(\omega t - \nu\alpha_S + {}^\nu\vartheta_\Delta\right) \tag{8.7}$$

$$F_Y(\alpha_S, t) = \frac{mzn_{cY}N_{wY}\hat{i}_Y}{\pi}\sum_\nu \frac{{}^\nu\xi_{wY}}{\nu}\sin\left[\left(\omega t - \frac{\pi}{6}\right) - \nu(\alpha_S - \varphi_{Y\Delta}) + {}^\nu\vartheta_Y\right] \tag{8.8}$$

where the corresponding spatial shift parameters ${}^{\nu}\vartheta_{\Delta}$ and ${}^{\nu}\vartheta_{Y}$ are defined as:

$$ {}^{\nu}\vartheta_{\Delta} = \nu(n_{c\Delta} - 1)\frac{\pi}{Q_S} + (n_{c\Delta} - 1)\frac{\pi}{2} \tag{8.9} $$

$$ {}^{\nu}\vartheta_{Y} = \nu(n_{cY} - 1)\frac{\pi}{Q_S} + (n_{cY} - 1)\frac{\pi}{2} \tag{8.10} $$

In the above MMF expressions, ${}^{\nu}\xi_{wY}$ and ${}^{\nu}\xi_{w\Delta}$ represent the respective winding factors of the winding sets. Since these parameters depend on the winding distribution, in particular on the number of coils per zone, they may be either equal or unequal, as indicated in Equations (8.3) and (8.4).

By applying the principle of superposition, the total MMF for the combined star-delta winding is obtained as the sum of the individual MMFs:

$$ \begin{aligned} F_{Y\Delta}(\alpha_S, t) &= F_{Y}(\alpha_S, t) \pm F_{\Delta}(\alpha_S, t) \\ &= \frac{mzN_{w,eq}\hat{i}}{\pi}\sum_{\nu}\frac{1}{\nu}\Big\{N^{*}_{w\Delta}\,{}^{\nu}\xi_{w\Delta}\sin\left(\omega t - \nu\alpha_S + {}^{\nu}\vartheta_{\Delta}\right) \\ &\quad \pm N^{*}_{wY}\,{}^{\nu}\xi_{wY}\sin\left[\left(\omega t - \frac{\pi}{6}\right) - \nu(\alpha_S - \varphi_{Y\Delta}) + {}^{\nu}\vartheta_{Y}\right]\Big\} \end{aligned} \tag{8.11} $$

In the above approach, $N_{w,eq}$ denotes the equivalent number of turns of the entire winding, and $N^{*}_{w\Delta}$ and N^{*}_{wY} represent the p.u. number of turns associated with the star and delta windings,

$$ N_{w,eq} = \left(n_{c\Delta}N_{w\Delta} + \sqrt{3}n_{cY}N_{wY}\right) \tag{8.12} $$

$$ N^{*}_{w\Delta} = \frac{n_{c\Delta}N_{w\Delta}}{N_{w,eq}} \tag{8.13} $$

$$ N^{*}_{wY} = \frac{\sqrt{3}n_{cY}N_{wY}}{N_{w,eq}} \tag{8.14} $$

The total MMF may also be expressed in closed form as:

$$ F_{Y\Delta}(\alpha_S, t) = \frac{mzN_{w,eq}\hat{i}}{\pi}\sum_{\nu}\frac{{}^{\nu}\xi_{w,eq}}{\nu}\sin\left(\omega t - \nu\alpha_S + {}^{\nu}\varphi_{1,2}\right) \tag{8.15} $$

The corresponding expressions for the equivalent winding factor and the resulting phase shift angle are given by:

$$ {}^{\nu}\xi_{w,eq} = \sqrt{\left(N^{*}_{w\Delta}\,{}^{\nu}\xi_{w\Delta}\right)^2 + \left(N^{*}_{wY}\,{}^{\nu}\xi_{wY}\right)^2 \pm 2N^{*}_{w\Delta}N^{*}_{wY}\,{}^{\nu}\xi_{w\Delta}\,{}^{\nu}\xi_{wY}\cos\left(\nu\varphi_{Y\Delta} - \frac{\pi}{6} + \left({}^{\nu}\vartheta_{Y} - {}^{\nu}\vartheta_{\Delta}\right)\right)} \tag{8.16} $$

$$ {}^{\nu}\varphi_{1,2} = a\tan\left(\frac{N^{*}_{w\Delta}\,{}^{\nu}\xi_{w\Delta}\sin\left({}^{\nu}\vartheta_{\Delta}\right) \pm N^{*}_{wY}\,{}^{\nu}\xi_{wY}\sin\left(\nu\varphi_{Y\Delta} - \frac{\pi}{6} + {}^{\nu}\vartheta_{Y}\right)}{N^{*}_{w\Delta}\,{}^{\nu}\xi_{w\Delta}\cos\left({}^{\nu}\vartheta_{\Delta}\right) \pm N^{*}_{wY}\,{}^{\nu}\xi_{wY}\cos\left(\nu\varphi_{Y\Delta} - \frac{\pi}{6} + {}^{\nu}\vartheta_{Y}\right)}\right) \tag{8.17} $$

In these relationships, the "+" sign corresponds to winding sets having the same polarity sequence, whereas the "−" sign indicates opposite polarity situation. An example of a star-delta winding with the same and opposite polarities winding sets is illustrated in Figure 8.6a,b.

8.2 Star-delta FSCW Topologies

Equations (8.1) and (8.2) indicate that, in a star-delta winding connection, the phase currents of winding sets are inherently shifted by a fixed phase angle of 30 degrees, i.e.,

$$\angle(i_Y, i_\Delta) = \frac{\pi}{6} \tag{8.18}$$

This phase displacement, however, should not be confused with the spatial displacement angle $\varphi_{Y\Delta}$ between the star- and delta-connected windings. In contrast to the fixed phase-current shift, $\varphi_{Y\Delta}$ is constrained by the star-delta conversion process and can only assume discrete values corresponding to integer multiples of the slot-pitch:

$$\varphi_{Y\Delta} = h\frac{2\pi}{Q_S} \tag{8.19}$$

where h is a positive integer.

For a given set of winding parameters, the number of stator slots for a three-phase DL-FSCW can be determined using Equation (8.20).

$$Q_S = m \cdot z \cdot n_c = 3zn_c \tag{8.20}$$

By combining Equations (8.19) and (8.20), it follows that the conversion of a conventional three-phase winding into a star-delta configuration yields two distinct winding groups:

1. Star-delta windings with equal phasor and space shift angles:

$$\begin{aligned} &\varphi_{Y\Delta} = \angle(i_Y, i_\Delta) \\ &\Rightarrow \varphi_{Y\Delta} = \frac{\pi}{6} \end{aligned} \tag{8.21}$$

2. Star-delta windings with unequal phasor and space shift angles:

$$\begin{aligned} &\varphi_{Y\Delta} \neq \angle(i_Y, i_\Delta) \\ &\Rightarrow \varphi_{Y\Delta} \neq \frac{\pi}{6} \end{aligned} \tag{8.22}$$

The corresponding winding arrangements are illustrated in Figure 8.3 and are referred to as symmetric (for case 1) and asymmetric (for case 2) star-delta windings, respectively. According to Figure 8.3b, the space shift angle for the general case can be written as:

$$\varphi_{Y\Delta} = \frac{\pi}{6} \pm \alpha_{Y\Delta} \tag{8.23}$$

Accordingly, when $\alpha_{Y\Delta} = 0°$, the winding is symmetric, whereas $\alpha_{Y\Delta} \neq 0°$ represents an asymmetric star-delta configuration. Moreover, for the asymmetric case, $\alpha_{Y\Delta}$ may assume both positive and negative values.

8.2.1 Symmetric Star-delta FSCWs

Considering first the symmetric star-delta winding category, characterized with equal phasor and spatial shift angles. From Equations (8.19) and (8.21), we have:

$$\varphi_{Y\Delta} = \frac{\pi}{6} \Rightarrow h\frac{2\pi}{Q_S} = \frac{\pi}{6} \tag{8.24}$$

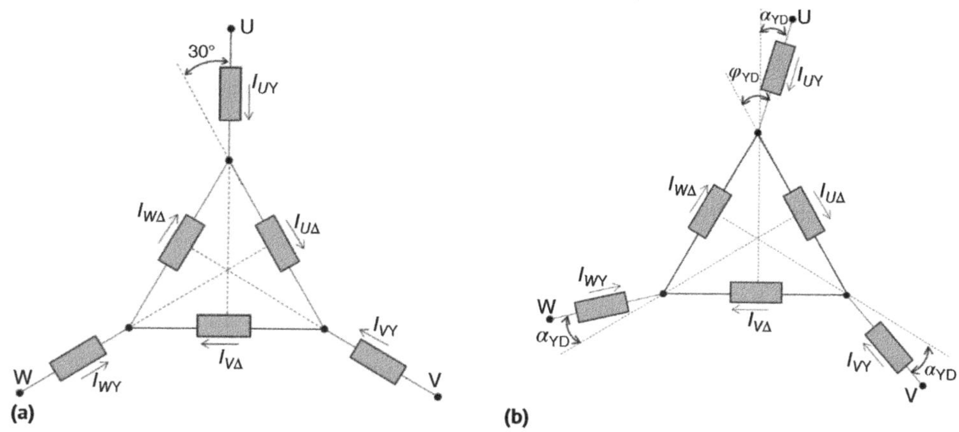

Figure 8.3 Star-delta winding with: (a) Equal phasor and space shift angles (symmetric winding). (b) Unequal phasor and space shift angles (asymmetric winding).

This condition means that the number of available stator slots for this winding group is limited and must satisfy:

$$Q_S = 12h, \quad h = 1, 2, 3, \ldots \tag{8.25}$$

Consequently, based on Equations (8.20) and (8.25), it follows that only specific three-phase FSCWs can be converted into symmetric star-delta configurations. The required number of coils per zone is given by:

$$n_c = \begin{cases} 4h, & \text{for one-zone winding } (z = 1) \\ 2h, & \text{for two-zone winding } (z = 2) \end{cases} \tag{8.26}$$

In particular, one-zone FSCWs suitable for conversion to symmetric star-delta windings must have $n_c = 4,8,12,\ldots$ coils per zone, whereas two-zone FSCWs require $n_c = 2,4,6,\ldots$ coils per zone. Moreover, although Equation (8.26) defines the feasible winding topologies for standard star-delta combinations, it does not explicitly specify the number of coils in the star or delta winding sets. According to Equations (8.3) and (8.4), the number of coils in each set may be either equal or unequal. The appropriate relationship between them, however, can be derived from the fundamental principles of star-delta transformation.

To illustrate this, let us consider a phase winding, or more specifically a winding zone, that is divided in two parts: one part assigned to the delta winding and the second part to the star winding, as illustrated in Figure 8.4. In this arrangement, the star winding is spatially shifted by $n_{c\Delta}$ slot-pitches relative to the delta winding, resulting in a certain displacement angle between the two winding sets:

$$\varphi_{Y\Delta} = n_{c\Delta}\frac{2\pi}{Q_S} \tag{8.27}$$

Combining Equations (8.21) and (8.27), the number of coils per zone for the delta winding is defined in Equation (8.28), while for the star winding in (8.29).

$$n_{c\Delta} = \frac{Q_S}{12} \tag{8.28}$$

$$n_{cY} = n_c - n_{c\Delta} \tag{8.29}$$

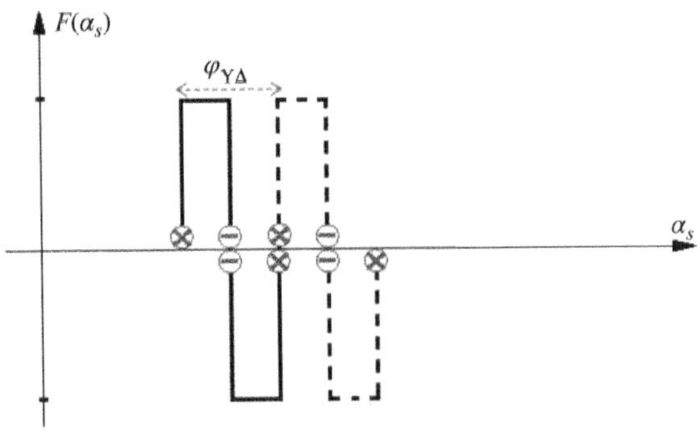

Figure 8.4 Winding layout (one phase) of a star-delta winding with $n_{c\Delta} = 2$ and $n_{cY} = 2$ coils per zone for the delta (in solid black) and star (in dashed black) winding sets.

Table 8.1 Available one-zone symmetric star-delta FSCWs and their respective parameters.

h	**1**	**2**	**3**	**4**
$n_c = 4h$	4	8	12	16
$n_{c\Delta} = \frac{Q_s}{12}$	1	2	3	4
$n_{cY} = n_c - n_{c\Delta}$	3	6	9	12
$Q_S = 3 \cdot z \cdot n_c$	12	24	36	48
$\varphi_{Y\Delta} = n_{c\Delta}\frac{2\pi}{Q_s}$	$\frac{\pi}{6}$	$\frac{\pi}{6}$	$\frac{\pi}{6}$	$\frac{\pi}{6}$

Table 8.2 Available two-zone symmetric star-delta FSCWs and their respective parameters.

h	**1**	**2**	**3**	**4**
$n_c = 2h$	2	4	6	8
$n_{c\Delta} = \frac{Q_s}{12}$	1	2	3	4
$n_{cY} = n_c - n_{c\Delta}$	1	2	3	4
$Q_S = 3 \cdot z \cdot n_c$	12	24	36	48
$\varphi_{Y\Delta} = n_{c\Delta}\frac{2\pi}{Q_s}$	$\frac{\pi}{6}$	$\frac{\pi}{6}$	$\frac{\pi}{6}$	$\frac{\pi}{6}$

As mentioned earlier, the number of feasible FSCW topologies for symmetric star-delta conversion is limited, and all valid configurations must satisfy the conditions specified in Equations (8.24)–(8.29). Based on these constraints, various suitable winding topologies for symmetric star-delta windings are summarized in Tables 8.1 and 8.2. Representative winding layouts and their associated MMF waveforms are shown in Figures 8.5 and 8.6. In these examples, the original winding zone containing n_c coils per zone is divided into two subsets. The coil MMF function for the delta winding (with $n_{c\Delta}$-coils) is represented by solid lines, while that of the star winding (with n_{cY}-coils) is indicated by dashed lines.

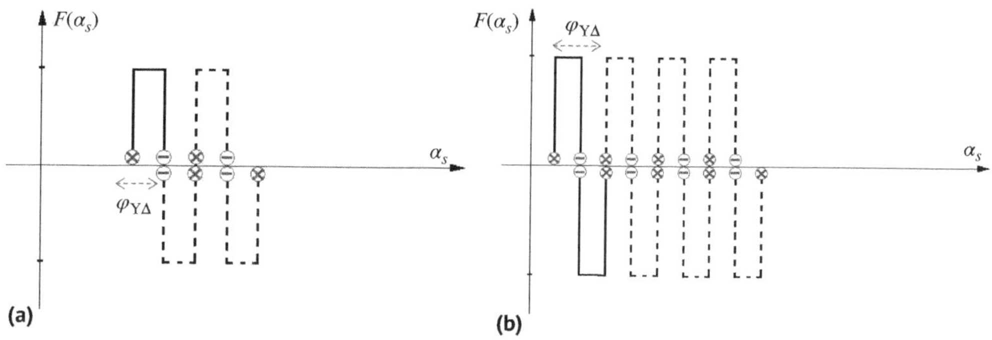

Figure 8.5 Winding layout (one phase) of a one-zone symmetric star-delta FSCW. Solid line for delta winding and dashed line for star winding. (a) FSCW with $Q_S = 12, z = 1, n_{c\Delta} = 1, n_{cY} = 3$, and $\varphi_{Y\Delta} = \pi/6$. (b) FSCW with $Q_S = 24, z = 1, n_{c\Delta} = 2, n_{cY} = 6$, and $\varphi_{Y\Delta} = \pi/6$.

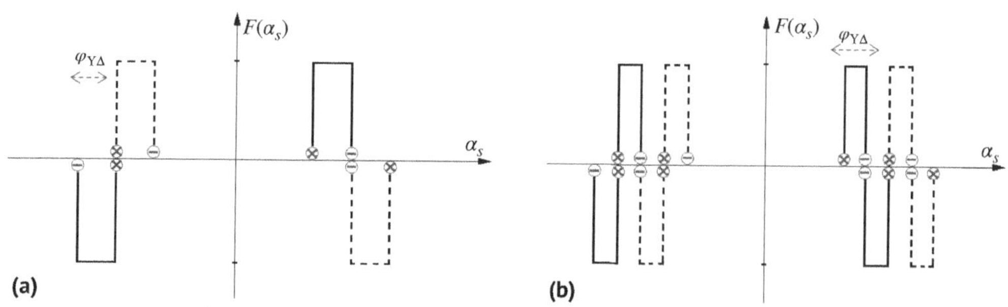

Figure 8.6 Winding layout (one phase) of a two-zone symmetric star-delta FSCW. Solid line for delta winding and dashed line for star winding. (a) FSCW with $Q_S = 12, z = 2, n_{c\Delta} = 1, n_{cY} = 1$, and $\varphi_{Y\Delta} = \pi/6$. (b) FSCW with $Q_S = 24, z = 2, n_{c\Delta} = 2, n_{cY} = 2$, and $\varphi_{Y\Delta} = \pi/6$.

8.2.2 Asymmetric Star-delta FSCWs

In contrast to the first category, the number of FSCW topologies suitable for asymmetric star-delta combinations is much higher. Essentially, most conventional FSCWs, except those with $n_c = 1$, can be converted to an asymmetric configuration, as summarized in Tables 8.3 and 8.4. In addition, certain winding types can be adapted for both symmetric and asymmetric star-delta FSCWs. In the tables below, the shift angles for symmetric configurations are highlighted in gray.

Figure 8.7 shows the winding arrangements (for one phase) and their MMF waveform for two possible asymmetric star-delta configurations, derived from the conversion of a conventional FSCW with parameters $Q_S = 9$, $z = 1$, and $n_c = 3$. Further examples in Figure 8.8 illustrate that topologies with a large number of coils per zone allow a wider range of asymmetric star-delta combinations, making it clear that the number of possible combinations increases with n_c. Additionally, Figure 8.9 provides a further example of a star-delta winding obtained from the conversion of a conventional two-zone FSCW with $n_c = 3$.

Table 8.3 Various one-zone asymmetric star-delta FSCWs and their respective parameters.

n_c	$n_{c\Delta}$	n_{cY}	$Q_S = 3 \cdot z \cdot n_c$	$\varphi_{Y\Delta} = n_{c\Delta}\dfrac{2\pi}{Q_s}$
1	—	—	3	—
2	1	1	6	$\pi/3$
3	1	2	9	$2\pi/9$
4	1, 2	3, 2	12	$\pi/6, \pi/3$
5	1, 2	4, 3	15	$2\pi/15, 4\pi/15$
6	1, 2, 3	5, 4, 3	18	$\pi/9, 2\pi/9, \pi/3$
7	1, 2, 3	6, 5, 4	21	$2\pi/21, 4\pi/21, 2\pi/7$
8	1, 2, 3, 4	7, 6, 5, 4	24	$\pi/12, \pi/6, \pi/4, \pi/3$

Table 8.4 Various two-zone asymmetric star-delta FSCWs and their respective parameters.

n_c	$n_{c\Delta}$	n_{cY}	$Q_S = 3 \cdot z \cdot n_c$	$\varphi_{Y\Delta} = n_{c\Delta}\dfrac{2\pi}{Q_s}$
1	—	—	6	—
2	1	1	12	$\pi/6$
3	1	2	18	$\pi/9$
4	1, 2	3, 2	24	$\pi/12, \pi/6$
5	1, 2	4, 3	30	$\pi/15, 2\pi/15$
6	1, 2, 3	5, 4, 3	36	$\pi/18, \pi/9, \pi/6$
7	1, 2, 3	6, 5, 4	48	$\pi/24, \pi/12, \pi/8$
8	1, 2, 3, 4	7, 6, 5	54	$\pi/22, \pi/11, 3\pi/22, 2\pi/11$

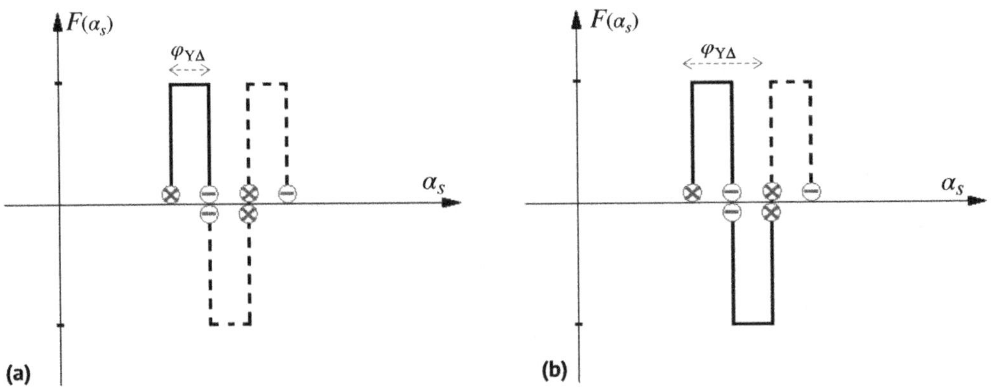

Figure 8.7 Winding layout (one phase) of an asymmetric star-delta FSCW with $Q_S = 9$ and $z = 1$. Solid line for delta winding and dashed line for star winding. (a) FSCW with $n_{c\Delta} = 1, n_{cY} = 2$, and $\varphi_{Y\Delta} = 2\pi/9$. (b) FSCW with $n_{c\Delta} = 2, n_{cY} = 1$, and $\varphi_{Y\Delta} = 4\pi/9$.

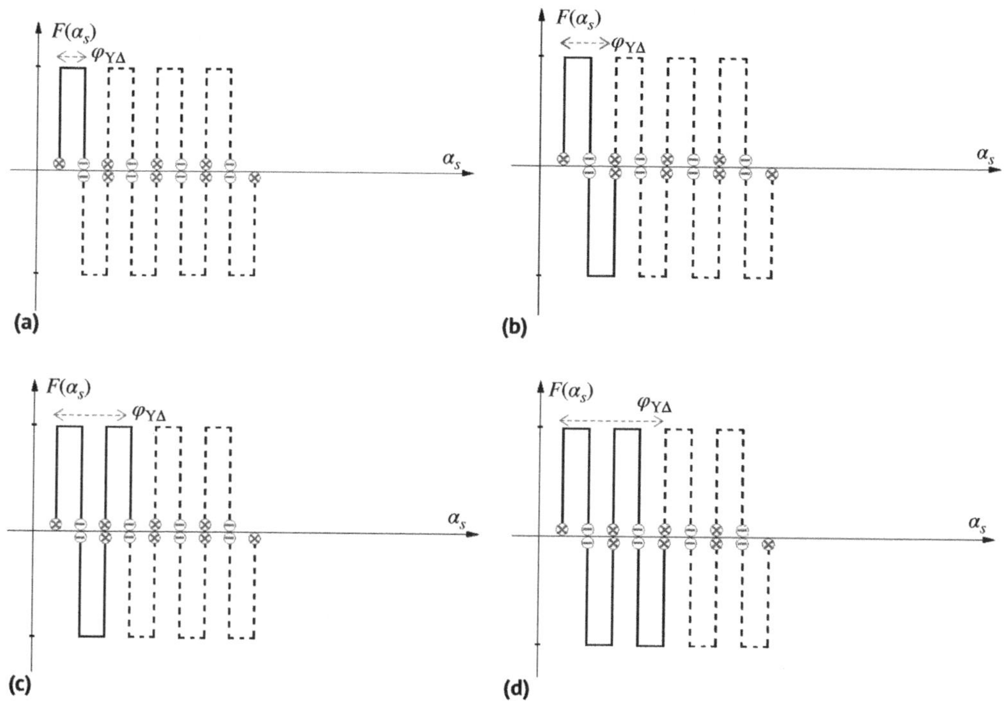

Figure 8.8 Winding layout (one phase) of an asymmetric star-delta FSCW with $Q_S = 24$ and $z = 1$. Solid line for delta winding and dashed line for star winding. (a) FSCW with $n_{c\Delta} = 1, n_{cY} = 7$, and $\varphi_{Y\Delta} = \pi/12$. (b) FSCW with $n_{c\Delta} = 2, n_{cY} = 6$, and $\varphi_{Y\Delta} = \pi/6$. (c) FSCW with $n_{c\Delta} = 3, n_{cY} = 5$, and $\varphi_{Y\Delta} = \pi/4$. (d) FSCW with $n_{c\Delta} = 4, n_{cY} = 4$, and $\varphi_{Y\Delta} = \pi/3$.

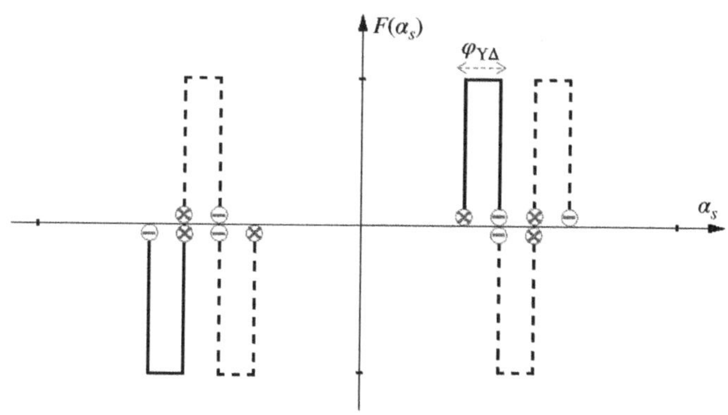

Figure 8.9 Winding layout (one phase) of an asymmetric star-delta FSCW with $Q_s = 18, z = 2, n_{c\Delta} = 1$, $n_{cY} = 2$, and $\varphi_{Y\Delta} = \pi/9$. Solid line for delta winding and dashed line for star winding.

8.3 Winding Analysis

As introduced in Chapter 2, the winding factor is defined as the product of the coil-pitch factor ${}^{\nu}\xi_p$, distribution factor ${}^{\nu}\xi_d$, and zone factor ${}^{\nu}\xi_z$. Since the star-delta transformation primarily modifies the number of coils per zone in winding sets by dividing the original winding, i.e.,

$$n_{c\Delta} < n_c$$
$$n_{cY} < n_c$$

as a result, the transformation specifically affects the distribution factors, while the coil-pitch and zone factors remain unchanged. Therefore, according to Equation (2.42), the winding factors for the star and delta winding sets may be written as:

$$^{\nu}\xi_{w\Delta} = {}^{\nu}\xi_{p}\,{}^{\nu}\xi_{d\Delta}\,{}^{\nu}\xi_{z} \tag{8.30}$$

$$^{\nu}\xi_{wY} = {}^{\nu}\xi_{p}\,{}^{\nu}\xi_{dY}\,{}^{\nu}\xi_{z} \tag{8.31}$$

Although the star-delta transformation can be applied to a wide range of FSCW topologies, the following analysis is restricted to DL-FSCWs with opposite coil polarity. For this class of windings, Equations (2.15), (2.31), and (2.38) yield the following expressions for the winding factors:

$$^{\nu}\xi_{w\Delta} = \sin\left(\nu\frac{\pi}{Q_S}\right)\frac{\sin\left(\nu n_{c\Delta}\frac{\pi}{Q_S} + n_{c\Delta}\frac{\pi}{2}\right)}{n_{c\Delta}\cdot\cos\left(\nu\frac{\pi}{Q_S}\right)}\sin\left(\nu\frac{\pi}{2}\right) \tag{8.32}$$

$$^{\nu}\xi_{wY} = \sin\left(\nu\frac{\pi}{Q_S}\right)\frac{\sin\left(\nu n_{cY}\frac{\pi}{Q_S} + n_{cY}\frac{\pi}{2}\right)}{n_{cY}\cdot\cos\left(\nu\frac{\pi}{Q_S}\right)}\sin\left(\nu\frac{\pi}{2}\right) \tag{8.33}$$

The above equations clearly show that a reduction in the number of coils per zone leads to an increase in the distribution factor for both the star and delta winding sets. Consequently, the winding factors for the operating harmonics are higher than those of the original winding. Moreover, since the star-delta transformation may lead to winding sets with either equal or unequal numbers of coils per zone, the resulting winding factors may be identical or different.

In terms of n_c parameter, star-delta windings can be classified into two distinct categories:

1. Balanced star-delta windings when $n_{c\Delta} = n_{cY}$, and
2. Unbalanced star-delta windings when $n_{c\Delta} \neq n_{cY}$.

8.3.1 Balanced Star-delta FSCWs

In a balanced star-delta winding, the number of coils per zone in the star and delta winding sets is identical, i.e.,

$$n_{c\Delta} = n_{cY} = n_c' \tag{8.34}$$

where,

$$n_c' = \frac{n_c}{2} \tag{8.35}$$

From Equation (8.32) and (8.33), it then follows directly that the winding factors of the two winding sets are equal:

$$^{\nu}\xi_{w\Delta} = {}^{\nu}\xi_{wY} = {}^{\nu}\xi_{w}' \tag{8.36}$$

that can be expressed as:

$$^{\nu}\xi_{w}' = \sin\left(\nu\frac{\pi}{Q_S}\right)\frac{\sin\left(\nu n_c'\frac{\pi}{Q_S} + n_c'\frac{\pi}{2}\right)}{n_c'\cdot\cos\left(\nu\frac{\pi}{Q_S}\right)}\sin\left(\nu\frac{\pi}{2}\right) \tag{8.37}$$

In addition to n_c' parameter, the spatial displacement angle between the star and delta winding sets also has a significant effect on the winding properties. For balanced star-delta windings, this angle can assume only two specific values. Using the expression for $\varphi_{Y\Delta}$ given in Equation (8.38), this parameter can be either $\pi/3$ or $\pi/6$, depending on the number of zones per phase as defined in Equation (8.39).

$$\varphi_{Y\Delta} = n_c' \frac{2\pi}{Q_S} = \frac{n_c}{2} \frac{2\pi}{mzn_c} = \frac{\pi}{3z} \tag{8.38}$$

$$\varphi_{Y\Delta} = \begin{cases} \dfrac{\pi}{3}, & \text{one-zone winding } (z = 1) \\ \dfrac{\pi}{6}, & \text{two-zone winding } (z = 2) \end{cases} \tag{8.39}$$

In this context, all common FSCW with an even number of coils per zone (e.g. $n_c = 2,4,6,\ldots$) fall into this category. In particular, the two-zone FSCWs listed in Table 8.2 represent symmetric balanced windings, whereas the one-zone FSCWs in Table 8.3 with $n_{c\Delta} = n_{cY}$ correspond to balanced asymmetric windings.

8.3.1.1 Balanced Star-delta Symmetric FSCWs

For the case when

$$n_{c\Delta} = n_{cY} = n_c'$$

and

$$\varphi_{Y\Delta} = \frac{\pi}{6}$$

the expressions for equivalent winding factor (8.16) and phase shift angle (8.17) simplify to:

$$^{\nu}\xi_{w,\mathrm{eq}} = \begin{cases} \sin\left((\nu - 1)\dfrac{\pi}{12}\right) \cdot {}^{\nu}\xi_{w}', & \text{opposite polarity winding sets} \\ \cos\left((\nu - 1)\dfrac{\pi}{12}\right) \cdot {}^{\nu}\xi_{w}', & \text{same polarity winding sets} \end{cases} \tag{8.40}$$

$$^{\nu}\varphi_{1,2} = (\nu - 1)\frac{\pi}{12} \tag{8.41}$$

It should also be noted that the equivalent MMFs generated by the star and delta coils are equal, i.e.,

$$N_{w\Delta}\hat{i}_{\Delta} = N_{wY}\hat{i}_{Y} = N_w'\hat{i} \tag{8.42}$$

Thus, the resulting MMF can be expressed as:

$$F(\alpha_S, t) = \frac{2mzn_c'N_w'\hat{i}}{\pi} \sum_{\nu} \frac{^{\nu}\xi_{w,\mathrm{eq}}}{\nu} \sin\left(\omega t - \nu\alpha_S - (\nu - 1)\frac{\pi}{12}\right) \tag{8.43}$$

Considering the winding factor defined in Equation (8.37), together with the fact that the number of coils per zone in the transformed star-delta winding is smaller than in the original winding:

$$n_{c\Delta} = n_{cY} < n_c \tag{8.44}$$

It follows that the winding factor of the star-delta winding is higher than that of the original winding:

$$^{\nu}\xi'_w > {}^{\nu}\xi_w \tag{8.45}$$

where $^{\nu}\xi_w$ denotes the winding factor for the original FSCW characterized by n_c coils per zone.

In addition to improved winding factor for the operating harmonic, as given in Equation (8.40), the star-delta combination in the balanced-symmetric case also results in the complete cancellation of certain MMF harmonics. Especially, for configurations with opposite winding set polarity, the MMF harmonics of order:

$$\begin{aligned} &\sin\left((\nu-1)\frac{\pi}{12}\right) = 0 \\ &\Rightarrow (\nu-1)\frac{\pi}{12} = k\pi \\ &\Rightarrow \nu = 12k+1, k = 0, \pm 1, \pm 2, \ldots \end{aligned} \tag{8.46}$$

are entirely eliminated. Similarly, for winding sets with the same polarity, the cancelled harmonic orders are given by:

$$\begin{aligned} &\cos\left((\nu-1)\frac{\pi}{12}\right) = 0 \\ &\Rightarrow (\nu-1)\frac{\pi}{12} = k\frac{\pi}{2} \\ &\Rightarrow \nu = 6k+1, k = \pm 1, \pm 2, \ldots \end{aligned} \tag{8.47}$$

Equation (8.46) indicates that the star-delta configuration with opposite winding set polarities completely cancels the 1st harmonic as well as the higher harmonics of order −11th, +13th, and so on. In contrast, for configurations with the same winding set polarity, Equation (8.47) shows that the −5th, +7th, −17th, +19th, and higher-order harmonics are eliminated. These conclusions are verified in the following examples.

Example 8.1 Star-delta FSCW with $Q_s = 12$, $z = 2$, and $n_{c\Delta} = n_{cY} = 1$
Figure 8.10 compares the MMF characteristics of the first example. The reference winding is a commonly used DL-FSCW with parameters $Q_s = 12$, $z = 2$, and $n_c = 2$. The corresponding star-delta winding retains the same number of stator slots Q_s and zones z, but uses $n_{c\Delta} = n_{cY} = 1$

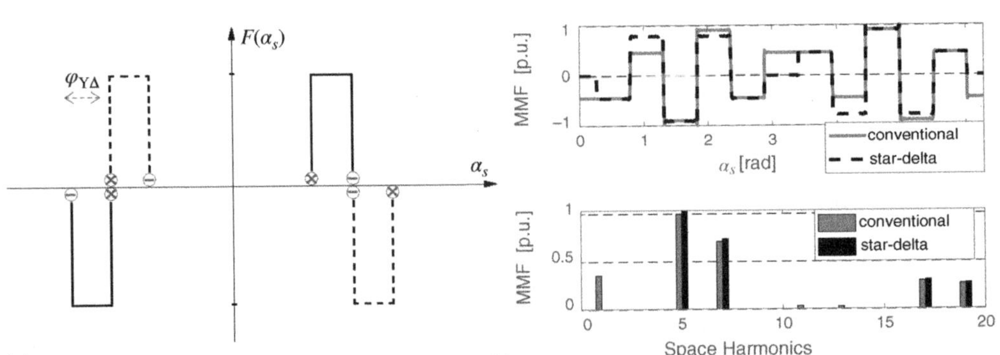

(a) **(b)**

Figure 8.10 (a) Winding layout (one phase) for the star-delta FSCW with $Q_s = 12, z = 2, n_{c\Delta} = n_{cY} = 1$, and $\varphi_{Y\Delta} = \pi/6$, and with opposite polarity winding sets. (b) Comparison of MMF characteristics with the conventional FSCW with $Q_s = 12, z = 2$, and $n_c = 2$.

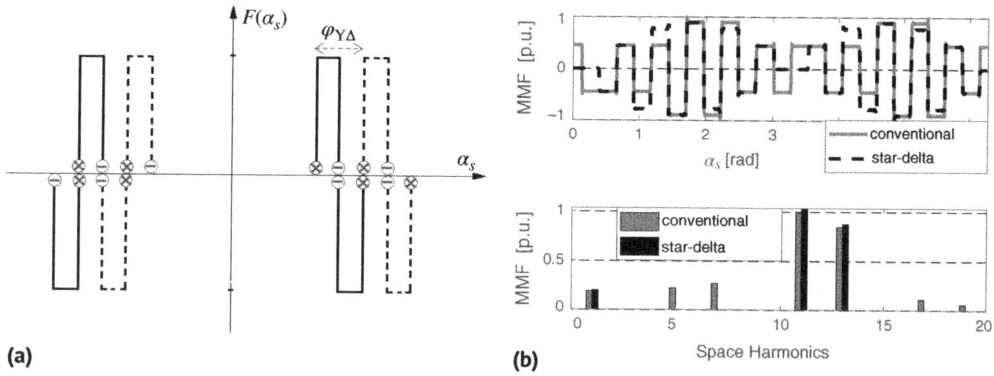

Figure 8.11 (a) Winding layout (one phase) for the star-delta FSCW with $Q_s = 24, z = 2, n_{c\Delta} = n_{cY} = 2$, $\varphi_{Y\Delta} = \pi/6$, and with same polarity winding sets. (b) Comparison of MMF characteristics with the conventional FSCW with $Q_s = 24, z = 2$, and $n_c = 4$.

coils per zone in both winding sets. This configuration represents a symmetric balanced star-delta winding with $\varphi_{Y\Delta} = \pi/6$ and opposite winding set polarity. A visual representation of the winding layout (one phase) is shown in Figure 8.10a. Notably, the star-delta configuration completely eliminates the first subharmonic while increasing the working harmonic by approximately 3.5%.

Example 8.2 Star-delta FSCW with $Q_s = 24$, $z = 2$, and $n_{c\Delta} = n_{cY} = 2$

Figure 8.11 examines the second example. The reference winding has $Q_s = 24$, $z = 2$, and $n_c = 4$, whereas the corresponding star-delta winding is configured with $n_{c\Delta} = n_{cY} = 2$, $\varphi_{Y\Delta} = \pi/6$, and winding sets with the same polarity, as illustrated in Figure 8.11a. In this case, the star-delta configuration improves the working harmonic by approximately 3.5% and completely cancels the 5th and 7th subharmonics and higher-order MMF harmonics.

8.3.1.2 Balanced Star-delta Asymmetric FSCWs

In addition to the two-zone windings listed in Table 8.2, which are classified as balanced and symmetric windings, certain one-zone FSCWs in Table 8.3 can also be arranged as a balanced star-delta configuration. These windings are characterized by a spatial displacement angle of $\pi/3$ between the winding sets and therefore belong to the asymmetric winding group. For

$$n_{c\Delta} = n_{cY} = n_c'$$

and

$$\varphi_{Y\Delta} = \frac{\pi}{3}$$

the equivalent winding factor (8.16) and phase shift angle (8.17) reduce to:

$$^{v}\xi_{w,eq} = \begin{cases} \sin\left((2v-1)\dfrac{\pi}{12}\right) \cdot {}^{v}\xi_w', & \text{opposite polarity winding sets} \\ \cos\left((2v-1)\dfrac{\pi}{12}\right) \cdot {}^{v}\xi_w', & \text{same polarity winding sets} \end{cases} \tag{8.48}$$

$$^{v}\varphi_{1,2} = (2v-1)\frac{\pi}{12} \tag{8.49}$$

The resulting MMF function for this winding category is:

$$F(\alpha_S, t) = \frac{2mzn_c'N_w'\hat{i}}{\pi} \sum_{\nu} \frac{^{\nu}\xi_{w,\text{eq}}}{\nu} \sin\left(\omega t - \nu\alpha_S - (2\nu - 1)\frac{\pi}{12}\right) \tag{8.50}$$

The impact of the star-delta connection on MMF harmonics is described by the equivalent winding factor in Equation (8.48). As in the symmetric case, the reduced number of coils per zone (i.e. $n_{c\Delta} = n_{cY} < n_c$) results in a higher winding factor than that of the corresponding conventional windings. However, unlike the symmetric case, the associated terms in (8.48), in particular:

$$\begin{aligned} &\sin\left((2\nu - 1)\frac{\pi}{12}\right) = 0 \\ &\Rightarrow (2\nu - 1)\frac{\pi}{12} = k\pi \\ &\Rightarrow \nu = 6k + \frac{1}{2} \end{aligned} \tag{8.51}$$

and

$$\begin{aligned} &\cos\left((2\nu - 1)\frac{\pi}{12}\right) = 0 \\ &\Rightarrow (2\nu - 1)\frac{\pi}{12} = k\frac{\pi}{2} \\ &\Rightarrow \nu = 3k + \frac{1}{2} \end{aligned} \tag{8.52}$$

indicate that no MMF harmonic is completely cancelled. Nevertheless, since the sin and cos terms for certain harmonic orders are notably smaller than unity, i.e.,

$$\sin\left((2\nu - 1)\frac{\pi}{12}\right) \ll 1 \quad (\text{e.g. for } \nu = 1)$$

or

$$\cos\left((2\nu - 1)\frac{\pi}{12}\right) \ll 1 \quad (\text{e.g. for } \nu = -2)$$

applying the star-delta technique to this winding group leads to a substantial reduction of specific MMF harmonics, and with this to a considerable improvement of the overall winding performance, as demonstrated in the next examples and summarized in Tables 8.5 to 8.11.

Example 8.3 Star-delta FSCW with $Q_s = 6$, $z = 1$, and $n_{c\Delta} = n_{cY} = 1$
Figure 8.12 examines a star-delta winding with $Q_s = 6$, $z = 1$, $n_{c\Delta} = n_{cY} = 1$, and $\varphi_{Y\Delta} = \pi/3$, as well as opposite winding set polarity. The reference winding is the conventional FSCW with $Q_s = 6$, $z = 1$, and $n_c = 2$. The results indicate that the star-delta combination increases the working harmonic (i.e. the 2nd harmonic) by approximately 11% and at the same time reduces the 1st subharmonic by more than 50%.

Example 8.4 Star-delta FSCW with $Q_s = 12$, $z = 1$, and $n_{c\Delta} = n_{cY} = 2$
Moving to Figure 8.13, the second example considers a winding topology with $Q_s = 12$, $z = 1$, $n_{c\Delta} = n_{cY} = 2$, $\varphi_{Y\Delta} = \pi/3$, but with the same winding set polarity. Here, the reference winding is the conventional FSCW with $Q_s = 12$, $z = 1$, and $n_c = 4$. As in the previous example, it can be observed that the star-delta combination increases the working harmonic (the 5th harmonic) by approximately 11%, while simultaneously reducing the 2nd and the 4th subharmonics by more than 50%.

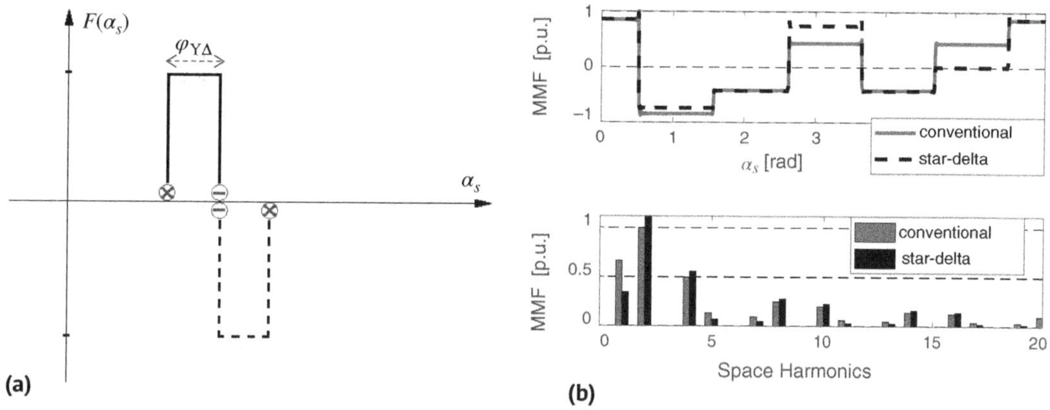

Figure 8.12 (a) Winding layout (one phase) for the star-delta FSCW with $Q_s = 6, z = 1, n_{c\Delta} = n_{cY} = 1$, $\varphi_{Y\Delta} = \pi/3$, and with opposite polarity winding sets. (b) Comparison of MMF characteristics with the conventional FSCW with $Q_s = 6, z = 1, n_c = 2$.

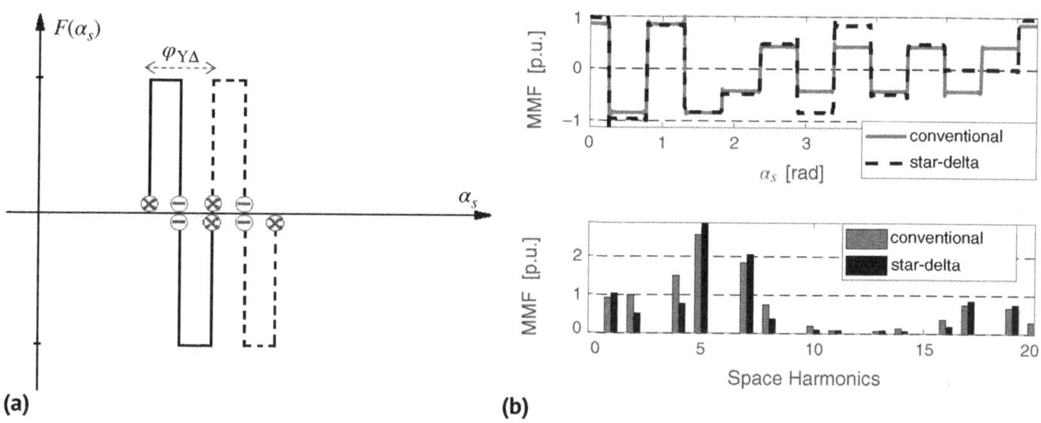

Figure 8.13 (a) Winding layout (one phase) for the star-delta FSCW with $Q_s = 12, z = 1, n_{c\Delta} = n_{cY} = 2$, $\varphi_{Y\Delta} = \pi/3$, and with the same polarity winding sets. (b) Comparison of MMF characteristics with the conventional FSCW with $Q_s = 12, z = 1, n_c = 4$.

Table 8.5 Winding factor (in %) for different balanced star-delta symmetric FSCWs. Comparison of results for conventional (rows in white) and star-delta (rows in gray) windings.

$n_{c\Delta}, n_{cY}$	Q_s	$v = 1$	2	3	4	5	6	7	8	9	10	11	12	13	14	15	16	17
1.1	12	6.7	—	—	—	**93.3**	—	**93.3**	—	—	—	6.7	—	6.7	—	—	—	**93.3**
		0.0	—	—	—	**96.6**	—	**96.6**	—	—	—	0	—	0	—	—	—	**96.6**
2.2	24	1.6	—	—	—	9.68	—	16.3	—	—	—	**94.9**	—	**94.9**	—	—	—	16.2
		1.7	—	—	—	0.0	—	0.0	—	—	—	**98.3**	—	**98.3**	—	—	—	0
3.3	36	0.73	—	—	—	3.9	—	5.8	—	—	—	11.9	—	17.9	—	—	—	**95.3**
		0.0	—	—	—	4.0	—	6.0	—	—	—	0.0	—	0	—	—	—	**98.6**

Table 8.6 Winding factor (in %) for different balanced star-delta asymmetric FSCWs. Comparison of results for conventional (rows in white) and star-delta (rows in gray) windings.

$n_{c\Delta}, n_{cY}$	Q_S	$v=1$	2	3	4	5	6	7	8	9	10	11	12	13	14	15	16
1.1	6	25.0	**75.0**	—	**75.0**	25.0	—	25.0	75.0	—	75.0	25.0	—	25.0	75.0	—	75.0
		12.9	**83.6**	—	**83.6**	12.9	—	12.9	83.6	—	83.6	12.9	—	12.9	83.6	—	83.6
2.2	18	5.8	12.5	—	37.5	**80.8**	—	**80.8**	37.5	—	12.5	5.8	—	5.8	12.5	—	37.5
		6.5	6.5	—	19.4	**90.1**	—	**90.1**	19.4	—	6.5	6.5	—	6.5	6.5	—	19.4
3.3	18	2.5	5.2	—	12.1	17.2	—	39.6	**81.8**	—	**81.8**	39.6	—	17.2	12.1	—	5.2
		1.3	5.8	—	13.5	8.9	—	20.5	**91.3**	—	**91.3**	20.5	—	8.9	13.5	—	5.8

8.3.2 Unbalanced Star-delta FSCWs

In an unbalanced star-delta winding, the number of coils per zone differs between the star and delta winding sets, as expressed by:

$$n_{c\Delta} \neq n_{cY} \tag{8.53}$$

As indicated by Equations (8.27) and (8.28), this unequal coil distribution leads to different winding factors for the two winding sets, i.e.,

$$^{v}\xi_{w\Delta} \neq {}^{v}\xi_{wY} \tag{8.54}$$

Unbalanced star-delta windings may be realized in either symmetric or asymmetric configurations, as summarized by the winding topologies listed in Tables 8.3 and 8.4. The spatial displacement angle between the winding sets can be:

$$\varphi_{Y\Delta} = \frac{\pi}{6}$$

or

$$\varphi_{Y\Delta} \neq \frac{\pi}{6}$$

In contrast to balanced star-delta windings, asymmetric configurations (i.e. $\varphi_{Y\Delta} \neq \pi/6$) allow for a wide range of spatial displacement angles, including values such as $\pi/24$, $\pi/18$, $\pi/15$, and others. Furthermore, the spatial displacement angle can be either positive or negative, as illustrated in Figure 8.14 and defined in Equation (8.55) or (8.56). Both cases may lead to improved MMF distribution.

$$\varphi_{Y\Delta} = n_{c\Delta}\frac{2\pi}{Q_S} \tag{8.55}$$

$$\varphi_{Y\Delta} = -n_{cY}\frac{2\pi}{Q_S} \tag{8.56}$$

However, it should be noted that, unlike balanced star-delta windings, the MMF functions and winding factors of unbalanced configurations generally cannot be expressed in simple closed form. Instead, they can be evaluated directly using the analytical formulae provided in Equations (8.15) and (8.16). The resulting MMF characteristics for various one- and also two-zone FSCWs are presented in Figures 8.15–8.21 and summarized in Tables 8.7 and 8.8.

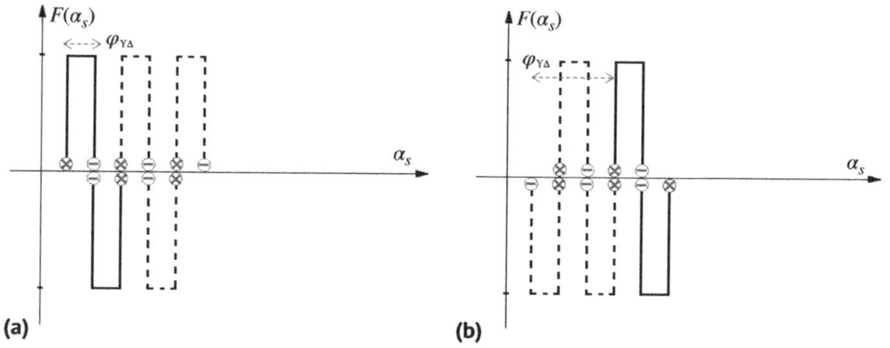

Figure 8.14 Winding layout (one phase) for the star-delta FSCW with $Q_s = 15, z = 1, n_{c\Delta} = 2$, and $n_{cY} = 3$. Solid line for delta winding and dashed line for star winding. (a) Winding configuration with $\varphi_{Y\Delta} = n_{c\Delta} \cdot 2\pi/15$. (b) Winding configuration with $\varphi_{Y\Delta} = -n_{cY} \cdot 2\pi/15$.

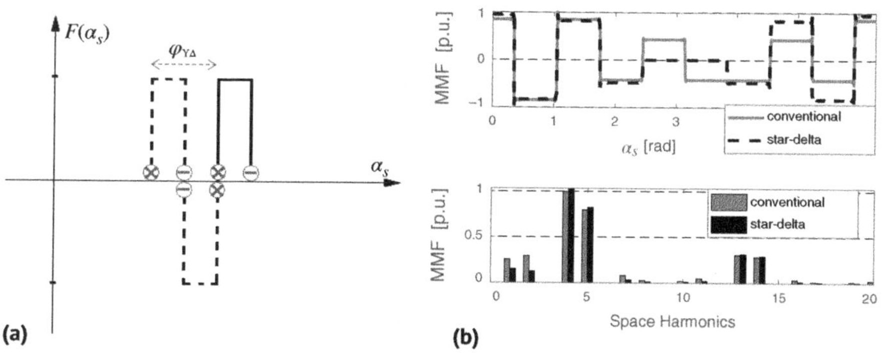

Figure 8.15 (a) Winding layout (one phase) for the star-delta FSCW with $Q_s = 9, z = 1, n_{c\Delta} = 1, n_{cY} = 2$, $\varphi_{Y\Delta} = -4\pi/9$, and with same polarity winding sets. (b) Comparison of MMF characteristics with the conventional FSCW with $Q_s = 9, z = 1$, and $n_c = 3$.

8.3.2.1 One-zone Star-delta FSCWs

Example 8.5 Star-delta FSCW with $Q_s = 9$, $z = 1$, $n_{c\Delta} = 1$, and $n_{cY} = 2$

The first examined winding is the FSCW with the following parameters: $Q_s = 9$, $z = 1$, and $n_c = 3$. In the star-delta combination shown in Figure 8.15a, the winding arrangement consists of $n_{c\Delta} = 1$ and $n_{cY} = 2$, with both winding sets having the same polarity. The shift angle between the winding sets is chosen to be negative, specifically $\varphi_{Y\Delta} = -4\pi/9$, that results from (8.56). A comparison of the MMF characteristics in Figure 8.15b shows a clear reduction of subharmonic components, accompanied by a slight increase in the working harmonics, namely the 4th and the 5th order components.

Example 8.6 Star-delta FSCW with $Q_s = 12$, $z = 1$, $n_{c\Delta} = 1$, and $n_{cY} = 3$

The next example considers the FSCW with $Q_s = 12$, $z = 1$, and $n_c = 4$. The corresponding star-delta winding layout, illustrated in Figure 8.16a, employs $n_{c\Delta} = 1$ and $n_{cY} = 3$, with the winding sets having opposite polarity. In this case, improved MMF characteristics are obtained by selecting a positive spatial displacement angle of $\varphi_{Y\Delta} = \pi/6$. The resulting MMF comparison is shown in Figure 8.16b.

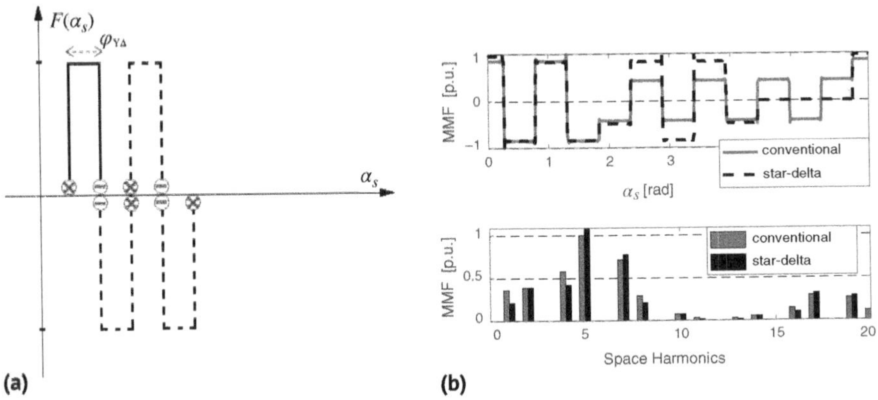

Figure 8.16 (a) Winding layout (one phase) for the star-delta FSCW with $Q_s = 12, z = 1, n_{c\Delta} = 1, n_{cY} = 3$, and $\varphi_{Y\Delta} = \pi/6$, and with opposite polarity winding sets. (b) Comparison of MMF characteristics with the conventional FSCW with $Q_s = 12, z = 1$, and $n_c = 4$.

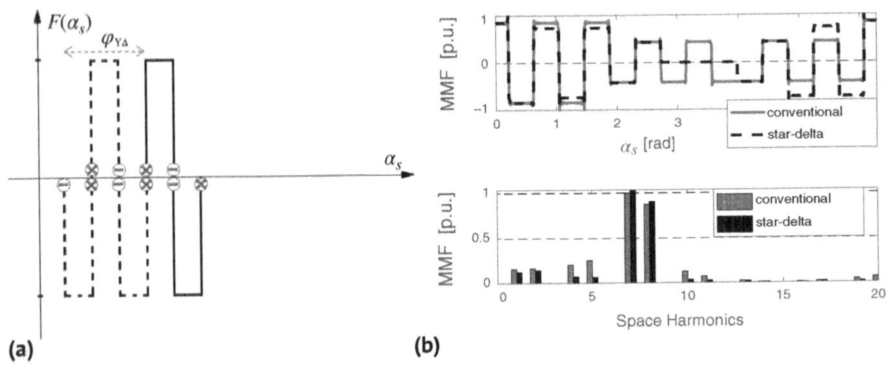

Figure 8.17 (a) Winding layout (one phase) for the star-delta FSCW with $Q_s = 15, z = 1, n_{c\Delta} = 2, n_{cY} = 3$, $\varphi_{Y\Delta} = -6\pi/15$, and with same polarity winding sets. (b) Comparison of MMF characteristics with the conventional FSCW with $Q_s = 15, z = 1$, and $n_c = 5$.

Example 8.7 Star-delta FSCW with $Q_s = 15, z = 1, n_{c\Delta} = 2$, and $n_{cY} = 3$

Another FSCW example with $Q_s = 15$, $z = 1$, and $n_c = 5$ is examined in Figure 8.17. The corresponding star-delta winding arrangement consists of $n_{c\Delta} = 2$ and $n_{cY} = 3$, and has negative winding set polarity, as illustrated in Figure 8.17a. Similar to the first example, a negative spatial displacement angle of $\varphi_{Y\Delta} = -6\pi/15$ is selected, resulting in a notable improvement of the MMF characteristics.

Example 8.8 Star-delta FSCW with $Q_s = 21, z = 1, n_{c\Delta} = 2(3)$, and $n_{cY} = 5(4)$

Finally, an FSCW with $Q_s = 21, z = 1$, and $n_c = 7$ is examined. For this winding type, two possible star-delta arrangements are studied. In Figure 8.18a, the case with $n_{c\Delta} = 2, n_{cY} = 5, \varphi_{Y\Delta} = -10\pi/21$, and with same winding sets polarity is analyzed. In Figure 8.18b, on the other hand, the alternative configuration with $n_{c\Delta} = 3, n_{cY} = 4, \varphi_{Y\Delta} = -8\pi/21$, and opposite winding sets polarity is examined. Also here, both arrangements lead to improved MMF characteristics compared to the conventional reference winding.

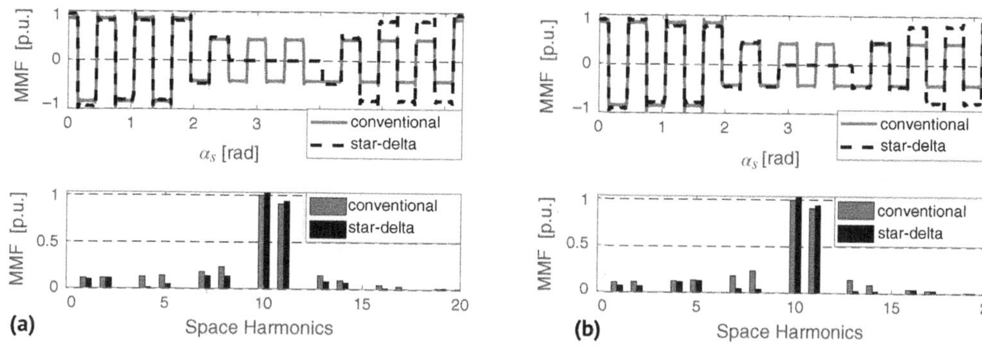

Figure 8.18 Comparison of MMF characteristics with the conventional FSCW with $Q_s = 21, z = 1$, and $n_c = 7$. (a) Star-delta FSCW with $Q_s = 21, z = 1, n_{c\Delta} = 2, n_{cY} = 5$, and $\varphi_{Y\Delta} = -10\pi/21$. (b) Star-delta FSCW with $Q_s = 21, z = 1, n_{c\Delta} = 3, n_{cY} = 4$, and $\varphi_{Y\Delta} = -8\pi/21$.

Table 8.7 Winding factor (in %) for different unbalanced star-delta one-zone FSCWs. Comparison of results for conventional (rows in white) and star-delta (rows in gray) windings.

$n_{c\Delta}, n_{cY}$	Q_s	$v=1$	2	3	4	5	6	7	8	9	10	11	12	13	14	15	16
1,2	9	6.1	13.9	—	**94.5**	**94.5**	—	13.9	6.1	—	6.1	13.9	—	94.5	94.5	—	13.9
		3.6	6.1	—	**97.5**	**97.5**	—	6.1	3.6	—	3.6	6.1	—	97.5	97.5	—	6.1
1,3	12	5.8	12.5	—	**37.5**	**80.8**	—	80.8	37.5	—	12.5	5.8	—	5.8	12.5	—	37.5
		3.3	12.5	—	**26.8**	**87.7**	—	87.7	26.8	—	12.5	3.3	—	3.3	12.5	—	26.8
2,3	15	2.1	4.5	—	11.1	17.3	—	**95.1**	**95.1**	—	17.3	11.1	—	4.5	2.1	—	2.1
		1.7	3.7	—	4.1	4.6	—	**98.4**	**98.4**	—	4.6	4.1	—	3.7	1.7	—	1.7
2,4	18	2.5	5.3	—	12.1	17.2	—	39.6	**81.8**	—	**81.8**	39.6	—	17.2	12.1	—	5.2
		2.8	3.2	—	9.9	14.1	—	24.5	**90.2**	—	**90.2**	24.5	—	14.1	9.9	—	3.2
3,4	21	1.1	2.2	—	4.9	6.6	—	12.4	18.2	—	**95.3**	**95.3**	—	18.2	12.4	—	6.6
		0.7	1.4	—	4.5	6.3	—	3.3	3.5	—	**98.6**	**98.6**	—	3.5	3.3	—	6.3

8.3.2.2 Two-zone Star-delta FSCWs

Example 8.9 Star-delta FSCW with $Q_s = 18$, $z = 2$, $n_{c\Delta} = 1$, and $n_{cY} = 2$

The first investigated two-zone FSCW is the topology with $Qs = 18$, $z = 2$, and $n_c = 3$. The corresponding star-delta winding layout, shown in Figure 8.19a, is characterized by $n_{c\Delta} = 1$, $n_{cY} = 2$, and the same winding set polarity. To enhance the MMF characteristics, a negative spatial displacement angle of $\varphi_{Y\Delta} = -2\pi/9$ is selected. A comparison of results is provided in Figure 8.19b.

Example 8.10 Star-delta FSCW with $Q_s = 24$, $z = 2$, $n_{c\Delta} = 1$, and $n_{cY} = 3$

The next example considers an FSCW with $Q_s = 24$, $z = 2$, and $n_c = 4$. The star-delta winding configuration, illustrated in Figure 8.20a, consists of $n_{c\Delta} = 1$, $n_{cY} = 3$, and negative winding set polarity. In this case, improved MMF characteristics are achieved by selecting a positive spatial displacement angle of $\varphi_{Y\Delta} = \pi/12$. The MMF characteristics and comparison with the reference winding are shown in Figure 8.20b.

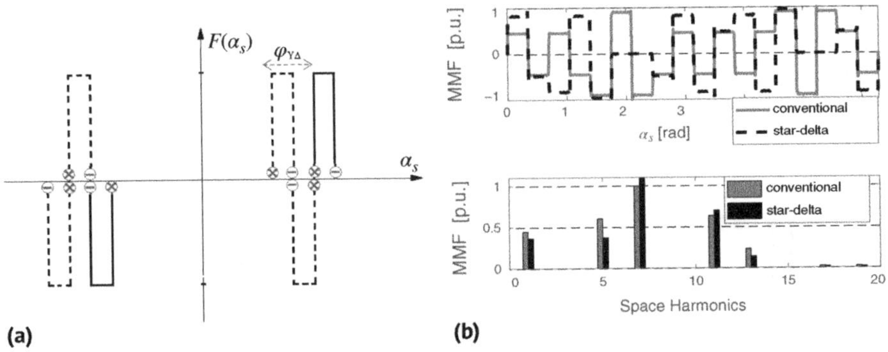

Figure 8.19 (a) Winding layout (one phase) for the star-delta FSCW with $Q_s = 18, z = 2, n_{c\Delta} = 1, n_{cY} = 2$, $\varphi_{Y\Delta} = -2\pi/9$, and with the same polarity winding sets. (b) Comparison of MMF characteristics with the conventional FSCW with $Q_s = 18, z = 2$, and $n_c = 3$.

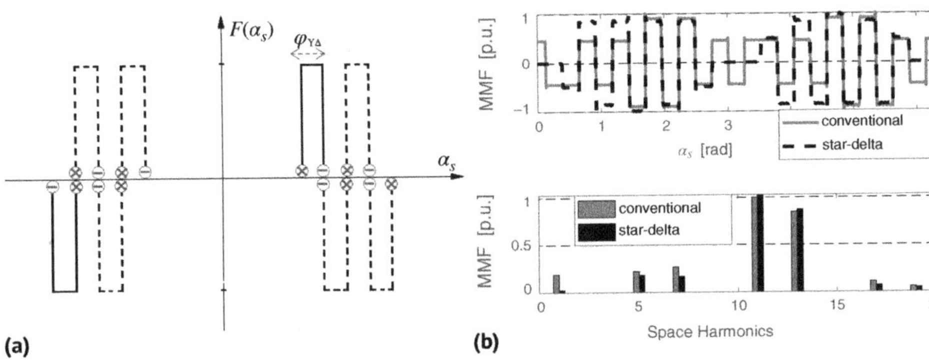

Figure 8.20 (a) Winding layout (one phase) for the star-delta FSCW with $Q_s = 24, z = 2, n_{c\Delta} = 1, n_{cY} = 3$, $\varphi_{Y\Delta} = \pi/12$, and with negative polarity winding sets. (b) Comparison of MMF characteristics with the conventional FSCW with $Q_s = 24, z = 2$, and $n_c = 4$.

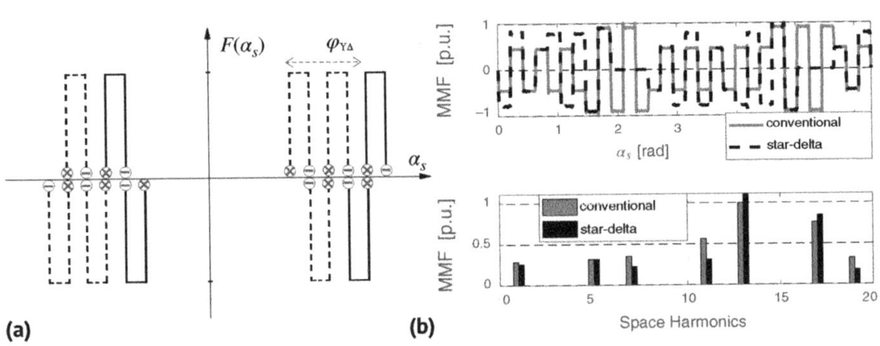

Figure 8.21 (a) Winding layout (one phase) for the star-delta FSCW with $Q_s = 30, z = 2, n_{c\Delta} = 2, n_{cY} = 3$, $\varphi_{Y\Delta} = -\pi/5$, and with opposite winding sets. (b) Comparison of MMF characteristics with the conventional FSCW with $Q_s = 30, z = 2$, and $n_c = 5$.

Example 8.11 Star-delta FSCW with $Q_s = 30$, $z = 2$, $n_{c\Delta} = 2$, and $n_{cY} = 3$
Similar trends are observed from the analysis of the FSCW with $Q_s = 30$, $z = 2$, and $n_c = 5$. The corresponding star-delta winding layout and its MMF properties are presented in Figure 8.21, which again confirms the effectiveness of this method for improving the MMF spectrum.

Table 8.8 Winding factor (in %) for different unbalanced star-delta two-zone FSCWs. Comparison of results for conventional (rows in white) and star-delta (rows in gray) windings.

$n_{c\Delta}, n_{cY}$	Q_S	$\nu=1$	2	3	4	5	6	7	8	9	10	11	12	13	14	15	16	17
1,2	18	5.1	—	—	—	34.4	—	**79.3**	—	—	—	**79.3**	—	34.4	—	—	—	5.1
		4.2	—	—	—	21.3	—	**87.4**	—	—	—	**87.4**	—	21.3	—	—	—	4.2
1,3	24	1.6	—	—	—	9.6	—	16.3	—	—	—	**94.9**	—	**94.9**	—	—	—	16.3
		0.2	—	—	—	7.9	—	10.3	—	—	—	**97.5**	—	**97.5**	—	—	—	10.3
2,3	30	1.8	—	—	—	10.0	—	15.6	—	—	—	38.9	—	**81.5**	—	—	—	**81.5**
		1.6	—	—	—	10.0	—	10.2	—	—	—	21.6	—	**90.5**	—	—	—	**90.5**

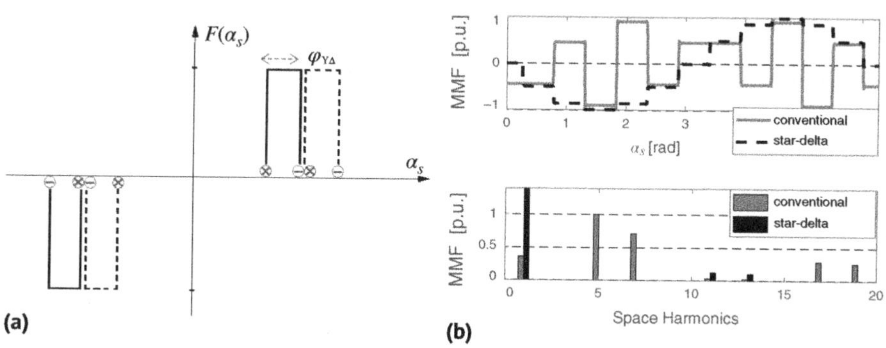

Figure 8.22 Star-delta winding with inversed coil polarity for star-connected winding set. (a) Winding layout (one phase) for an FSCW with $Q_S = 12, z = 2, n_{c\Delta} = n_{cY} = 1$, and $\varphi_{Y\Delta} = \pi/6$. (b) Comparison of MMF characteristics with the conventional FSCW with $Q_S = 12, z = 2$, and $n_c = 2$.

8.3.3 Star-delta FSCWs with Sinusoidal MMF Distribution

In the previously discussed winding examples, the transformation from a standard FSCW to a star-delta combination involves dividing the original winding (the phase zone) into two sets, while maintaining the coil polarity consistent with the original winding. However, if the coil polarity of one set is reversed (e.g. in the star winding), this modification in certain cases may result in a high-quality MMF distribution.

To illustrate this effect, a typical FSCW with $Q_s = 12$, $z = 2$, and $n_c = 2$ is considered. The standard star-delta winding arrangement for this topology was presented earlier in Figure 8.10. An alternative star-delta configuration is obtained by reversing the polarity of one winding set, for example, the star-connected winding, as illustrated in Figure 8.22a. Since both winding sets now share the same polarity, the corresponding MMF waveform can be determined using Equation (8.43). The equivalent winding factor, defined in Equation (8.40), is then given by:

$$^{\nu}\xi_{w,\mathrm{eq}} = \cos\left((\nu - 1)\frac{\pi}{12}\right) \cdot {}^{\nu}\xi'_{w} \tag{8.57}$$

The MMF results for this example are presented in Figure 8.22b and summarized in Table 8.9. It is important to note that this particular winding combination produces an almost sinusoidal MMF waveform that is very similar to that of a distributed winding with $q = 2$. The main drawback of this topology, however, is the reduced winding factor for the operating harmonic, particularly the

Table 8.9 Comparison of winding factors.

	Star-delta winding			Conventional
ν	$\cos((\nu-1)\pi/12)$	${}^{\nu}\xi'_w$ (%)	${}^{\nu}\xi_{w,\mathrm{eq}}$ (%)	${}^{\nu}\xi'_w$ (%)
+1	1	25.88	**25.88**	6.70
5	0	96.6	0	**93.3**
+7	0	96.6	0	93.3
−11	1	25.88	25.88	6.70
+13	1	25.88	25.88	6.70

first harmonic. Nevertheless, its significantly shorter end-winding length and simple structure, compared to distributed windings, make this particular star-delta configuration with tooth-concentrated coils an attractive alternative for applications where the available axial space is limited.

8.3.4 Concluding Remarks

The preceding analysis shows that the star-delta technique is a practical and effective method for a wide range of FSCW topologies. The MMF characteristics are strongly influenced by the choice of coil-group parameters and their spatial displacement. It follows that:

- Depending on the selected number of coils per zone in each winding set, star-delta FSCWs may be classified as balanced ($n_{c\Delta} = n_{cY}$) or unbalanced ($n_{c\Delta} \neq n_{cY}$).
- If the phase-current shift angle and the spatial displacement between the winding sets are equal (i.e. $\varphi_{Y\Delta} = \pi/6$), the winding is referred to as a symmetric star-delta winding. Otherwise, it is asymmetric.
- In a star-delta connection, the spatial displacement angle $\varphi_{Y\Delta}$ may assume either positive or negative values; however, only certain values lead to optimal MMF performance.
- In the balanced and symmetric star-delta windings, a large number of lower- and higher-order MMF harmonics are reduced, while the operating harmonic is enhanced. In this case, the optimal spatial displacement angle is always positive (i.e. $\varphi_{Y\Delta} > 0$).
- In asymmetric star-delta windings, both lower- and higher-order MMF harmonics are also significantly reduced, and the operating harmonic remains enhanced. The optimal spatial displacement angle is again positive (i.e. $\varphi_{Y\Delta} > 0$).
- For unbalanced star-delta windings, the technique remains effective in suppressing MMF unwanted harmonics and enhancing the working harmonic. In contrast to balanced configurations, both positive and negative spatial displacement angles may lead to improved MMF performance.

8.4 FSCWs with Sequential Star-delta Coil Arrangements

In addition to the standard star-delta windings, which consist of two coil-group sequences, one belonging to the star-connected winding and the other to the delta-connected winding, Dajaku and Gerling [116] proposed an alternative coil arrangement consisting of star and delta coil sequences (e.g. $U_{1,\Delta}$ $U_{1,Y}$ $U_{2,\Delta}$ $U_{2,Y}$, ...), as illustrated in Figure 8.23. In this approach, additional wound teeth are initially inserted between adjacent coil pairs within a zone of the reference winding during the star-delta conversion process. Then, the original coils are connected in delta, while the newly added coils are in star. Two illustrative examples of this transformation are provided in Figures 8.24 and 8.25.

Compared with the original winding layout, the total number of stator slots is increased according to:

$$Q_{SY} = m \cdot z \cdot n_{cY} \tag{8.58}$$

that corresponds to the slots associated with the star-connected winding coils.

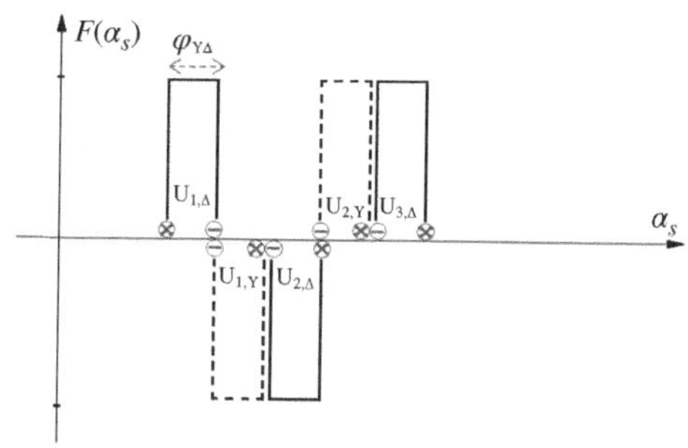

Figure 8.23 Star-delta combination with sequential star and delta coil arrangement and with winding parameters $Q_s = 15, z = 1, n_{c\Delta} = 3$, and $n_{cY} = 2$. Solid line for delta winding and dashed line for star winding.

Similar to conventional star-delta windings, the number of coils per zone in the star and delta winding sets may be either the same or different, as defined in Equation (8.59).

$$\begin{cases} n_{c\Delta} = n_{cY} = \dfrac{n_c}{2}, & \text{if } n_c \text{ even} \\ n_{c\Delta} = n_{cY} \pm 1, & \text{if } n_c \text{ odd} \end{cases} \tag{8.59}$$

From an MMF standpoint, this configuration is particularly effective when the winding sets satisfy the conditions:

$$n_{c\Delta} = n_{cY} + 1 \tag{8.60}$$

and

$$\angle(i_Y, i_\Delta) = \frac{\pi}{2} \tag{8.61}$$

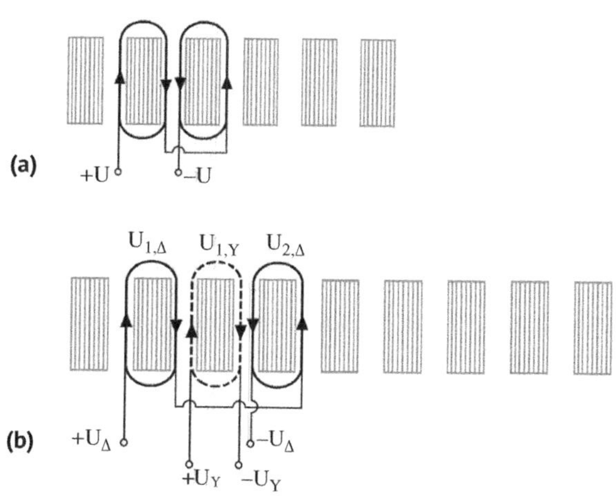

Figure 8.24 (a) Conventional FSCW with $Q_s = 6, z = 1$, and $n_c = 2$. (b) Star-delta FSCW with sequential star and delta coils and with winding parameters $Q_s = 9, z = 1, n_{c\Delta} = 2$, and $n_{cY} = 1$.

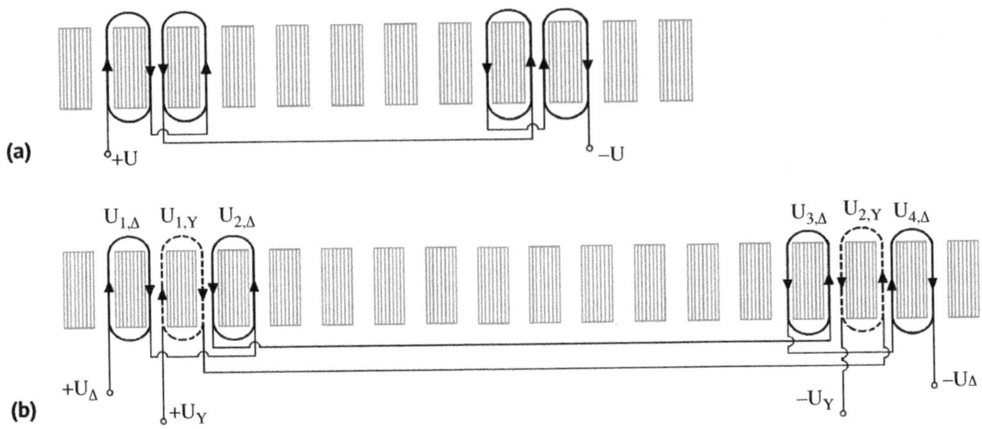

Figure 8.25 (a) Conventional FSCW with $Q_s = 12, z = 2$ and $n_c = 2$. (b) Star-delta FSCW with sequential star and delta coils and with winding parameters $Q_s = 18, z = 2, n_{c\Delta} = 2$, and $n_{cY} = 1$.

Additionally, as shown in Figure 8.23, the spatial displacement between winding sets is always equal to one slot pitch, i.e.,

$$\varphi_{Y\Delta} = \frac{2\pi}{Q_S} \tag{8.62}$$

Furthermore, with regard to the winding set polarity, both polarity arrangements can be applied, depending on the specific design objectives. The example in Figure 2.23 corresponds to the case of opposite polarity.

8.4.1 MMF Winding Function

In order to fulfill the condition defined in Equation (8.61), the star-delta combination illustrated in Figure 8.26 is required. According to Kirchhoff's law, the phase currents in the star-connected winding are given by:

$$\begin{aligned}
i_{u,Y}(t) &= i_{w,\Delta}(t) - i_{v,\Delta}(t) = \sqrt{3}\hat{i} \cdot \cos\left(\omega t + \frac{\pi}{2}\right)\\
i_{v,Y}(t) &= i_{u,\Delta}(t) - i_{w,\Delta}(t) = \sqrt{3}\hat{i} \cdot \cos\left(\omega t + \frac{\pi}{2} - \frac{2\pi}{3}\right)\\
i_{w,Y}(t) &= i_{v,\Delta}(t) - i_{u,\Delta}(t) = \sqrt{3}\hat{i} \cdot \cos\left(\omega t + \frac{\pi}{2} + \frac{2\pi}{3}\right)
\end{aligned} \tag{8.63}$$

The MMF expressions for the individual winding sets are analogous to those presented in Equations (8.7) and (8.8). Accordingly, when the condition:

$$\angle(i_Y, i_\Delta) = \pi/2$$

is fulfilled, the MMF contributions can be expressed as:

$$F_\Delta(\alpha_S, t) = \frac{mzn_{c\Delta}N_{w\Delta}\hat{i}_\Delta}{\pi} \sum_\nu \frac{{}^\nu\xi_{w\Delta}}{\nu} \sin\left(\omega t - \nu\alpha_S + {}^\nu\vartheta_\Delta\right) \tag{8.64}$$

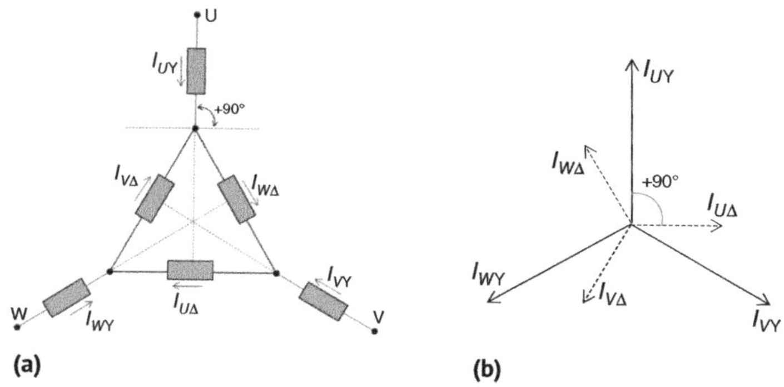

Figure 8.26 (a) Star-delta connection for $\pi/2$ phase shift case. (b) Respective current phasors for the three-phase star-delta-connected winding.

$$F_{\mathrm{Y}}(\alpha_S, t) = \frac{m z n_{c\mathrm{Y}} N_{w\mathrm{Y}} \hat{i}_{\mathrm{Y}}}{\pi} \sum_{\nu} \frac{{}^{\nu}\xi_{w\mathrm{Y}}}{\nu} \sin\left[\left(\omega t + \frac{\pi}{2}\right) - \nu(\alpha_S - \varphi_{\mathrm{Y}\Delta}) + {}^{\nu}\vartheta_{\mathrm{Y}}\right] \tag{8.65}$$

Consequently, the resulting MMF winding function can be derived in closed analytical form as follows:

$$F_{\mathrm{Y}\Delta}(\alpha_S, t) = \frac{m z N_{w,\mathrm{eq}} \hat{i}}{\pi} \sum_{\nu} \frac{{}^{\nu}\xi_{w,\mathrm{eq}}}{\nu} \sin\left(\omega t - \nu\alpha_S + {}^{\nu}\varphi_{1,2}\right) \tag{8.66}$$

$${}^{\nu}\xi_{w,\mathrm{eq}} = \sqrt{\left(N^{*}_{w\Delta}\,{}^{\nu}\xi_{w\Delta}\right)^2 + \left(N^{*}_{w\mathrm{Y}}\,{}^{\nu}\xi_{w\mathrm{Y}}\right)^2 \pm 2N^{*}_{w\Delta}N^{*}_{w\mathrm{Y}}\,{}^{\nu}\xi_{w\Delta}\,{}^{\nu}\xi_{w\mathrm{Y}} \cos\left(\nu\varphi_{\mathrm{Y}\Delta} + \frac{\pi}{2} + \left({}^{\nu}\vartheta_{\mathrm{Y}} - {}^{\nu}\vartheta_{\Delta}\right)\right)} \tag{8.67}$$

$${}^{\nu}\varphi_{1,2} = a\tan\left(\frac{N^{*}_{w\Delta}\,{}^{\nu}\xi_{w\Delta}\sin\left({}^{\nu}\vartheta_{\Delta}\right) \pm N^{*}_{w\mathrm{Y}}\,{}^{\nu}\xi_{w\mathrm{Y}}\sin\left(\nu\varphi_{\mathrm{Y}\Delta} + \frac{\pi}{2} + {}^{\nu}\vartheta_{\mathrm{Y}}\right)}{N^{*}_{w\Delta}\,{}^{\nu}\xi_{w\Delta}\cos\left({}^{\nu}\vartheta_{\Delta}\right) \pm N^{*}_{w\mathrm{Y}}\,{}^{\nu}\xi_{w\mathrm{Y}}\cos\left(\nu\varphi_{\mathrm{Y}\Delta} + \frac{\pi}{2} + {}^{\nu}\vartheta_{\mathrm{Y}}\right)}\right) \tag{8.68}$$

From Figure 8.24b, it is evident that each winding set, when considered separately, resembles a SL-FSCWs. The coresponding winding factors, according to Equations (2.15), (2.28) and (2.31), are therefore:

$${}^{\nu}\xi_{w\Delta} = \sin\left(\nu\frac{\pi}{Q_S}\right) \frac{\sin\left(\nu n_{c\Delta}\frac{2\pi}{Q_S} + n_{c\Delta}\frac{\pi}{2}\right)}{n_{c\Delta}\cdot\cos\left(\nu\frac{2\pi}{Q_S}\right)} \sin\left(\nu\frac{\pi}{2}\right) \tag{8.69}$$

$${}^{\nu}\xi_{w\mathrm{Y}} = \sin\left(\nu\frac{\pi}{Q_S}\right) \frac{\sin\left(\nu n_{c\mathrm{Y}}\frac{2\pi}{Q_S} + n_{c\mathrm{Y}}\frac{\pi}{2}\right)}{n_{c\mathrm{Y}}\cdot\cos\left(\nu\frac{2\pi}{Q_S}\right)} \sin\left(\nu\frac{\pi}{2}\right) \tag{8.70}$$

Whereas the respective parameters ${}^{\nu}\vartheta_{\Delta}$ and ${}^{\nu}\vartheta_{Y}$ in the MMF function are analogous to those defined in (8.9) and (8.10) but adapted for the single-layer case according to Equation (2.27),

$$
\begin{aligned}
{}^{\nu}\vartheta_{\Delta} &= \nu(n_{c\Delta} - 1)\frac{2\pi}{Q_S} + (n_{c\Delta} - 1)\frac{\pi}{2} \\
{}^{\nu}\vartheta_{Y} &= \nu(n_{cY} - 1)\frac{2\pi}{Q_S} + (n_{cY} - 1)\frac{\pi}{2}
\end{aligned}
\tag{8.71}
$$

When the relationship between the number of coils in the winding sets satisfies:

$$n_{c\Delta} = n_{cY} + 1$$

the argument of the trigonometric term in Equations (8.67) reduce to:

$$
\begin{aligned}
&\nu\varphi_{Y\Delta} + \frac{\pi}{2} + \left({}^{\nu}\vartheta_{Y} - {}^{\nu}\vartheta_{\Delta}\right) \\
&= \nu\frac{2\pi}{Q_S} + \frac{\pi}{2} - \nu\frac{2\pi}{Q_S} - \frac{\pi}{2} = 0
\end{aligned}
\tag{8.72}
$$

Consequently, the resulting equivalent winding factor and phase angle simplified to:

$$
{}^{\nu}\xi_{w,eq} = N^{*}_{w\Delta}\,{}^{\nu}\xi_{w\Delta} \pm N^{*}_{wY}\,{}^{\nu}\xi_{wY} = \frac{n_{c\Delta}{}^{\nu}\xi_{w\Delta} \pm (n_{c\Delta} - 1)\,{}^{\nu}\xi_{wY}}{2n_{c\Delta} - 1}
\tag{8.73}
$$

$$
{}^{\nu}\varphi_{1,2} = {}^{\nu}\vartheta_{\Delta} = \nu(n_{c\Delta} - 1)\frac{2\pi}{Q_S} + (n_{c\Delta} - 1)\frac{\pi}{2}
\tag{8.74}
$$

From Equation (8.60), it follows that this winding group belongs to the category of unbalanced star-delta windings,

$$n_{c\Delta} \neq n_{cY}$$

yielding different winding factors for the individual winding sets, i.e.,

$${}^{\nu}\xi_{w\Delta} \neq {}^{\nu}\xi_{wY}$$

Moreover, according to (8.61) and (8.62), the resulting star-delta configuration is also asymmetric, as illustrated in Figure 8.27.

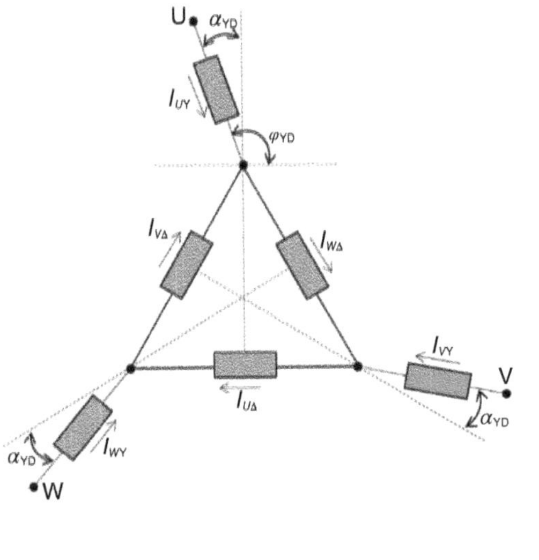

Figure 8.27 Asymmetric star-delta winding connection with $\angle(i_Y, i_\Delta) = \pi/2$ and $\alpha_{Y\Delta} > 0$.

8.4.2 FSCW Examples with Sequential Star-delta Coil Arrangements

8.4.2.1 One-zone FSCWs

Example 8.12 Star-delta FSCW with $Q_s = 9, z = 1, n_{c\Delta} = 2$, and $n_{cY} = 1$

Starting from the conventional FSCW with $Q_s = 6$, $z = 1$, and $n_c = 2$, one additional wound tooth (per zone and per phase) is inserted between the coil pairs of the initial winding, as illustrated in Figure 8.28a. The original coils are connected in delta, while the added coils are in star, resulting in a corresponding star-delta winding with $Q_s = 9$, $z = 1$, $n_{c\Delta} = 2$, and $n_{cY} = n_{c\Delta} - 1 = 1$. The winding sets have the same polarity and a spatial displacement of $\varphi_{Y\Delta} = 2\pi/9$. This transformation approach leads to an effective reduction of both lower- and higher-order harmonics; however, a notable decrease of approximately 15% is also observed in working harmonic. The MMF characteristics for both the reference and resulting star-delta winding are provided in Figure 8.28b and Table 8.10.

Example 8.13 Star-delta FSCW with $Q_s = 15, z = 1, n_{c\Delta} = 3$, and $n_{cY} = 2$

Applying the same procedure to an FSCW with $Q_s = 9$, $z = 1$, and $n_c = 3$ yields a star-delta configuration with $Q_s = 15$, $z = 1$, $n_{c\Delta} = 3$, and $n_{cY} = 2$. In this case, opposite winding set polarity is selected in order to improve the MMF characteristics (Figure 8.29a), whereas, as before, the spatial displacement angle is one-slot pitch, i.e. $\varphi_{Y\Delta} = 2\pi/15$. The MMF comparison in Figure 8.29b confirms the effective suppression of the targeted harmonics, accompanied by only a slight reduction in operating harmonics.

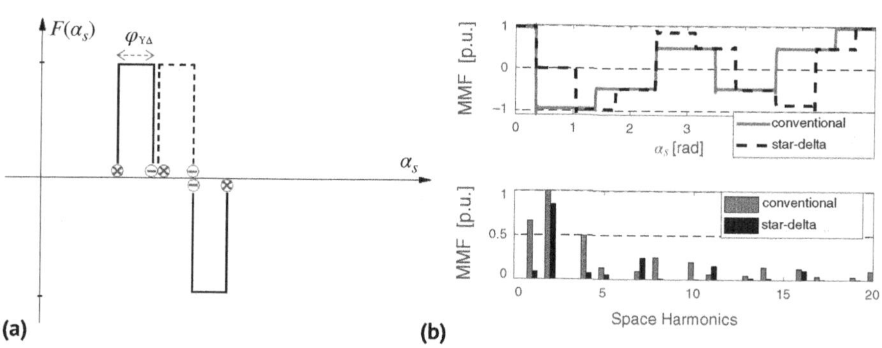

Figure 8.28 (a) Winding layout (one phase) for the star-delta FSCW with $Q_s = 9, z = 1, n_{c\Delta} = 2, n_{cY} = 1$, $\varphi_{Y\Delta} = 2\pi/9$, and with same polarity winding sets. (b) Comparison of MMF characteristics with the conventional FSCW with $Q_s = 6, z = 1$, and $n_c = 2$.

Table 8.10 Winding factor (in %) for different sequential star-delta coils one-zone FSCWs. Comparison of the results for conventional (rows in white) and corresponding star-delta (rows in gray) windings.

$n_{c\Delta}, n_{cY}$	Q_S	$v = 1$	2	3	4	5	6	7	8	9	10	11	12	13	14	15	16
2,1	6	25.0	**75.0**	—	**75.0**	25.0	—	25.0	75.0	—	75.0	25.0	—	25.0	75.0	—	75.0
	9	3.25	**63.6**	—	10.8	10.8	—	63.6	3.25	—	3.25	63.6	—	10.4	10.4	—	63.6
3,2	9	6.1	13.9	—	**94.5**	**94.5**	—	13.9	6.1	—	6.1	13.9	—	94.5	94.5	—	13.9
	15	1.9	2.3	—	**73.5**	4.6	—	8.2	8.2	—	4.6	73.5	—	2.3	1.9	—	1.9

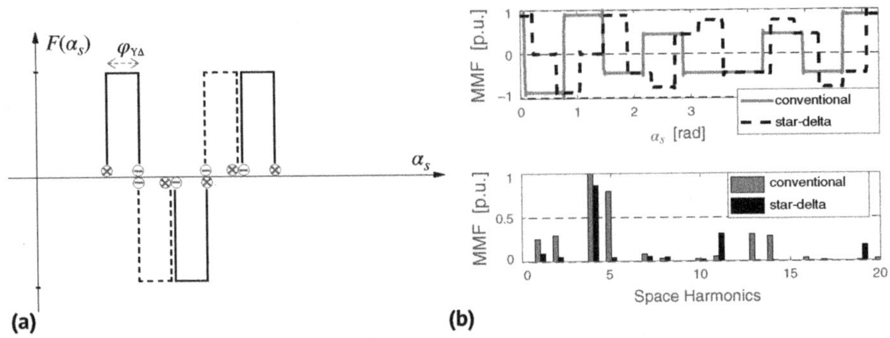

Figure 8.29 (a) Winding layout (one phase) for the star-delta FSCW with $Q_s = 15, z = 1, n_{c\Delta} = 3, n_{cY} = 2$, $\varphi_{Y\Delta} = 2\pi/15$ and with opposite polarity winding sets. (b) Comparison of MMF characteristics with the conventional FSCW with $Q_s = 9, z = 1$, and $n_c = 3$.

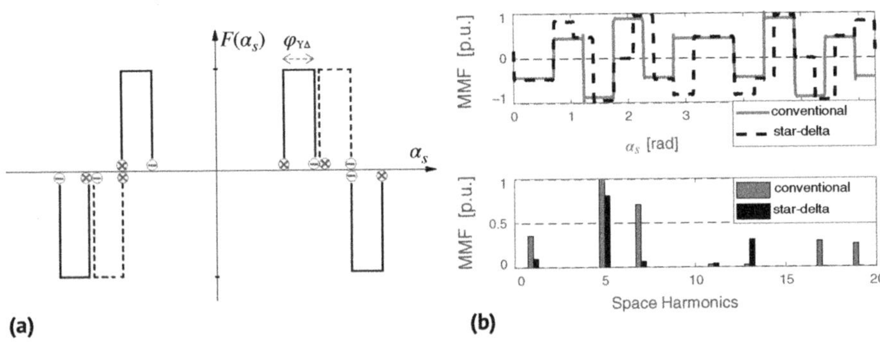

Figure 8.30 (a) Winding layout (one phase) for the star-delta FSCW with $Q_s = 18, z = 2, n_{c\Delta} = 2, n_{cY} = 1$, and $\varphi_{Y\Delta} = \pi/9$, and with the same polarity winding sets. (b) Comparison of MMF characteristics with the conventional FSCW with $Q_s = 12, z = 2$, and $n_c = 2$.

8.4.2.2 Two-zone FSCWs

Example 8.14 Star-delta FSCW with $Qs = 18$, $z = 2$, $n_{c\Delta} = 2$, and $n_{cY} = 1$

Starting from an FSCW with $Qs = 12$, $z = 2$, and $n_c = 2$, the star-delta transformation yields a configuration with $Qs = 18$, $z = 2$, $n_{c\Delta} = 2$, and $n_{cY} = 1$, as illustrated in Figure 8.30a. The winding sets have the same polarity and a spatial displacement angle of $\varphi_{Y\Delta} = \pi/9$. The MMF results in Figure 8.30b demonstrate significant improvements in the reduction of harmonic content. However, similar to the one-zone windings, the working harmonic is also affected.

Example 8.15 Star-delta FSCW with $Q_s = 30$, $z = 2$, $n_{c\Delta} = 3$, and $n_{cY} = 2$

Next, for an FSCW with $Qs = 18$, $z = 2$, and $n_c = 3$, the star-delta transformation leads to a winding with $Q_s = 30$, $z = 2$, $n_{c\Delta} = 3$, and $n_{cY} = 2$. To ensure high-quality MMF characteristics, an opposite winding set polarity was selected. The MMF results shown in Figure 8.31a exhibit a substantial improvement in the harmonic spectrum compared to the reference case.

Example 8.16 Star-delta FSCW with $Q_s = 42$, $z = 2$, $n_{c\Delta} = 4$, and $n_{cY} = 3$

Finally, starting from an FSCW with $Qs = 24$, $z = 2$, and $n_c = 4$, the star-delta transformation yields the winding with $Q_s = 42$, $z = 2$, $n_{c\Delta} = 4$, and $n_{cY} = 3$. As shown in Figure 8.31b and Table 8.11, the

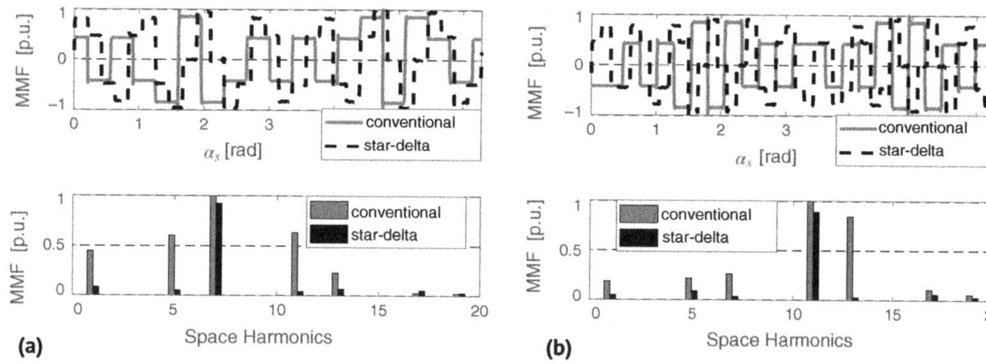

Figure 8.31 Comparison of MMF characteristics. (a) Star-delta FSCW with $Q_s = 30, z = 2, n_{c\Delta} = 3, n_{cY} = 2, \varphi_{Y\Delta} = \pi/15$, and with opposite polarity winding sets; conventional FSCW with $Q_s = 18, z = 2$, and $n_c = 3$. (b) Star-delta FSCW with $Q_s = 42, z = 2, n_{c\Delta} = 4, n_{cY} = 3, \varphi_{Y\Delta} = \pi/21$, and with the same polarity winding sets; conventional FSCW with $Q_s = 24, z = 2$, and $n_c = 4$.

Table 8.11 Winding factor for different sequential star-delta coils two-zone FSCWs. Comparison of the results for conventional (rows in white) and corresponding star-delta (rows in gray) windings.

$n_{c\Delta}$, n_{cY}	Q_S	$v=1$	2	3	4	5	6	7	8	9	10	11	12	13	14	15	16	17
2,1	12	6.7	—	—	—	**93.3**	—	**93.3**	—	—	—	6.7	—	6.7	—	—	—	**93.3**
	18	1.8	—	—	—	**75.8**	—	8.9	—	—	—	8.9	—	75.8	—	—	—	1.8
3,2	18	5.1	—	—	—	34.4	—	**79.3**	—	—	—	**79.3**	—	34.4	—	—	—	5.1
	30	0.8	—	—	—	2.6	—	**66.2**	—	—	—	5.1	—	9.3	—	—	—	9.3
4,3	24	1.6	—	—	—	9.6	—	16.3	—	—	—	**94.9**	—	**94.9**	—	—	—	16.3
	42	0.4	—	—	—	3.4	—	1.9	—	—	—	**72.5**	—	3.1	—	—	—	7.5

use of the star-delta connection with the same winding set polarity significantly reduces the MMF harmonic content, with only slight attenuation of the operating harmonic.

8.4.2.3 Key Observations and Conclusions

Based on the analysis of both one-zone and two-zone windings, the following conclusions can be made:

1. Winding sets polarity:
   - If $n_{c\Delta}$ is even, employing the same winding set polarity generally results in superior MMF performance.
   - If $n_{c\Delta}$ is odd, the opposite polarity configuration is preferred.
2. Harmonic reduction:
   - For both one-zone and two-zone windings, the star-delta transformation proves to be highly effective in reducing both lower- and higher-order MMF harmonics.
   - The improvement in harmonic content is accompanied by a slight reduction in the amplitude of the working harmonic.

8.5 Star-delta FSCWs with a Dual Slot-layer Stator

FSCWs are typically characterized by a non-sinusoidal MMF waveform and a high harmonic content in the MMF spectrum. Consequently, such windings are often unsuitable for certain applications, particularly asynchronous machines, where higher-order space harmonics give rise to parasitic torque components and increased rotor losses. For these machine types, a stator winding with a high-quality sinusoidal MMF waveform is therefore preferred, as introduced in section 8.3.3. However, the main drawback of this configuration is the low winding factor for the working harmonic (approximately 25.5%).

An alternative approach to achieve a near-sinusoidal MMF waveform while maintaining an FSCW topology with improved winding factor is to combine a star-delta winding with a stator arrangement with dual slot layers (Figure 8.32). This concept, initially proposed by Dajaku et al. [118], is illustrated in Figure 8.32a. The configuration consists of two different winding sets, each with unique winding layouts and coil-pitches. The first winding set is placed in the upper stator slot layer, has a shorter coil-pitch of $2\pi/12$, and is defined by the parameters $Q_{S\Delta} = 12$, $z = 2$, and $n_{c\Delta} = 2$. The second winding set is arranged in the lower stator slot layer and has a larger coil-pitch of $2\pi/6$, with parameters $Q_{SY} = 6$, $z = 2$, and $n_{cY} = 1$. In addition, the star and delta winding sets have opposite polarities and are spatially shifted by:

$$\varphi_{Y\Delta} = \frac{\pi}{Q_{S\Delta}} = \frac{\pi}{12} \tag{8.75}$$

In order to achieve a high-quality MMF distribution, the winding sets are connected according to the star-delta combination shown in Figure 8.26, which provides a phase shift of $\pi/2$ (90 degrees) between the two current systems.

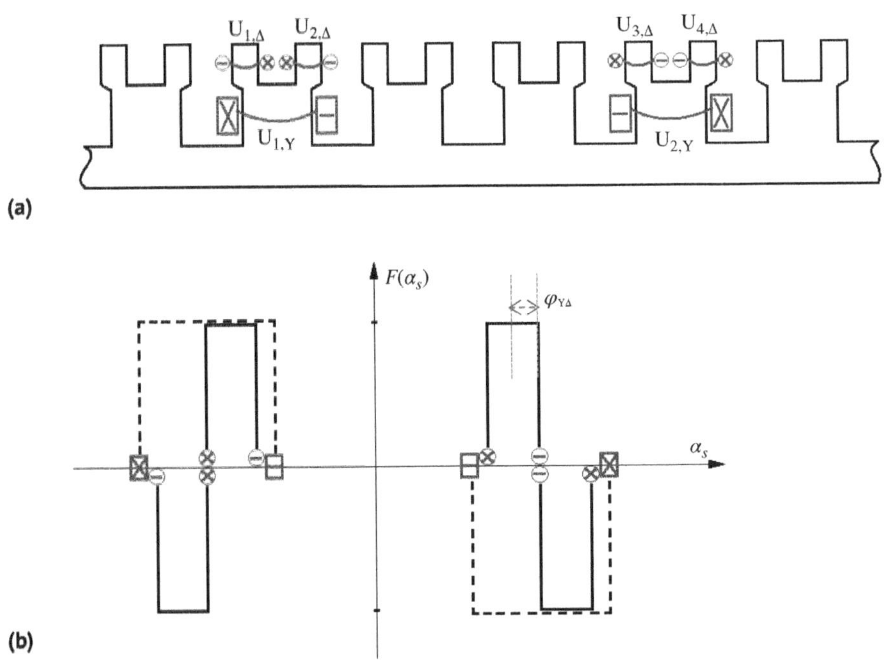

Figure 8.32 Star-delta FSCW with dual stator slot-layer configuration. (a) Winding layout (one phase). (b) Respective MMF distribution (solid line for delta winding and dashed line for star winding).

Based on the winding specifications, i.e.,

$$n_{c\Delta} = 2, \quad Q_{S\Delta} = 12$$

$$n_{cY} = 1, \quad Q_{SY} = 6$$

$$\varphi_{Y\Delta} = \frac{\pi}{12} \text{ and } \angle(i_Y, i_\Delta) = \pi/2$$

the corresponding MMF winding function parameters are given by:

$$^{\nu}\vartheta_\Delta = \nu(n_{c\Delta} - 1)\frac{\pi}{Q_{S\Delta}} + (n_{c\Delta} - 1)\frac{\pi}{2} = \nu\frac{\pi}{12} + \frac{\pi}{2}$$

$$^{\nu}\vartheta_Y = \nu(n_{cY} - 1)\frac{\pi}{Q_{SY}} + (n_{cY} - 1)\frac{\pi}{2} = 0$$

Consequently, using Equations (8.64) and (8.65), the MMF functions for the individual winding sets can be expressed as:

$$F_\Delta(\alpha_S, t) = \frac{2mzN_{w\Delta}\hat{i}_\Delta}{\pi}\sum_\nu \frac{^{\nu}\xi_{w\Delta}}{\nu}\sin\left(\omega t - \nu\alpha_S + \nu\frac{\pi}{12} + \frac{\pi}{2}\right) \tag{8.76}$$

$$F_Y(\alpha_S, t) = \frac{mzN_{wY}\hat{i}_Y}{\pi}\sum_\nu \frac{^{\nu}\xi_{wY}}{\nu}\sin\left[\omega t + \frac{\pi}{2} - \nu\left(\alpha_S - \frac{\pi}{12}\right)\right] \tag{8.77}$$

where, according to Equations (8.32) and (8.33), the respective winding factors are:

$$^{\nu}\xi_{w\Delta} = \sin^2\left(\nu\frac{\pi}{12}\right)\sin\left(\nu\frac{\pi}{2}\right) \tag{8.78}$$

$$^{\nu}\xi_{wY} = \sin\left(\nu\frac{\pi}{6}\right)\sin\left(\nu\frac{\pi}{2}\right) \tag{8.79}$$

The resulting combined MMF function is therefore obtained as:

$$F_{Y\Delta}(\alpha_S, t) = \frac{mzN_{w,\text{eq}}\hat{i}}{\pi}\sum_\nu \frac{^{\nu}\xi_{w,\text{eq}}}{\nu}\sin\left(\omega t - \nu\left(\alpha_S - \frac{\pi}{12}\right) - \frac{\pi}{2}\right) \tag{8.80}$$

where the equivalent winding factor is defined by:

$$^{\nu}\xi_{w,\text{eq}} = N^*_{w\Delta}\,{}^{\nu}\xi_{w\Delta} - N^*_{wY}\,{}^{\nu}\xi_{wY} = \frac{2N_{w\Delta}{}^{\nu}\xi_{w\Delta} - \sqrt{3}N_{wY}{}^{\nu}\xi_{wY}}{2N_{w\Delta} + \sqrt{3}N_{wY}} \tag{8.81}$$

The corresponding number of turns per phase for this particular winding type is defined in Equations (8.79)–(8.81).

$$N_{w,\text{eq}} = \left(n_{c\Delta}N_{w\Delta} + \sqrt{3}n_{cY}N_{wY}\right) = 2N_{w\Delta} + \sqrt{3}N_{wY} \tag{8.82}$$

$$N^*_{w\Delta} = \frac{n_{c\Delta}N_{w\Delta}}{N_{w,\text{eq}}} = \frac{2N_{w\Delta}}{2N_{w\Delta} + \sqrt{3}N_{wY}} \tag{8.83}$$

$$N^*_{wY} = \frac{\sqrt{3}n_{cY}N_{wY}}{N_{w,\text{eq}}} = \frac{\sqrt{3}N_{wY}}{2N_{w\Delta} + \sqrt{3}N_{wY}} \tag{8.84}$$

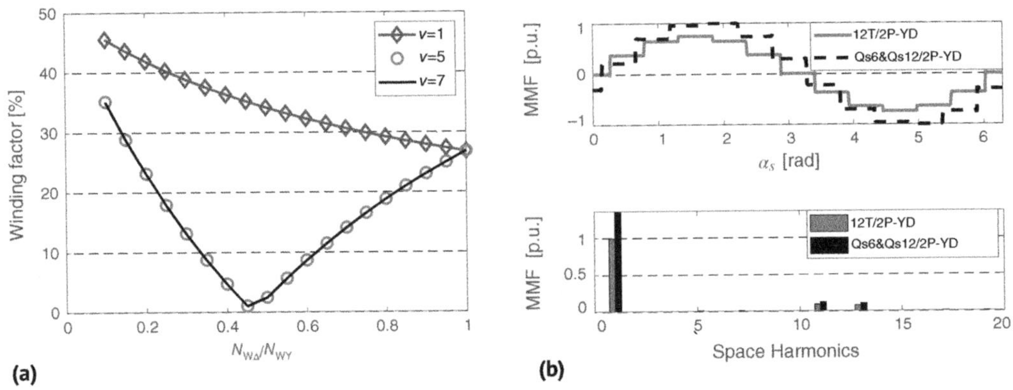

Figure 8.33 Star-delta winding with dual slot-layer stator. (a) Variation of winding factors as function of turns ratio $N_{w\Delta}/N_{wY}$. (b) Comparison of MMF characteristics with the topology in Figure 8.22.

A key feature of this configuration is the ratio between the number of turns of the star- and delta-connected winding sets. In contrast to the winding topologies investigated previously, which are characterize by:

$$N_{w\Delta} = \sqrt{3}N_{wY}$$

the winding factor results presented in Figure 8.33a show that a complete cancellation of the fifth and seventh order MMF harmonics is achieved when the turns ratio satisfies:

$$N_{w\Delta} = (4/9)N_{wY}$$

which is considerably smaller than that of standard star-delta windings.

A comparison of the MMF characteristics between this configuration and the two-poles FSCW topology shown in Figure 8.22 is presented in Figure 8.33b. The results indicate that, while the overall MMF waveform remains similar, the winding factor associated with the fundamental harmonic is approximately 44% higher for the configuration with the dual slot-layer stator.

8.6 Exemplary Star-delta FSCW Machines

As demonstrated in earlier sections, the conventional star-delta transformation is not restricted to a particular FSCW group but can be effectively applied to all FSCW configurations. In addition, several advanced winding concepts have been introduced that further expand the design range of star-delta FSCWs. These include topologies featuring sequential star and delta coil arrangements, as well as configurations combining two winding sets implemented on a dual slot-layer stator.

In the following, a representative set of star-delta FSCW topologies for PM machines is presented and discussed. For each example, the winding layout, key design parameters, and the resulting MMF characteristics are examined in detail.

Example 8.17 6-slot/4-pole star-delta machine
The first example, illustrated in Figure 8.34, considers a machine with six stator slots and four rotor poles, equipped with a standard star-delta winding combination. The winding parameters

are $Q_s = 6$, $z = 1$, $n_{c\Delta} = n_{cY} = 1$, and $\varphi_{Y\Delta} = 2\pi/6$. The star and delta winding sets are connected according to the schema in Figure 8.3, yielding a $\pi/6$ phase shift between the two current systems. Under these conditions, the 2nd MMF harmonic, with a winding factor of 83.6%, is selected as the working harmonic.

Example 8.18 9-slot/8-pole star-delta machine
The second example, depicted in Figure 8.35, is a 9-slot, 8-pole machine equipped with a star-delta winding defined by $Q_s = 9$, $z = 1$, $n_{c\Delta} = 1$, $n_{cY} = 2$, and $\varphi_{Y\Delta} = -4\pi/9$. As in the previous case, the current systems are phase-shifted by $\pi/6$. For this winding, the 4th MMF harmonic has a winding factor of 97.5% and is therefore selected as the operating harmonic.

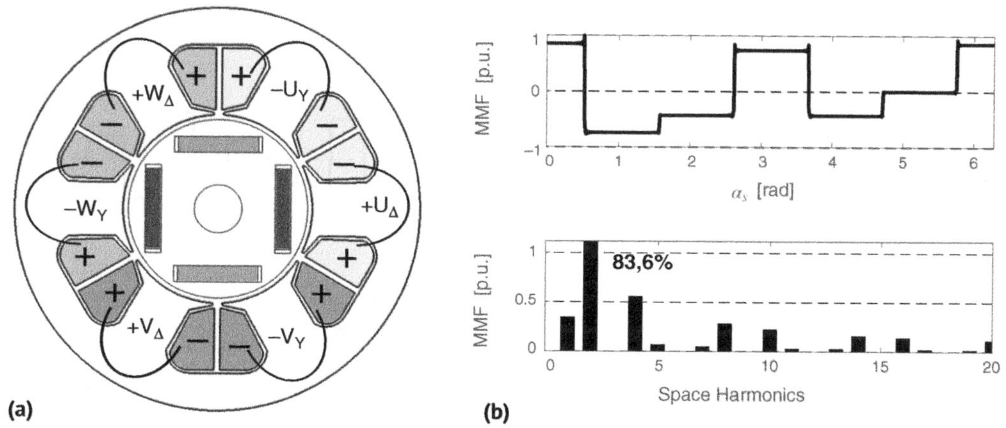

Figure 8.34 (a) Winding layout for the star-delta FSCW with $Q_s = 6, z = 1, n_{c\Delta} = n_{cY} = 1, \varphi_{Y\Delta} = \pi/3$, and with opposite polarity winding sets. (b) MMF characteristics.

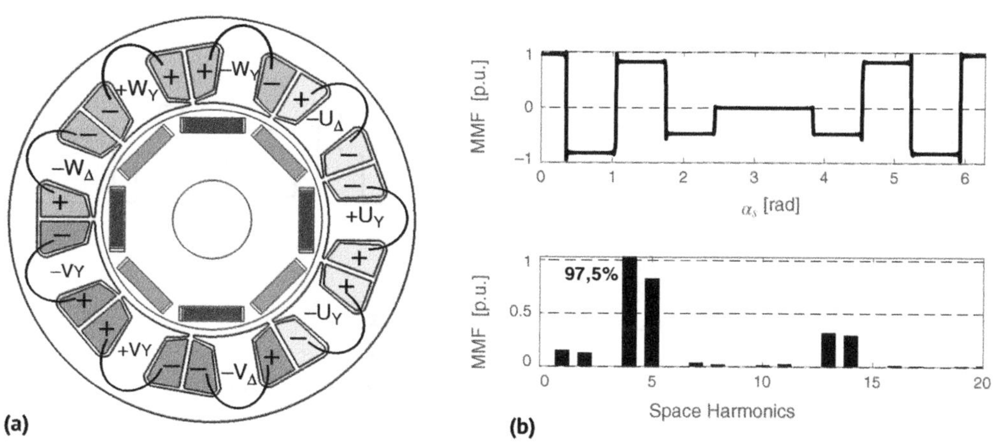

Figure 8.35 (a) Winding layout for the star-delta FSCW with $Q_s = 9, z = 1, n_{c\Delta} = 1, n_{cY} = 2, \varphi_{Y\Delta} = -4\pi/9$, and with the same polarity winding sets. (b) MMF characteristics.

Example 8.19 12-slot/10-pole star-delta machine

Figure 8.36 illustrates another star-delta FSCW with a $\pi/6$ phase shift between the star and delta current. The winding parameters are $Q_s = 12$, $z = 2$, $n_{c\Delta} = n_{cY} = 1$, and $\varphi_{Y\Delta} = \pi/6$. In this case, the 5th MMF harmonic, with a winding factor of 96.6%, is selected as the working harmonic, leading to a 10-pole PM machine design.

Example 8.20 18-slot/10-pole star-delta machine

An example featuring sequential star and delta coil combinations and a current phase shift of $\pi/2$ is presented in Figure 8.37. The winding parameters are $Q_s = 18, z = 2, n_{c\Delta} = 2, n_{cY} = 1,$ and $\varphi_{Y\Delta} = \pi/9$. Compared to the previous configuration, this winding produces an MMF waveform with significantly reduced harmonic content. The 5th harmonic is again used as the working harmonic, with a winding factor of 75.8%.

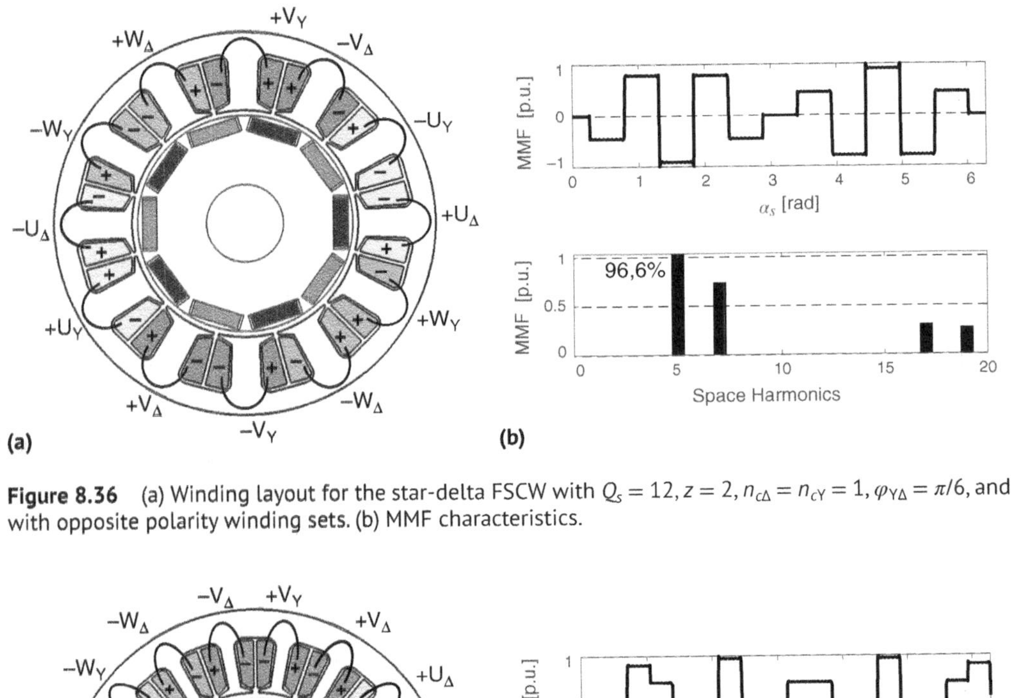

Figure 8.36 (a) Winding layout for the star-delta FSCW with $Q_s = 12, z = 2, n_{c\Delta} = n_{cY} = 1, \varphi_{Y\Delta} = \pi/6$, and with opposite polarity winding sets. (b) MMF characteristics.

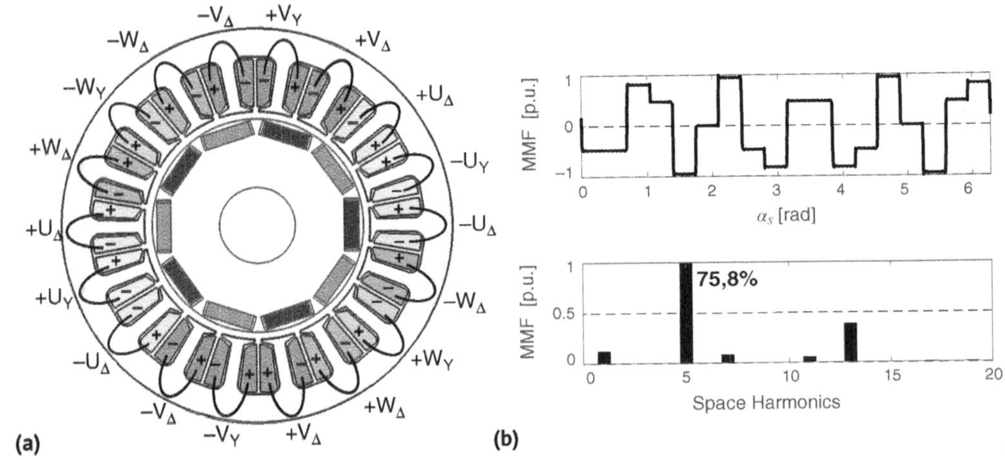

Figure 8.37 (a) Winding layout for the star-delta FSCW with $Q_s = 18, z = 2, n_{c\Delta} = 2, n_{cY} = 1, \varphi_{Y\Delta} = \pi/9$, and with same polarity winding sets. (b) MMF characteristics.

Example 8.21 12-slot/2-pole star-delta machine
For low pole-number applications, a particular star-delta FSCW with same-polarity winding sets is shown in Figure 8.38. The winding parameters are $Qs = 12$, $z = 2$, and $n_{c\Delta} = n_{cY} = 1$, with a spatial displacement angle $\varphi_{Y\Delta} = \pi/6$ and current systems likewise phase-shifted by $\pi/6$. This configuration produces an MMF distribution that is nearly sinusoidal, with the fundamental (1st harmonic) serving as the operating wave. The corresponding winding factor is 25.8%.

Example 8.22 12-slot/2-pole star-delta machine with dual slot-layer stator
To further enhance the winding factor for the operating harmonics while maintaining a sinusoidal MMF waveform, Figure 8.39 introduces a star-delta winding implemented on a dual slot-layer stator. The delta winding is characterized by $Q_{S\Delta} = 12$, $z = 2$, and $n_{c\Delta} = 2$, while the star winding uses $Q_{SY} = 6$, $z = 2$, and $n_{cY} = 1$. The two winding sets are shifted in space by $\varphi_{Y\Delta} = \pi/12$, and their current systems are phase-shifted by $\pi/2$. As a result, the winding factor of the working harmonic is increased by approximately 44% compared to the low-pole FSCW presented in the previous example.

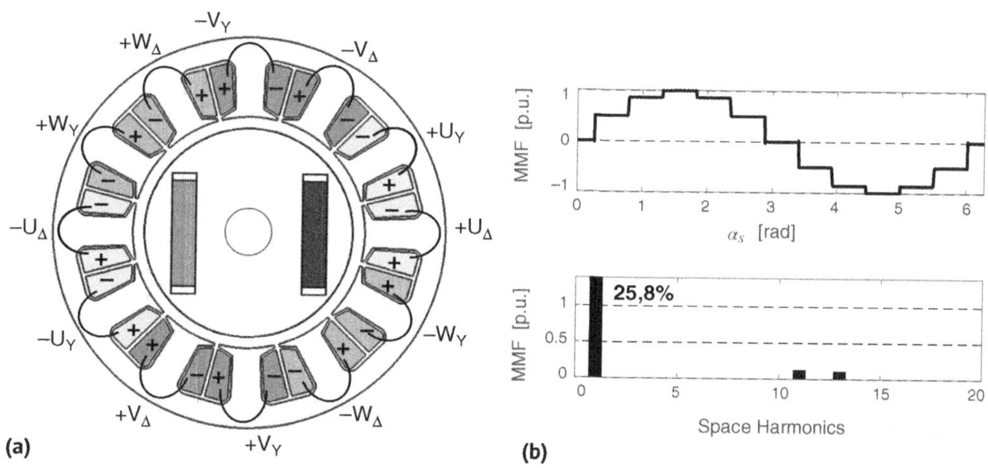

Figure 8.38 (a) Winding layout for the star-delta FSCW with $Q_s = 12, z = 2, n_{c\Delta} = n_{cY} = 1, \varphi_{Y\Delta} = \pi/6$, and with the same polarity winding sets. (b) MMF characteristics.

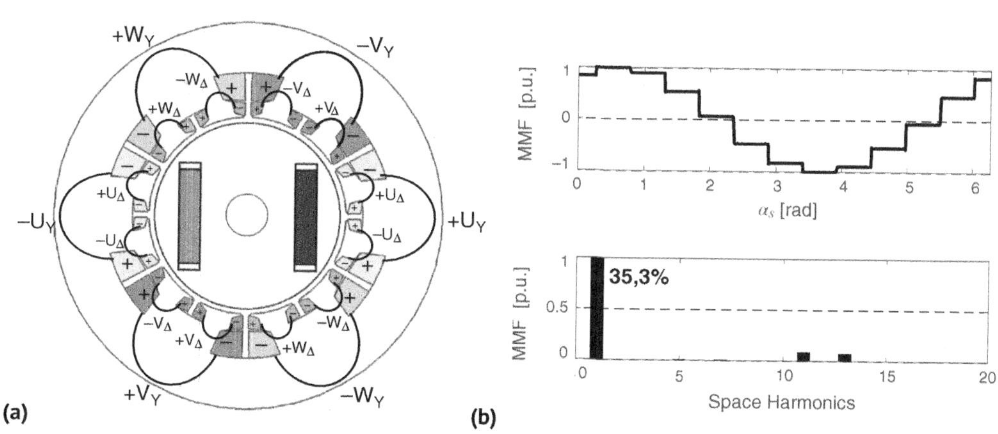

Figure 8.39 Star-delta FSCW with dual stator slot-layer configuration. (a) Winding layout. (b) MMF characteristics.

9

Fractional Slot Concentrated Windings for Axial Flux Machines

The application of concentrated windings is not restricted to a specific topology of electric machines. In the preceding chapters, these windings were mainly examined in the context of radial flux machines (RFMs) in order to explain their fundamental characteristics and performance aspects. However, the same winding principles can be directly applied to axial flux machines (AFMs) without any conceptual modifications. This machine type represents an attractive alternative to conventional RFMs, particularly in applications requiring high torque density and short axial length [119–122]. Moreover, owing to their simple structure and compact design, AFMs with concentrated windings have become a preferred solution in a wide range of practical applications, as reported by Eastham et al. [123], Marignetti et al. [124], Di Gerlando et al. [125], and Woolmer and McCulloch [126]. From an electromagnetic standpoint, the magnetomotive force (MMF) distribution generated by a given fractional slot concentrated winding (FSCW) is exclusively determined by the spatial placement of the conductors and the associated current loading and therefore is independent of whether the machine employs a radial or axial flux path. As a result, the space-harmonic content introduced by concentrated windings is an inherent characteristic of the winding topology rather than the machine geometry. Therefore, the same MMF harmonic phenomena observed in RFMs are also present in AFMs. Accordingly, the performance limitations associated with MMF harmonics, as well as the optimization strategies and harmonic mitigation techniques discussed in the preceding chapters, are equally applicable to AFMs and therefore do not require further general investigations. However, an exception among various optimization approaches is the stator-shifting technique introduced in Chapter 7. When applied to AFMs, this method represents a special case owing to the unique geometric construction. The following discussion therefore focuses on the practical implementation of this technique and the evaluation of the resulting performance improvements in AFMs. Since this approach is particularly suited to a specific machine configuration, particularly the stator-rotor-stator arrangement, this topology is considered as the reference structure for investigating various FSCW types in AFMs.

9.1 AFMs – Construction and Realization

In general, electric machines may be classified into RFMs and AFMs, as illustrated in Figure 9.1. The fundamental difference between these two constructions lies in the direction of the main magnetic flux and the corresponding rotor-stator arrangement. In RFMs, the magnetic flux is predominantly radial, flowing from the rotor toward the stator in a direction perpendicular to the

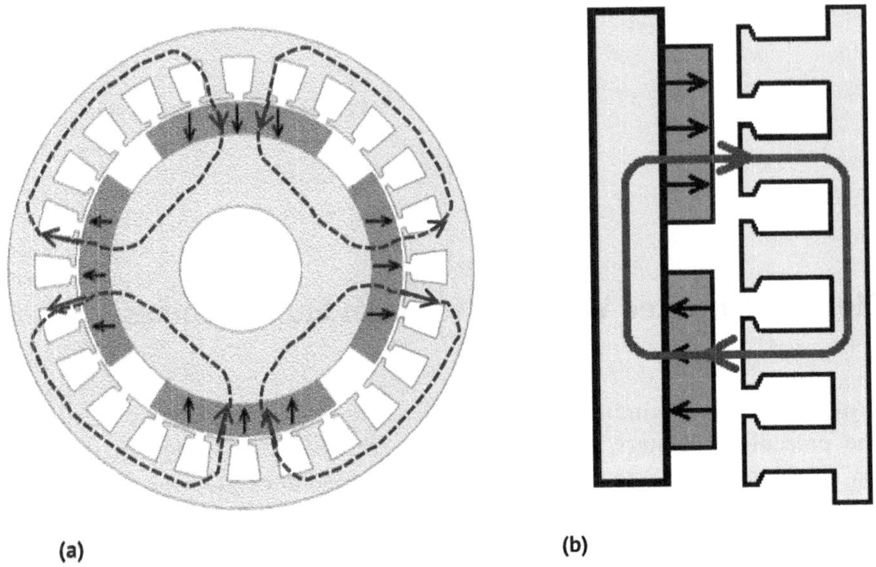

Figure 9.1 (a) Radial flux machine. (b) Axial flux machine.

shaft. In contrast, AFMs are characterized by an axial magnetic flux that flows parallel to the shaft, from one face of the rotor to the opposing face. This difference in flux orientation leads to distinct mechanical arrangements. In RFMs, the rotor is typically cylindrical and rotates either inside or outside a concentric stator. By comparison, AFMs employ disc-shaped rotors and stators that are arranged face to face, forming a compact axial structure.

From a practical standpoint, RFMs benefit from well-established manufacturing processes and are generally more cost-effective. Consequently, they dominate a wide range of applications, including household appliances, fans, pumps, industrial drives, and electric vehicles. AFMs, while typically are more complex and costly to manufacture, offer advantages in terms of power density, torque density, and compactness. As a result, they are particularly well suited for applications in which constraints on space and weight, as well as high efficiency, are of primary importance, such as aircraft, robotics, and high-performance electric vehicle drivetrains.

AFMs can be realized in several structural configurations, as illustrated in Figure 9.2. The most common arrangements include [127]:

1. single-sided configurations, consisting of one stator-rotor pair,
2. double-sided rotor-stator-rotor configurations, and
3. double-sided stator-rotor-stator configurations.

The rotor and stator components of an AFM require specific structural designs which depend on the selected configuration. In single-sided AFMs, the rotor has a magnetic yoke to provide a return path for the magnetic flux, as shown in Figure 9.2a. In rotor-internal axial flux configurations, the rotors may be realized without a magnetic yoke, as illustrated in Figure 9.2b. Similarly, the stators of stator-internal AFMs can be designed either without a yoke (yokeless configuration), Figure 9.2c, or with a magnetic yoke to complete the flux path, Figure 9.2d. A three-dimensional exploded view of a double-sided stator-rotor-stator AFM is shown in Figure 9.3.

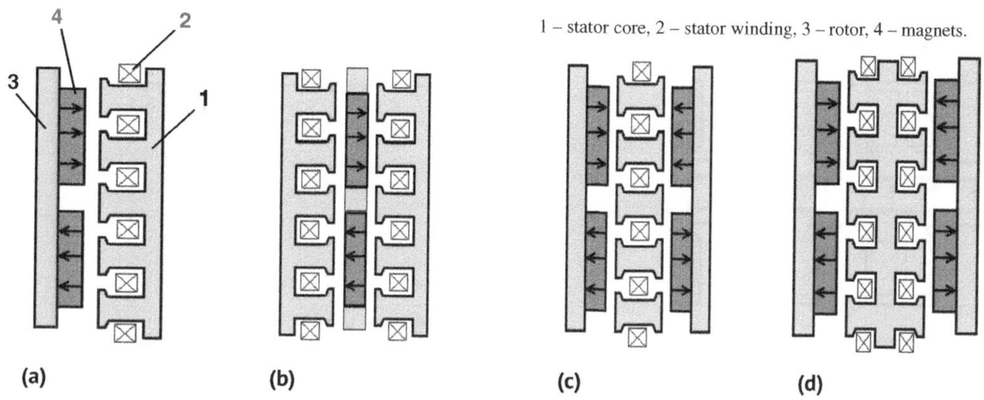

Figure 9.2 Different AFM topologies. (a) Single-sided. (b) Double-sided stator-rotor-stator. (c) Double-sided rotor-stator-rotor (yokeless stator). (d) Double-sided rotor-stator-rotor (stator with magnetic yoke).

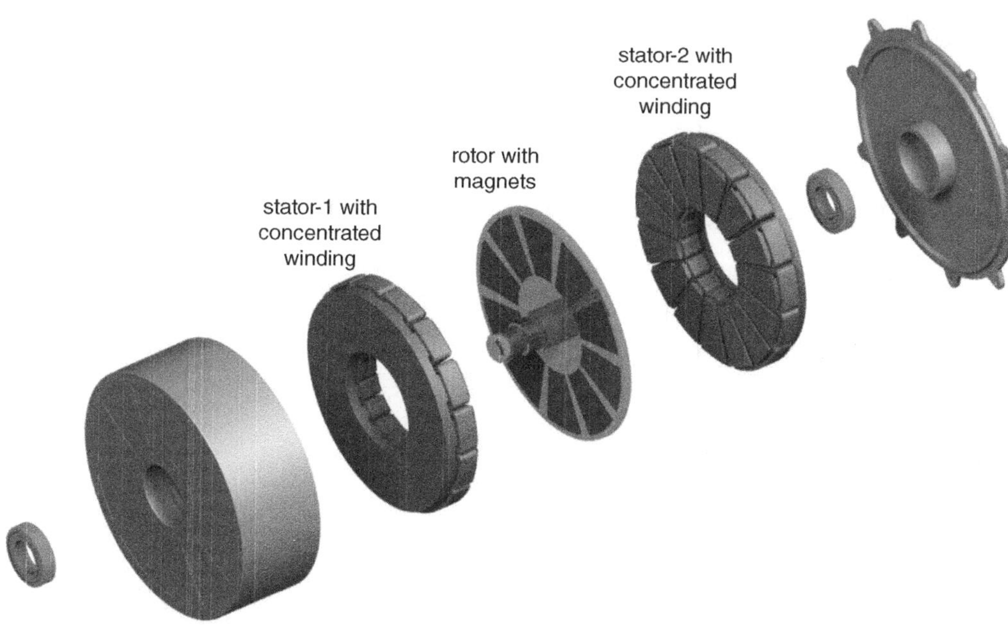

Figure 9.3 Explored view of an AFM with double-sided stator-rotor-stator configuration.

9.2 AFMs – Stator Winding MMF Harmonics

A widely used winding type for the AFMs is the tooth-concentrated winding. This configuration provides a compact and cost-effective solution; however, its principal drawback is the relatively high MMF harmonic content, which can lead to several adverse effects on the machine performance.

Figure 9.4a, illustrates the basic structure of a double-sided AFM employing a concentrated winding and an inner permanent-magnet rotor. The stator utilizes a conventional FSCW with a winding sequence +U+V+W, where the symbols "+" denotes positive coil polarities (other winding topologies contain both "+" and "−," i.e. positive and negative coil polarities).

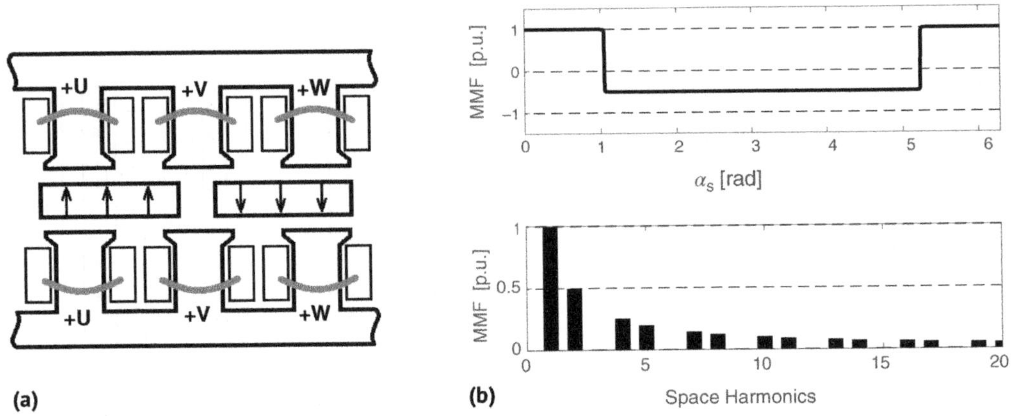

Figure 9.4 (a) Double-sided stator-rotor-stator AFM with +U+V+W concentrated winding. (b) MMF characteristics of the stator winding.

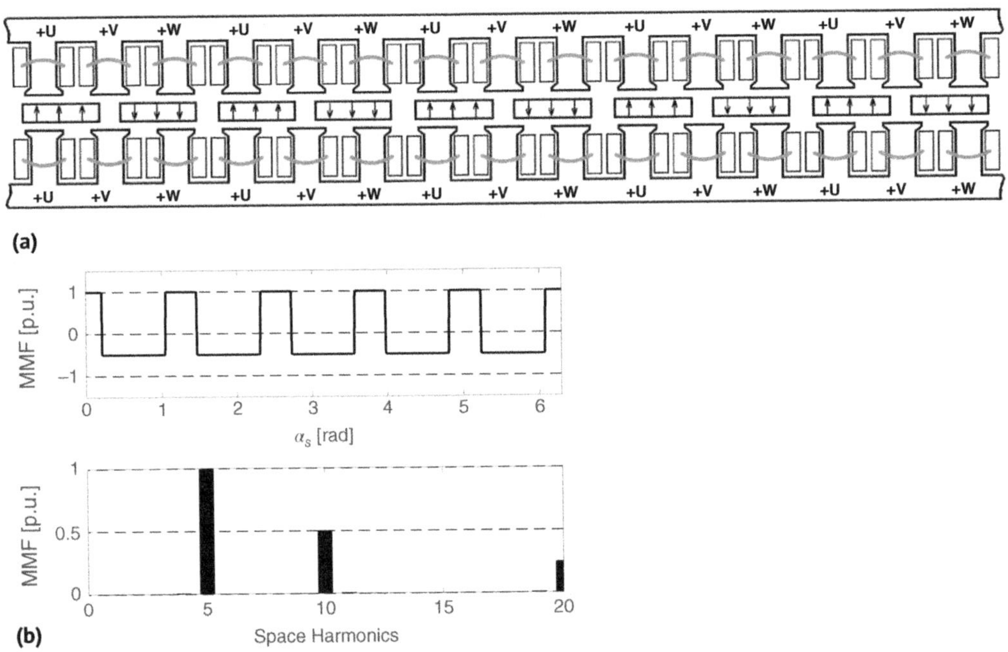

Figure 9.5 (a) 10-pole AFM with a +U+V+W concentrated winding and a periodicity factor of five. (b) Resulting MMF characteristics of the stator winding.

This winding topology produces the MMF distribution shown in Figure 9.4b. The resulting MMF spectrum is dominated by the 1st, 2nd, 4th, 5th, and higher-order harmonics. As discussed in Chapter 2, this winding configuration is unique among FSCWs in that it utilizes the fundamental harmonic, i.e., the 1st harmonic, as the working harmonic. A key attribute of this winding type is its flexibility. Any desired pole-pair number can be achieved by repeating the base winding pattern p-times around the stator circumference. Figure 9.5a depicts the machine configuration obtained when the base winding is repeated five times, yielding a 15-slot/10-pole AFM topology. In this case, the winding periodicity factor is five, i.e. $\wp = 5$, the MMF waveform repeats five times along the stator circumference, and the corresponding MMF harmonic occurs at integer multiples of five (Figure 9.5b). Consequently, for higher pole-number application employing this winding type, the MMF spectrum is free of subharmonic components, which is an important advantage in terms of rotor losses and additional saturation of magnetic path.

Another commonly used concentrated winding employs the +U−U+U, +V−V+V, +W−W+W coil sequence. This configuration generates two MMF harmonics suitable for machine operation, namely the 4th and the 5th harmonic order. The example machine shown in Figure 9.6a corresponds to a 9-slot/10-pole AFM and uses the 5th harmonic as a working harmonic. In addition, a 10-pole PM machine can also be designed using another concentrated winding with the layout +U−U, −V+V, +W−W, −U+U, +V−V, −W+W, as illustrated in Figure 9.6b. This winding type is commonly referred to as the 12-slot/10-pole FSCW.

As demonstrated, a 10-pole PM machine can be realized using different FSCW configurations leading to different slot/pole combinations, as well as different MMF characteristics and winding factors associated with the operating harmonics. The corresponding MMF spectra and harmonic winding factors are presented in Figure 9.7 and summarized in Table 9.1. The following

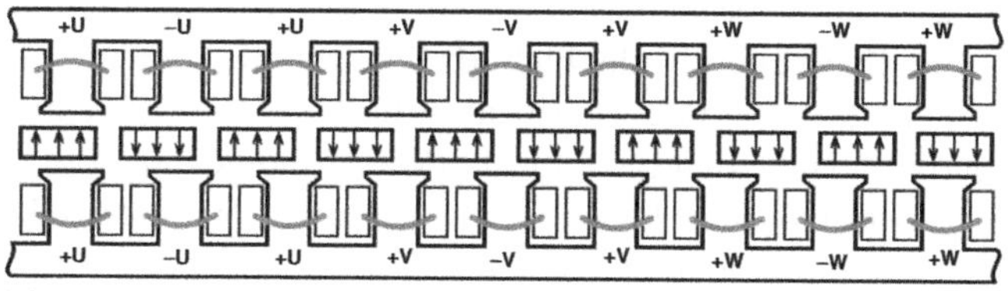

(a)

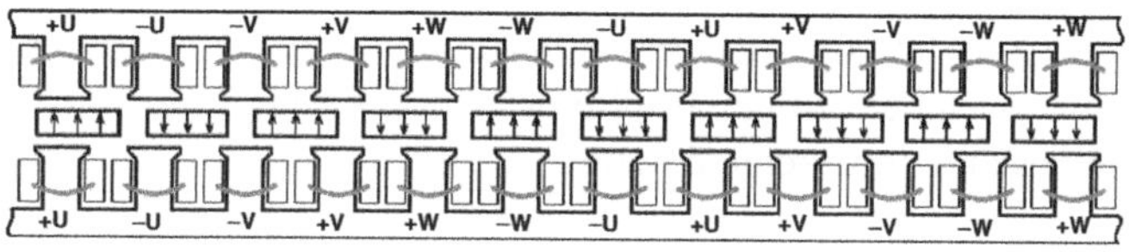

(b)

Figure 9.6 10-pole AFMs with concentrated windings. (a) AFM with a 9-slot/10-pole FSCW. (b) AFM with a 12-slot/10-pole FSCW.

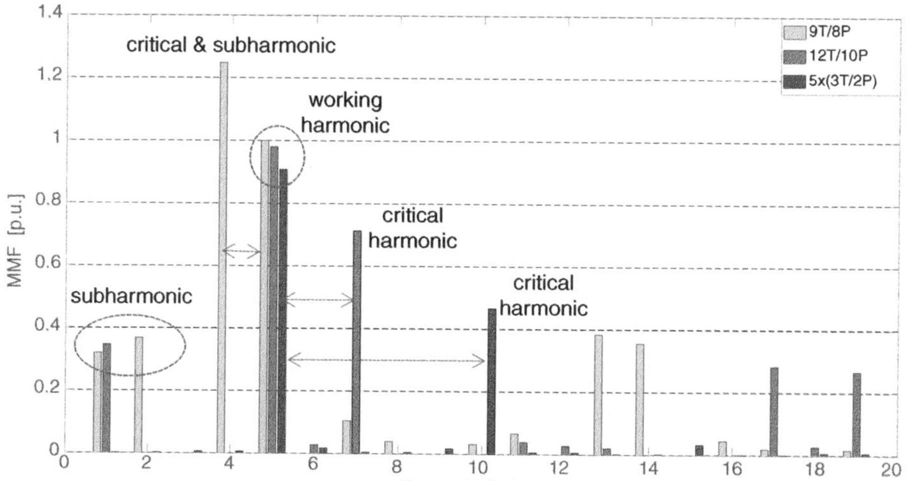

Figure 9.7 Comparison of MMF harmonic content for different FSCWs for application in 10-pole machines.

Table 9.1 Winding factor (in %) for different DL-FSCWs for 10-pole machines.

Winding type	$v = 1$	2	3	4	5	6	7	8	9	10	11	12	13	14	15	16	17
9T/10P	6.1	14.0	—	**94.5**	**94.5**	—	14.0	6.1	—	6.1	14.0	—	**94.5**	**94.5**	—	14.0	6.1
12T/10	6.7	—	—	—	**93.3**	—	**93.3**	—	—	—	6.7	—	6.7	—	—	—	93.3
15T/10P	—	—	—	—	**86.6**	—	—	—	—	**86.6**	—	—	—	—	—	—	—

Note: Bold represents highest winding factors.

abbreviations are used through the comparisons: the 9-slot/10-pole = 9T/10P, the 12-slot/10-pole = 12T/10P, and the 15-slot/10-pole = 5 × (3T/2P).

Based on these results, the following conclusions can be made:

- *9-slot/10-pole FSCW:*
  This winding type contains a large number of subharmonics and higher-order harmonics in the MMF spectrum. The winding factor for working harmonic (5th) is 94.5%, while the first critical harmonic is the 4th harmonic. Consequently, the high MMF harmonic content, and especially the 4th harmonic with its large amplitude, is critical in terms of rotor (magnet) losses and acoustic noise. It is important to note that the critical MMF harmonics adjacent to the operating harmonic are not only critical for rotor losses but also generate low-order radial force components (modes) that are a major source of machine noises.
- *12-slot/10-pole FSCW:*
  This winding type also produces a high harmonic content in the MMF spectrum. The winding factor for working harmonic (5th) is 93.3% and the first critical harmonic is the 7th harmonic. Similar to the 9-slot/10-pole case, the presence of high-amplitude critical harmonic (7th) can lead to increased rotor (magnet) losses and noise related issues.
- *15-slot/10-pole FSCW:*
  When the common +U+V+W winding is applied to high-pole electric machines, the resulting MMF spectrum is free of subharmonic components. The winding factor for working harmonic (5th) is 86.6%. The first critical harmonic occurs at an order of $2p$, which is located far away from the working harmonic and is therefore less critical than in the previous examples. Although the fundamental winding factor is lower compared to the 9-slot/10-pole or the 12-slot/10-pole configurations, the absence of subharmonics and critical harmonics provides clear advantages, particularly in terms of reduced rotor losses and improved acoustic behavior. As a result, this winding type has seen extensive use in a variety of practical high-pole machine applications.

9.3 Stator-shifting Technique for AFMs with Concentrated Windings

The optimization of FSCWs using the stator-shifting method has been discussed in Chapter 7 and demonstrated for a variety of RFMs with concentrated windings. The fundamental principle involves integrating two identical winding systems in one stator core, with one system spatially shifted relative to the other by a defined angle. In order to achieve an effective reduction of critical harmonics, the second winding set must be displaced by an angle corresponding to the midpoint of the stator tooth (i.e. coil-pitch) of the original winding. Accordingly, additional slots are introduced at the center of each stator tooth, and the second winding is placed in these newly created slots. As a consequence, in RFMs each stator coil spans two adjacent stator teeth, leading to a new winding configuration referred to as a two-tooth FSW.

In AFMs, however, the stator-shifting concept can be implemented in a fundamentally different manner due to their double-sided structural design. Since the machine comprises two physically separate stator components, the corresponding winding systems can be spatially shifted with respect to each other by any desired angle without requiring any modification to the stator geometry. Consequently, this approach enables enhanced performance while maintaining the same machine structure and manufacturing effort as conventional AFMs.

In general, the stator shifting technique can be applied to any AFM with concentrated windings through the following procedure:

- Two identical stator components are used, analogous to a conventional AFM.
- The first stator component is shifted (in rotation direction) relative to the second stator component by a specific angle α_τ.
- Additionally, the winding of the first stator is displaced relative to the winding of the second stator by a specific angle α_w.
- The two winding systems may be configured with either the same or opposite coil polarities.

Using this approach, various AFM topologies employing different FSCWs are investigated. Their corresponding winding layouts are presented, and the resulting MMF characteristics are analyzed and compared with those of their conventional counterparts.

9.3.1 AFMs with Double-layer FSCWs

Example 9.1 AFM with DL-FSCW with $Q_S = 3$, $z = 1$, and $n_c = 1$
Figure 9.8 presents the optimized version of the 3-slot/2-pole AFM employing a concentrated winding. This winding type produces several dominant harmonics, such as the 1st, 2nd, 4th, 5th, and further higher-order MMF harmonics. Consequently, the practical use of this winding requires careful suppression of critical harmonics. To achieve this objective, the two stator components are circumferentially displaced by one-half of a slot-pitch, defined as:

$$\alpha_\tau = \pi / Q_S$$

In addition, the corresponding winding sets are shifted by a second displacement angle α_w. According to Equation (7.10), the optimal value of this angle is given by:

$$\alpha_w = 3\alpha_\tau = 3\pi / Q_S$$

Furthermore, depending on the relative polarity of the two winding sets, either the first or the second critical MMF harmonic can be selectively suppressed. This leads in two possible machine configurations, each optimized with respect to a specific working harmonic component.

Design-1: 3-slot/2-pole AFM

When the above displacement angles are applied to both stator components, and opposite winding set polarities are employed, the AFM configuration shown in Figure 9.8a is obtained. The corresponding MMF results, presented in Figure 9.8b, demonstrate effective suppression of the 2nd, 4th, 8th, and 10th spatial harmonics, while the 1st harmonic remains essentially unchanged. This behavior makes the proposed winding arrangement particularly well suited for 2-pole machines and more generally, for machines whose pole numbers are integer multiples of two.

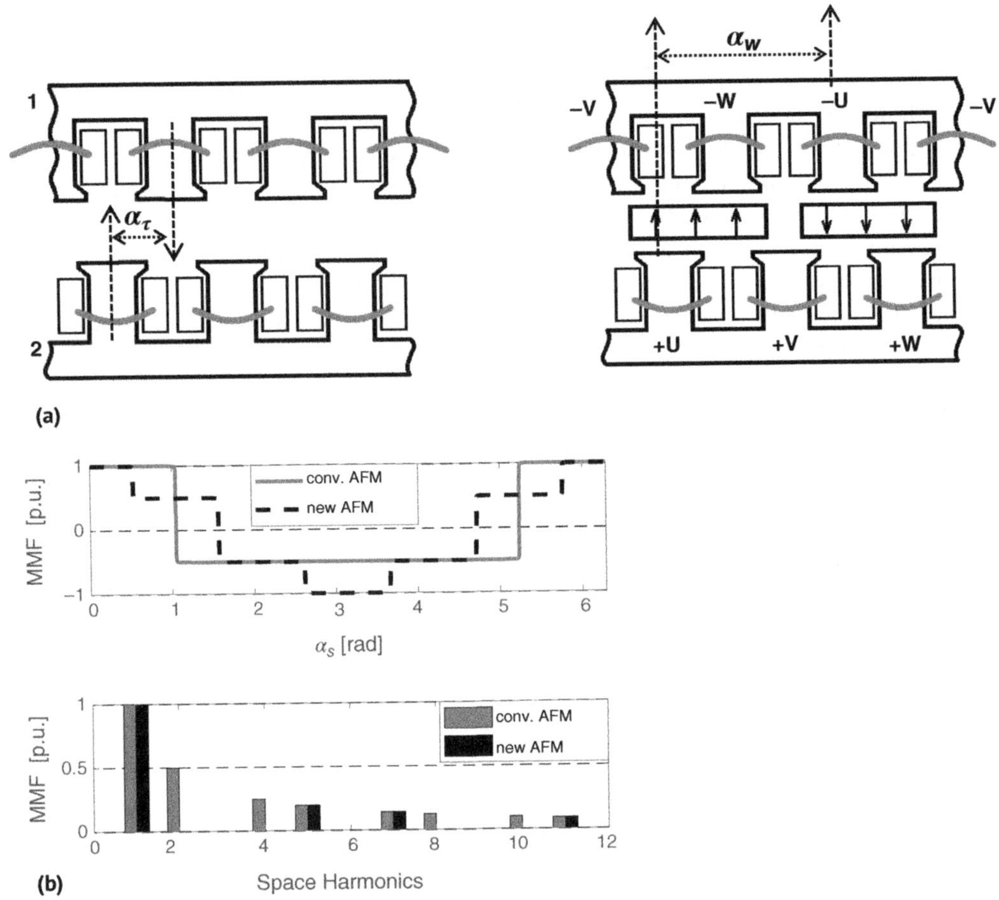

Figure 9.8 (a) Optimized AFM with a 3-slot/2-pole DL-FSCW and opposite winding set polarity. (b) Comparison of the MMF characteristics with the conventional design.

Design-2: 3-slot/4-pole AFM

In the second configuration, the displacement angles α_τ and α_w are retained, while the winding polarity is chosen to be the same for both stator components, as shown in Figure 9.9. This modification leads to substantial attenuation of the 1st, 5th, 7th, and higher-order MMF harmonics, whereas the 2nd harmonic remains unaffected. As a result, this winding arrangement is suitable for 4-pole AFM designs.

Example 9.2 AFM with DL-FSCW with $Q_S = 9$, $z = 1$, and $n_c = 3$

Another widely used concentrated winding is the FSCW with $Q_S = 9$, $z = 1$, and $n_c = 3$. An example AFM employing this winding configuration is illustrated in Figure 9.6a. For this topology, the dominant operating harmonic components are typically the 4th or the 5th harmonic order. To suppress the associated critical harmonics, the displacement angles are selected as follows:

- stator displacement angle: $\alpha_\tau = \pi/Q_S$,
- winding-set displacement angle: $\alpha_w = 9\alpha_\tau = 9\pi/Q_S$.

As in the previous example, appropriate selection of winding sets polarity allows selective suppression of either the first or the second critical MMF harmonic, leading to two alternative machine designs.

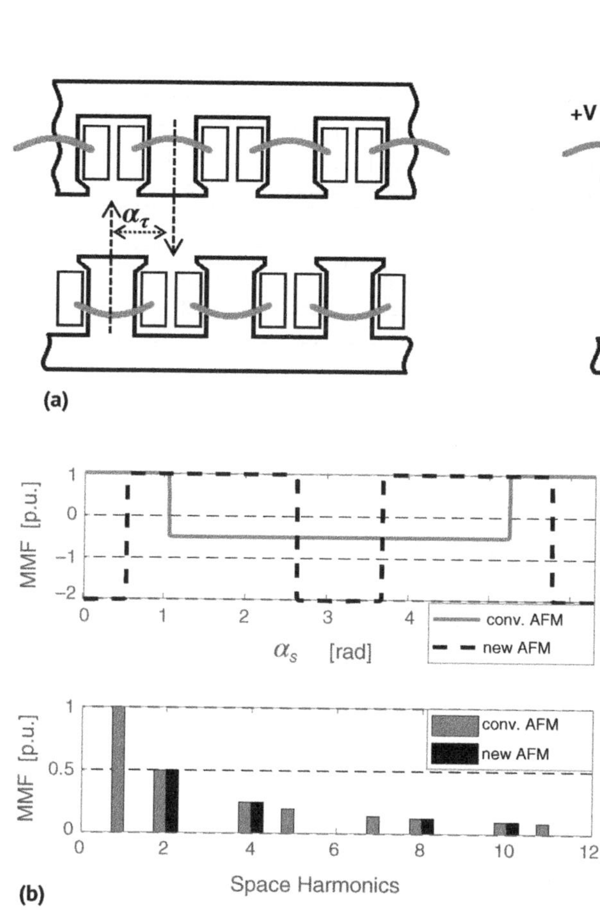

Figure 9.9 (a) Optimized AFM with a 3-slot/2-pole DL-FSCW and same winding set polarity. (b) Comparison of the MMF characteristics with the conventional design.

Design-1: 9-slot/10-pole AFM

The AFM configuration with opposite winding set polarity is presented in Figure 9.10a, with the corresponding MMF results compared to the reference machine in Figure 9.10b. This arrangement significantly reduces the 2nd, 4th, 8th, and higher-order harmonics, while leaving the 5th harmonic essentially unaffected. The final result is an efficient 10-pole AFM solution with improved performance.

Design-2: 9-slot/8-pole AFM

By retaining the same displacement angles α_τ and α_w and reversing the winding sets polarity, a second AFM configuration suitable for 8-pole machines is obtained, as illustrated in Figure 9.11. This winding arrangement is highly effective in suppressing the most critical MMF harmonics, notably the 1st, 5th, and 7th order.

Example 9.3 AFM with DL-FSCW with $Q_S = 12$, $z = 2$, and $n_c = 2$

A further AFM example uses a DL-FSCW topology with $Q_S = 12$, $z = 2$, and $n_c = 2$, as illustrated in Figure 9.6b for the case with a 10-pole rotor. The typical operating harmonic here is either the

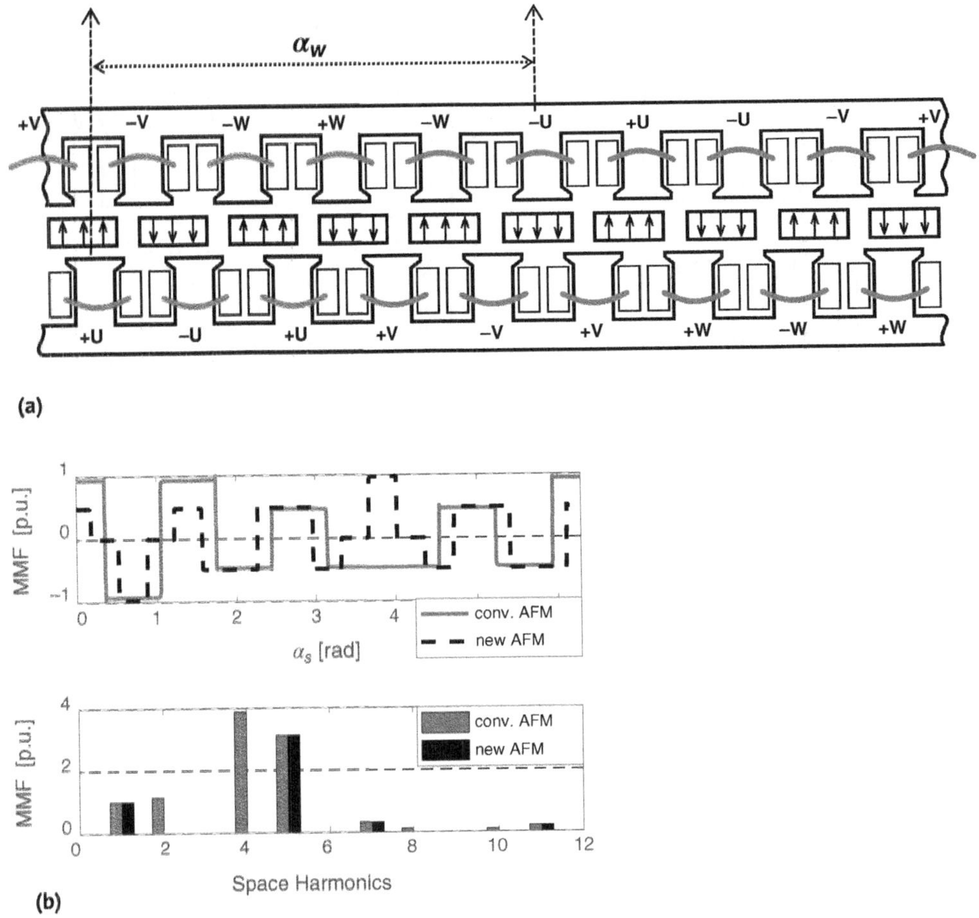

Figure 9.10 (a) Optimized AFM with a 9-slot/10-pole DL- FSCW and opposite winding set polarity. (b) Comparison of the MMF characteristics with the conventional design.

5th or the 7th MMF harmonics. In order to suppress the corresponding critical harmonics, the displacement angles are selected as:

- stator displacement angle: $\alpha_\tau = \pi/Q_S$,
- winding-set displacement angle: $\alpha_w = 5\alpha_\tau = 5\pi/Q_S$.

Once again, the selection of winding set polarity determines which critical harmonic will be suppressed, resulting in two alternative AFM designs.

Design-1: 12-Slot/14-Pole AFM

Figure 9.12 presents the winding arrangement and corresponding MMF results for the configuration with opposite winding set polarity. With the proposed displacement angles, the 1st and 5th winding MMF harmonics are significantly reduced, while the 7th harmonic remains largely unaffected. This configuration is therefore well suited for 14-pole AFM applications.

Design-2: 12-Slot/10-Pole AFM

An alternative realization is shown in Figure 9.13, where the winding polarity is modified while maintaining the same displacement angles. The resulting MMF distribution indicates effective

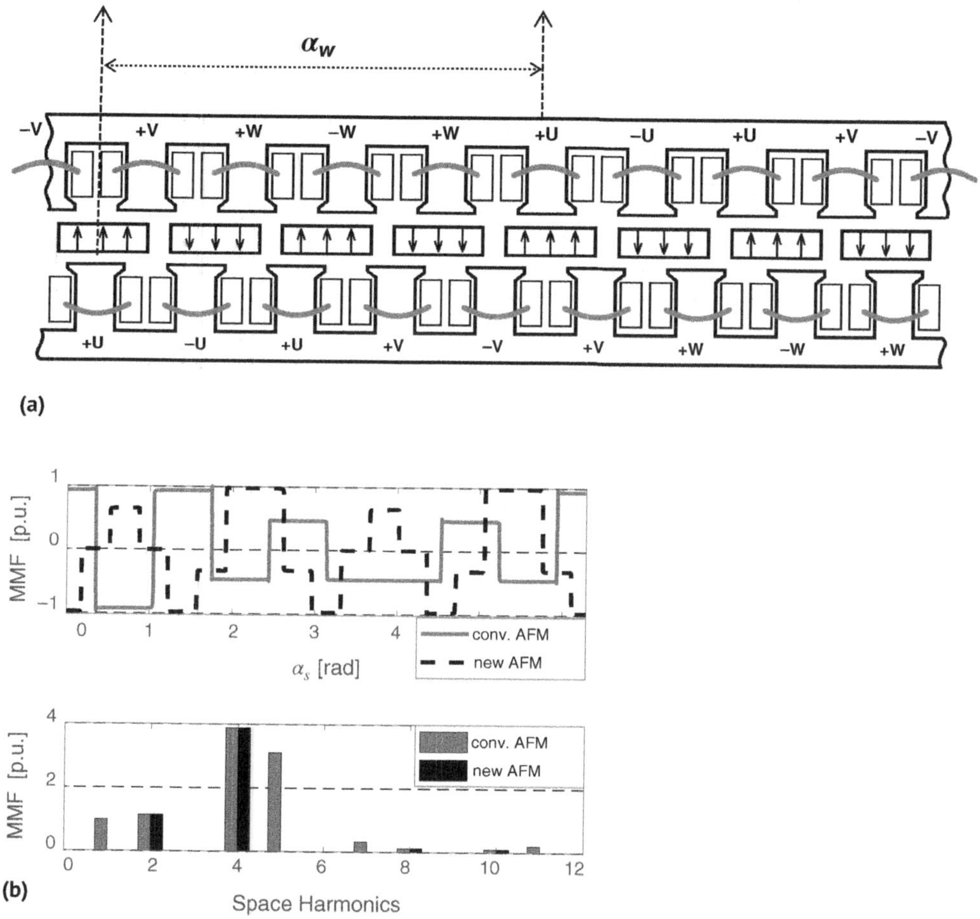

Figure 9.11 (a) Optimized AFM with a 9-slot/8-pole DL- FSCW and same winding set polarity. (b) Comparison of the MMF characteristics with the conventional design.

suppression of the 1st and 7th harmonics, whereas the 5th harmonic remains unaffected. This makes the proposed arrangement an effective solution for 10-pole AFMs.

9.3.2 AFMs with Single-layer FSCWs

The example machines discussed above employ double-layer concentrated windings, with each stator tooth carrying a coil. However, AFMs can also be realized using single-layer concentrated windings, where coils are wound on every second stator tooth. An example of this configuration is illustrated in Figure 9.14, corresponding to a 12-slot/10-pole SL-FSCW topology. This winding exhibits a high winding factor for the working harmonic, i.e. the 5th or the 7th, reaching 96.6%. Nevertheless, it also contains a subharmonic component of relatively large amplitude. The displacement angles required to suppress the associated critical harmonics are:

- stator displacement angle: $\alpha_\tau = \pi/Q_S$,
- winding-set displacement angle: $\alpha_w = 5\alpha_\tau = 5\pi/Q_S$.

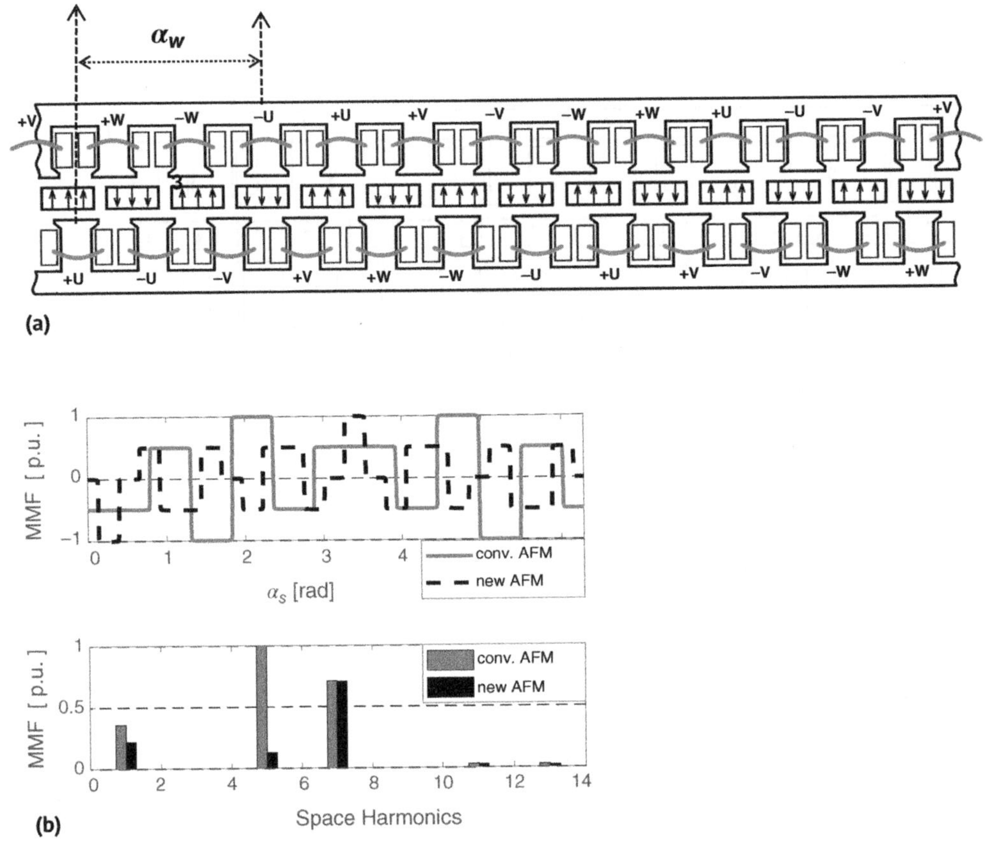

Figure 9.12 (a) Optimized AFM with a 12-slot/14-pole DL- FSCW and opposite winding set polarity. (b) Comparison of the MMF characteristics with the conventional design.

Whereby, depending on the relative polarity of winding sets, two effective AFM topologies can be realized.

9.3.2.1 Design-1: 12-slot/14-pole AFM

The winding arrangement corresponding to this machine configuration is illustrated in Figure 9.15a. In this design, the two stator components are displaced by the angles α_τ and α_w, and opposite winding-set polarities are employed. The resulting MMF characteristics, shown in Figure 9.15b, demonstrate effective reduction of the 1st and 5th MMF harmonics, while the 7th-order harmonic remains essentially unaffected. This configuration is therefore well suited for 14-pole AFM applications.

9.3.2.2 Design-2: 12-slot/10-pole AFM

An alternative configuration is presented in Figure 9.16a. Here, the same stator and winding displacement angles α_τ and α_w are retained; however, unlike the previous case, the winding sets are selected to have the same polarity. The corresponding MMF spectra, shown in Figure 9.16b, indicate substantial suppression of the 1st and 7th harmonics, while the 5th harmonic remains unaffected, making this configuration particularly suitable for 10-pole AFMs.

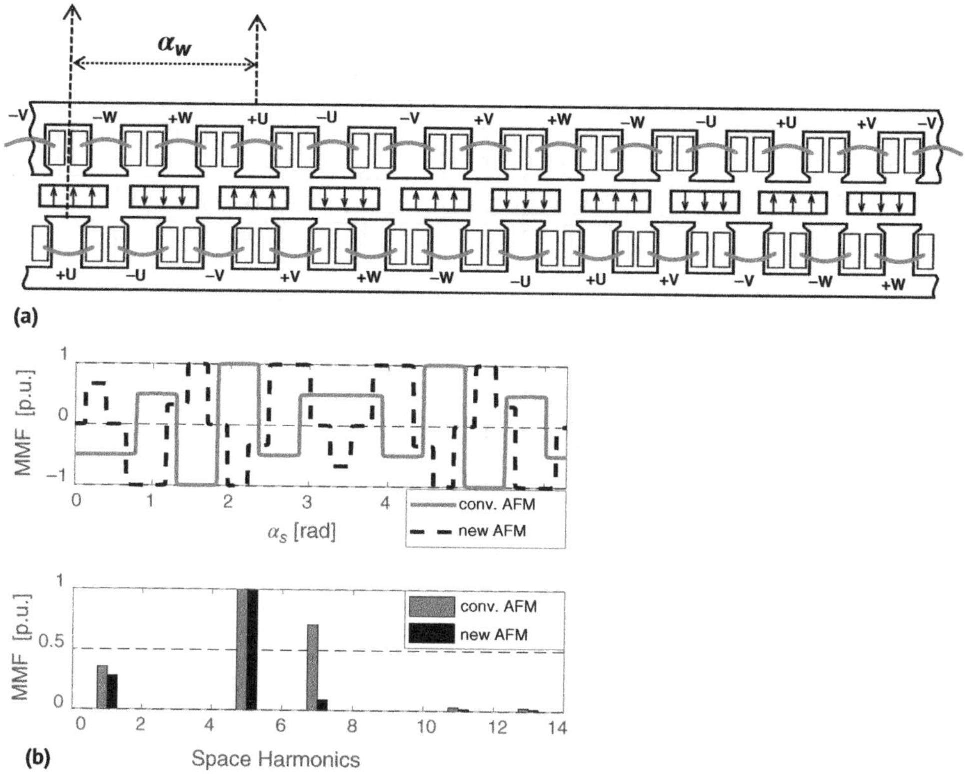

Figure 9.13 (a) Optimized AFM with a 12-slot/10-pole DL- FSCW and the same winding set polarity. (b) Comparison of the MMF characteristics with the conventional design.

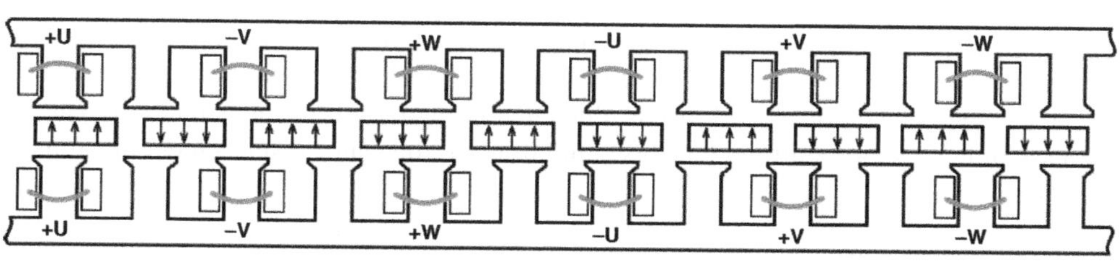

Figure 9.14 Standard AFM with the 12-slot/10-pole SL-FSCW.

9.4 AFMs with Asymmetric Stator Shifting

In the example machines examined earlier, the circumferential displacement between the stator components is selected to be exactly one-half of a slot-pitch angles, defined as:

$$\alpha_{\tau 1} = \pi / Q_S$$

A machine employing this displacement is referred to as an AFM with symmetric stator shifting, as illustrated in Figure 9.17a.

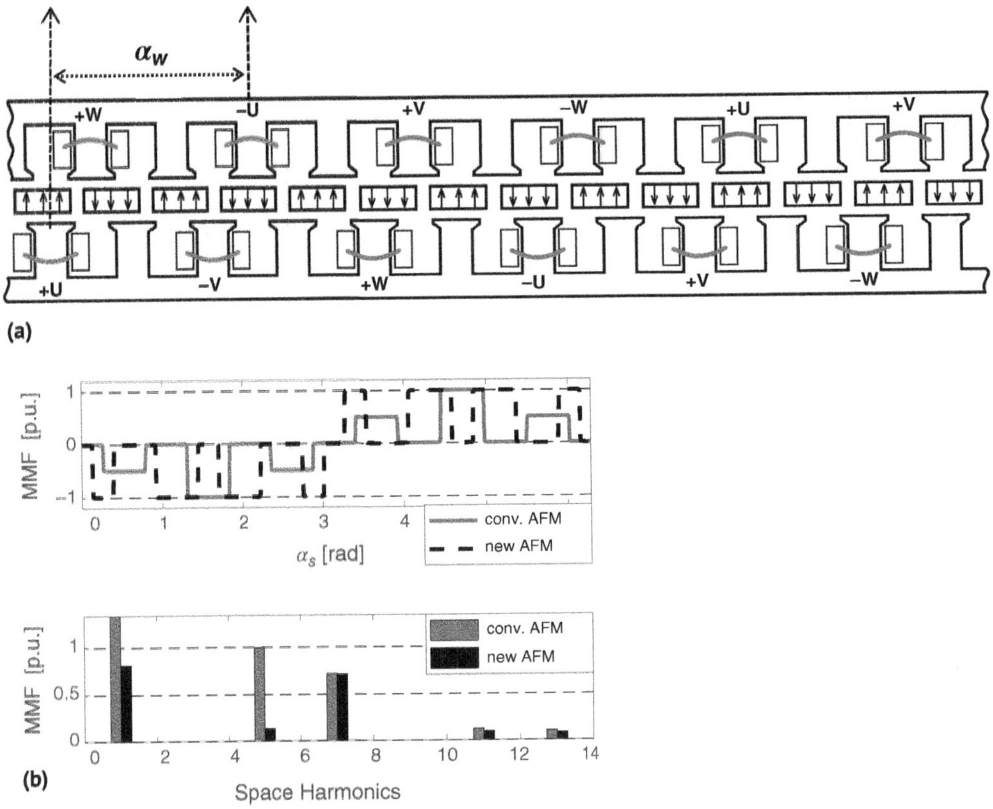

Figure 9.15 (a) Optimized AFM with a 12-slot/14-pole SL-FSCW and opposite winding set polarity. (b) Comparison of the MMF characteristics with the conventional design.

An important advantage of the double-sided stator AFM topology, however, is the flexibility to select stator displacement angles that are either equal to or different from an integer of the half slot-pitch angle. This more general case is illustrated in Figure 9.17b, and is referred to as an AFM with asymmetric stator shifting. For this configuration, the stator displacement angle may be expressed as:

$$\alpha_{\tau 2} = \pi / Q_S \pm \Delta\alpha_x, \text{where } \Delta\alpha_x > 0$$

Accordingly, the displacement angle of the winding systems is given by:

$$\alpha_{w2} = k \cdot \alpha_{\tau 2}$$

where k is an integer number $(k = 2,3,4,\ldots)$, selected based on the specific critical harmonic targeted for suppression.

By introducing the additional angular offset $\Delta\alpha_x > 0$, a fine-tuning degree of freedom is obtained that allows further optimization of the MMF spectrum. In particular, the appropriate selection of $\Delta\alpha_x$ parameter enables complete attenuation of specific undesired MMF harmonics that cannot be fully suppressed using symmetrical stator shifting method. An alternative AFM based on the machine configuration shown in Figure 9.9b, but employing an asymmetric stator displacement is illustrated in Figure 9.18.

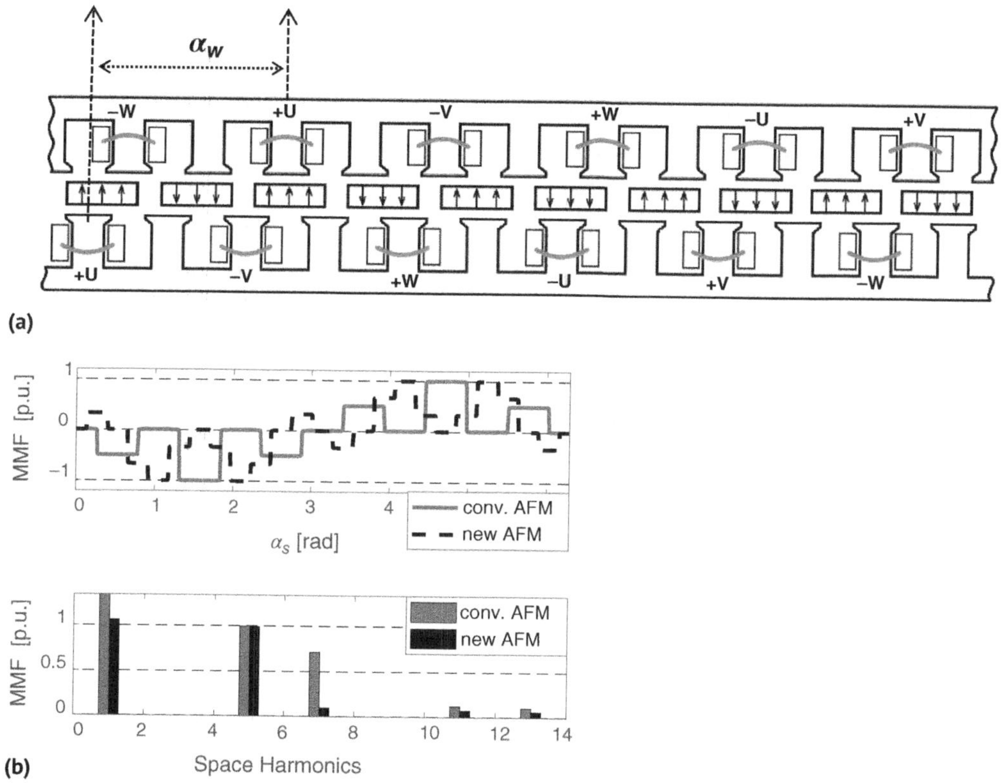

Figure 9.16 (a) Optimized AFM with a 12-slot/10-pole SL-FSCW and same winding set polarity. (b) Comparison of the MMF characteristics with the conventional design.

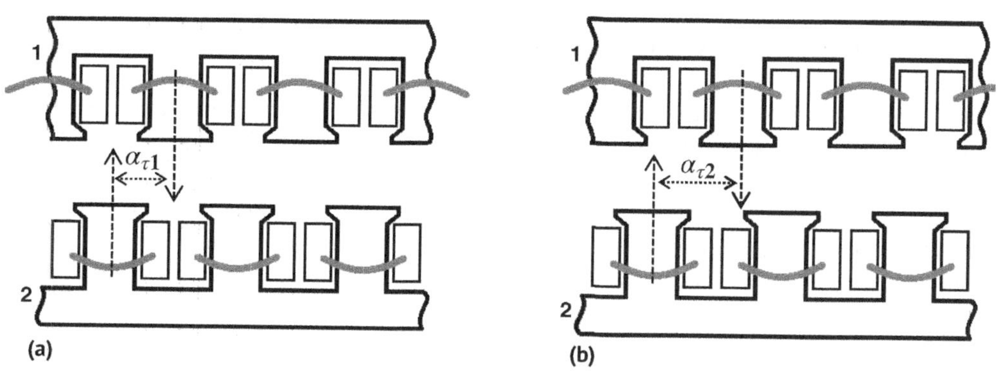

Figure 9.17 (a) AFM with symmetric stator shifting. (b) AFM with asymmetric stator shifting.

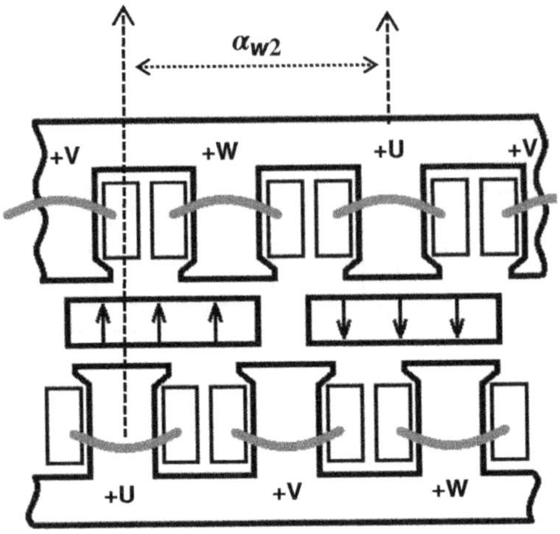

Figure 9.18 The 3-slot/2-pole AFM with asymmetric stator shifting configuration.

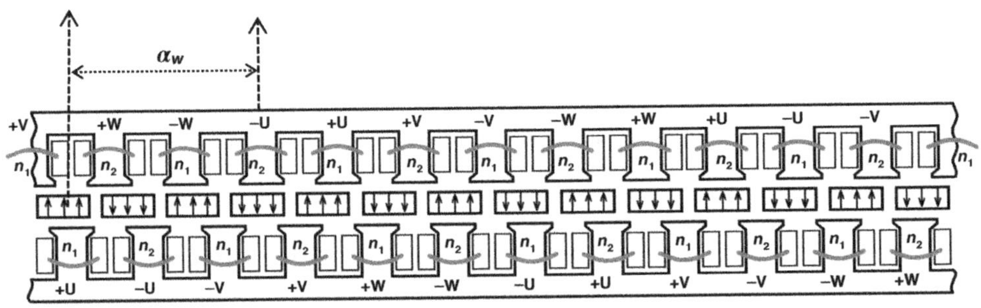

Figure 9.19 AFM with a 12-slot/14-pole DL-FSCW and unequal turns per coil.

9.5 AFMs with Unequal Turns per Coil

In an alternative realization, the new AFM concept can be implemented using concentrated windings with unequal turns per coil, as illustrated in Figure 9.19. In this configuration, n_1 denotes the number of turns of the first coil of a given winding phase, while n_2 represents the number of turns on the adjacent coil belonging to the same phase. As discussed in Chapter 3, the use of unequal turn coil combination provides an additional degree of freedom in modifying the MMF distribution, leading to further improvement in the harmonic content, particularly through the reduction of winding subharmonics.

9.6 AFMs with Flux-barrier Stator

In terms of the winding factor of the operating harmonic, and consequently the torque density of electric machine, SL-FSCWs outperform other concentrated winding topologies. However, the presence of high-amplitude subharmonic components, along with critical MMF harmonics, leads to undesirable electromagnetic effects, which have often limited the practical application of this winding group.

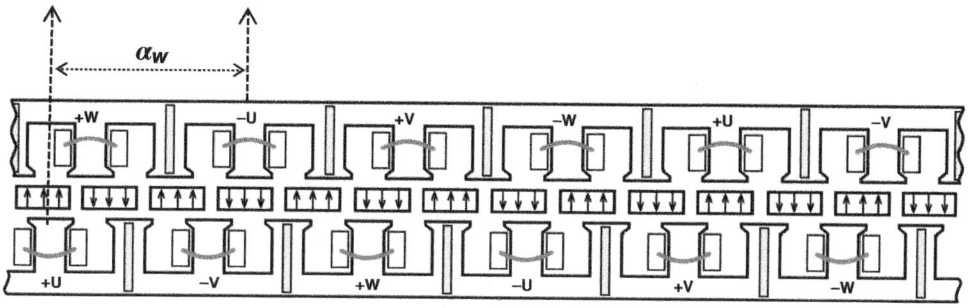

Figure 9.20 AFM with a 12-slot/14-pole SL-FSCW and flux-barrier stator.

The results in Section 9.3.2 show that the stator-shifting technique is highly effective in reducing critical MMF harmonics in single-layer windings. However, despite this significant improvement in the MMF spectrum, a substantial subharmonic component remains, which can continue to have a negative effect on machine performance. Therefore, further optimization is necessary for this winding configuration. As explained in Chapter 6, the flux-barrier technique provides an effective means of reducing subharmonics in SL-FSCWs and could therefore be combined with the stator-shifting technique to further improve the winding MMF performance. Specifically, the stator-shifting approach can be first employed to suppress the critical harmonics. Subsequently, the implementation of flux-barriers, particularly within the stator-tooth regions, serves to reduce the subharmonic component while simultaneously enhancing the operating harmonic. Figure 9.20 illustrates the resulting machine configuration, featuring an SL-FSCW with flux-barriers introduced in the unwound stator teeth. This combined approach represents a highly efficient solution for 14-pole AFMs.

9.7 AFMs Topologies

Figure 9.21 shows a representative AFM example designed for a practical traction application in electric vehicles (EVs) in a three-dimensional view. The machine features a 15-slot/10-pole topology with a standard FSCW, arranged in sequence +U+V+W, and repeated five times around the stator circumference. Figure 9.22 provides a comparison with an alternative realization that includes the stator-shifting technique, which is employed to reduce critical MMF harmonics. For clarity, the sketches depict only the minimum symmetry required to represent the winding layout and spatial phase relationships.

As discussed in Section 9.3.1, the optimization of MMF characteristics for this winding type is achieved by circumferentially displacing the two stator components by $\alpha_\tau = \pi/Q_S$, while the corresponding winding sets are shifted by $\alpha_w = 3\alpha_\tau = 3\pi/Q_S$. To utilize the fundamental (1st) MMF harmonic as the operating wave, winding sets of the same polarity are selected. The resulting machine performance (p.u.), including torque capability, rotor magnet losses, and critical force modes, is summarized in Table 9.2. The finite element analysis (FEA) performed in a real AFM proves that the stator-shifting technique effectively reduces magnet losses without compromising the machine's torque capability. Moreover, the targeted suppression of critical force harmonics (modes) contributes to reduced electromagnetic noise, making this approach particularly suitable for low-noise and high-efficiency applications.

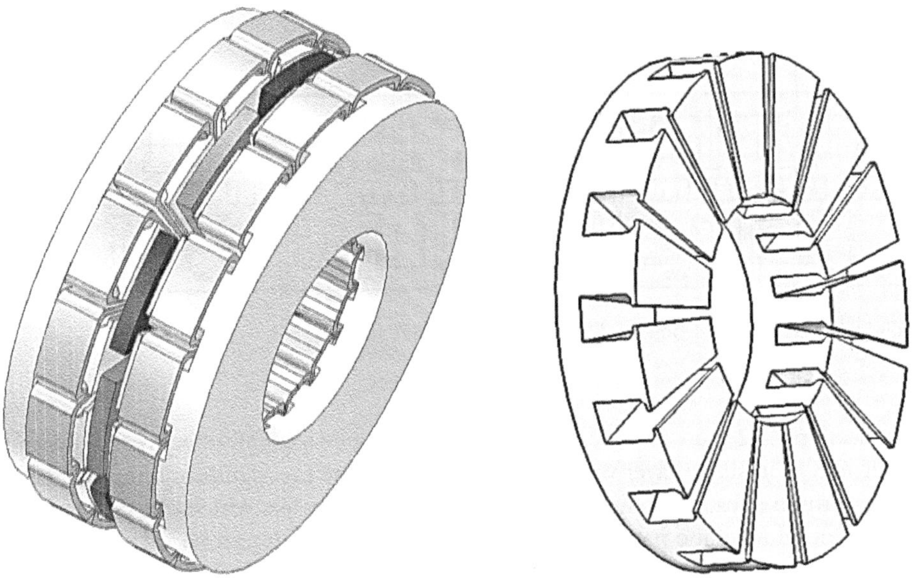

(a) (b)

Figure 9.21 Conventional 15-slot/10-pole AFM with concentrated winding. (a) Complete view of the machine. (b) Stator core geometry.

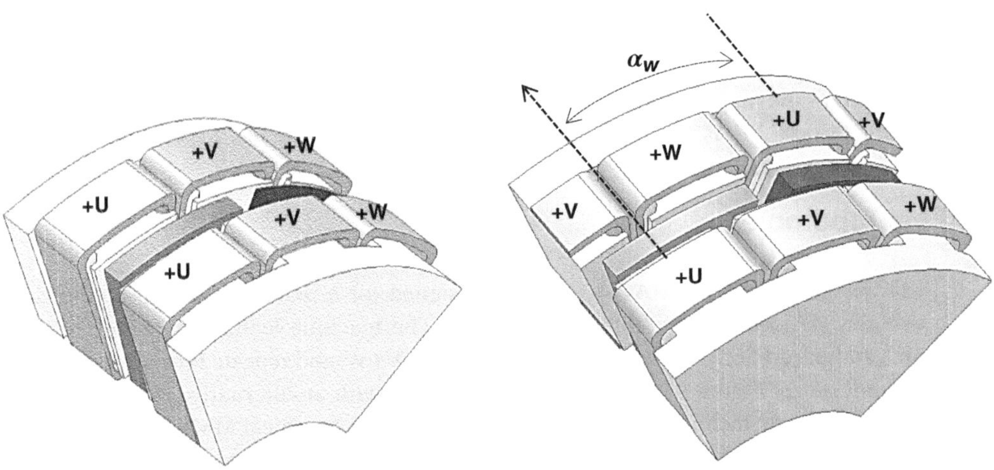

(a) (b)

Figure 9.22 Minimal symmetry of the 15-slot/10-pole AFM. (a) Standard design. (b) Stator-shifting design.

Table 9.2 Comparison of electromagnetic performance for a 15-slot/10-pole AFM.

15-Slot/10-pole AFM	Electromagnetic torque (%)	Magnet losses (%)	Critical radial force mode (%)
Standard design	100	100	100
Stator-shifting	100	18.5	0

References

1 Arnold E (1891) Die Ankerwicklungen und Ankerkonstruktionen der Gleichstrom-Dynamomaschinen. Julius Springer, Berlin/München, Germany.
2 Arnold E (1904) Die Wechselstromtechnik III – Die Wicklungen der Wechselstrommaschinen. Springer, Berlin Heidelberg, Germany.
3 Richter R (1920) Ankerwicklungen für Gleich- und Wechselstrommaschinen, Julius Springer, Berlin, Germany.
4 Kauders W (1932) Systematik der Drehstromwicklungen. Elektrotechnik und Maschinenbau 50.
5 Heller F, Kauders W (1935) Das Görgessche Durchflutungspolygon, Arch Elektrotech 29, pp. 599–616.
6 Liwschitz M (1943) Distribution factor and pitch factors of the harmonics of a fractional slot winding. AIEE, pp. 664–666.
7 Veinott C (1950) Theory and Design of Small Induction Motors, McGraw-Hill, New York, USA.
8 Liwschitz M (1950) Alternating-Current Machines – A Book for Winders, Repairmen, and Designers of Electric Machines, Van Nostrand, New York, USA
9 Sequenz H (1950) Die Wicklungen elektrische Maschinen, Springer, Vienna, Austria.
10 Richter R (1952) Lehrbuch der Wicklungen elektrischer Maschinen, Wissenschaftliche Bücherei, Verlag und Druck G Braun, Karlsruhe, Germany.
11 Heiles F (1953) Wicklungen Elektrischer Maschinen und Ihre Herstellung. Springer, Berlin, Heidelberg, Germany.
12 Say M G (1976) Alternating Current Machines, Pitman, New York, USA.
13 Heller B, Hamata F (1977) Harmonic Field Effects in Induction Machines, Elsevier, Amsterdam, The Netherlands.
14 Bianchi N, Dai Pre M (2006) Use of the star of slots in designing fractional slot single-layer synchronous motors. IEE Proc Electr Power Appl 153 (3), pp. 459–466.
15 Müller G, Vogt K, Ponick B (2008) Berechnung elektrischer Maschinen, Wiley-VCH, Berlin, Germany.
16 Hague B (1917) The mathematical treatment of the magnetomotive forces of armature windings. J Inst Electr Eng 55 (268), pp. 489–514.
17 Graham Q (1927) The MMF wave of polyphase windings with special reference to sub synchronous harmonics. AIEE 46, pp. 19–28.
18 Calvert J F (1938) Amplitude of magnetomotive force harmonics for fractional slot windings-I. AIEE 57 (12), pp. 777–785.
19 Schuisky W (1957) Induktionsmaschinen, Springer, Vienna, Austria.
20 Krause P C (1986) Analysis of Electric Machinery, McGraw-Hill, New York, USA.

21 Toliyat H A, Lipo T A, White J C (1991) Analysis of a concentrated winding induction machine for adjustable speed drive applications, IEEE Trans Energy Conver 6 (4), pp. 679–683.

22 Sadarangani C (2000) Electrical machines - Design and Analysis of Induction and Permanent Magnet Motors. KTH, Stockholm, Sweden.

23 Lipo T A (2004) Introduction to AC Machine Design. Wisconsin Power Electronics Research Center, University of Wisconsin, USA.

24 Ouyang W, El-Antably A, Huang S, and Lipo T A (2005) Analysis of optimal stator concentric winding patterns design, International Conference on Electrical Machines and Systems ICEMS, Nanjing, China, pp. 94–98.

25 Binder A (2012) Elektrische Maschinen und Antriebe. Springer-Verlag, Berlin Heidelberg, Germany.

26 Pyrhönen J, Jokinen T, Hrabovcova V (2014) Design of Rotating Electrical Machines. John Wiley & Sons, Ltd.

27 Gerling D (2015) Electrical Machines – Mathematical Fundamentals of Machine Topologies. Springer, Berlin Heidelberg, Germany.

28 Cros J, Viarouge P (2002) Synthesis of high performance PM motors with concentrated windings. IEEE Trana Energy Convers 17 (2), pp. 248–253.

29 Koch T, Binder A (2002) Permanent magnet motors with fractional slot winding for electric traction, International Conference in Electric Motors, Brugge, Belgium, pp. 1–6.

30 Magnussen F, Sadarangani C (2003) Winding factors and Joule losses of permanent magnet machines with concentrated windings. IEEE International Electric Machines and Drives Conference IEMDC, Madison Wisconsin, USA, pp. 333–339.

31 Libert F, Soulard J (2004) Investigation on pole-slot combinations for permanent-magnet machines with concentrated windings, International Conference on Electrical Machines ICEM, Cracow, Poland, pp. 1–6.

32 Ishak D, Zhu Z Q, Howe D (2004) Comparative study of permanent magnet brushless motors with all teeth and alternative teeth windings, IEE Int Conf PEMD (2), pp. 834–839.

33 Ishak D, Zhu Z Q, Howe D (2005) Permanent magnet brushless machines with unequal tooth widths and similar slot and pole numbers, IEEE Trans Ind Appl 41 (2), pp. 584–590.

34 Ishak D, Zhu Z Q, Howe D (2006) Comparison of PM brushless motors having either all teeth or alternate teeth wound. IEEE Trans Energy Convers 21 (1), pp. 95–103.

35 Bianchi N, Bolognani S, Dai Pre M, Grezzani G (2006) Design considerations for fractional slot winding configurations of synchronous machines, IEEE Trans Ind Appl 42 (4), pp. 997–1007.

36 Bianchi N, Dai Pré M, Alberti L (2007) Theory and Design of Fractional Slot PM Machines. CLEUP, Padova, Italy.

37 EL-Refaie A M, Shah M R, Qu R, Kern J M (2007) Effect of number of phases on losses in conducting sleeves of high speed surface PM machine rotors, IEEE Ind Appl Annul Meeting IAS, New Orleans, USA, pp. 1522–1529.

38 El-Refaie A M (2010) Fractional-slot concentrated-windings synchronous permanent magnet machines: opportunities and challenges, IEEE Trans Ind Electron 57 (1), pp. 107–121.

39 EL-Refaie A M, Jahns T M (2006) Impact of winding layer number and magnet type on synchronous surface PM machines designed for wide constant-power speed range operation, IEEE IAS Annu Meeting, Tampa, FL, USA, pp. 1486–1493.

40 Meier F, Soulard J (2009) PMSMs with non-overlapping concentrated windings: design guidelines and model references, Ecologic vehicles-renewable energy EVER, Monaco, pp. 1–9.

41 Ito K, Naka K, Nakano M, Kobayashi M (2009) Electric machine. US Patent US7 605 514.

42 Cistelecan M V, Ferriera F J T E. (2010) Three phase tooth concentrated multiple-layer fractional windings with low space harmonic content, IEEE Energy Convers Congress and Exposition ECCE, pp. 1399–1405.
43 Reddy P B, EL-Refaie A M, Huh K K (2011) Effect of number of layers on performance of fractional slot concentrated-windings interior permanent magnet machines, International Conference on Power Electronics – ECCE Asia, The Shilla Jeju, Korea, pp. 1921–1928.
44 Alberti L, Bianchi N (2013) Theory and design of fractional slot multilayer windings. IEEE Trans Ind Appl 49 (2), pp. 841–849.
45 Bianchi N, Alberti L, Barcaro M (2015) Design and tests of a four-layer fractional slot interior permanent magnet motor, IEEE Workshop on Electrical Machines Design, Control and Diagnosis WEMDCD 2015, Torino, Italy.
46 Fornasiero E, Alberti L, Bianchi N, Bolognani S (2013) Consideration on selecting fractional slot nonoverlapped coil winding, IEEE Trans Ind Appl 49 (3), pp. 1316–1324.
47 Dinca Ch (2016) Motor Design for Maximum Material Exploitation and Magnetization Procedure with In-line Quality Check for Mass Production, Ph. D. Thesis, Technische Universität Berlin.
48 Stenzel P (2017) Großserientaugliche Nadelwickeltechnik für verteilte Wicklungen im Anwendungsfall der E-Traktionsantriebe, Ph. D. Thesis, Technischen Fakultät der Friedrich-Alexander-Universität Erlangen-Nürnberg, Germany.
49 Canders W R, Hülsmann D (2011) Analysis and determination of symmetrical three-phase windings with focus on tooth coil windings, International Symposium on Electromagnetic Fields in Mechatronics ISEF, Funchal, Madeira, pp. 1–8.
50 Kometani H, Asao Y, Adachi K (2000) Dynamo-electric machine, US Patent 6, 166, 471.
51 Dajaku G, Gerling D (2011) Eddy current loss minimization in rotor magnets of PM machines using high-efficiency 12-teeth/10-poles winding topology, International Conference on Electrical Machines and Systems, Beijing, China, pp. 1–6.
52 Dajaku G, Gerling D (2011) Novel 24-slots/10-poles winding topology for electric machines, IEEE International Electric Machines and Drives Conference IEMDC-2011, Niagara Falls, Kanada, pp. 65–70.
53 Dotz B, Gerling D (2017) Winding factor analysis of star-delta windings, IEEE Transportation Electrification Conference and Expo ITEC, Harbin, China, pp. 579–585.
54 Dotz B, Gerling D (2017) Windings with various numbers of turns per phasor, IEEE International Electric Machines and Drives Conference IEMDC, Miami, USA, pp. 1–7.
55 Ward E, Härer H (1969) Preliminary investigation of an inverter-fed 5-phase induction motor, Proc. Institution Electr Engineers 116 (6), pp. 980–984.
56 Nelson R H, Krause P C (1974) Induction machine analysis for arbitrary displacement between multiple winding sets, IEEE Trans Power App Syst PAS-93, pp. 841–848.
57 Jahns T M (1980) Improved reliability in solid-state AC drives by means of multiple independent phase drive units, IEEE Trans Ind Appl IA16 (3), pp. 321–331.
58 Klingshirn E A (1983) High phase order induction motors, Part I: Description and theoretical considerations, IEEE Trans Power App Syst PAS-102, pp. 47–53.
59 Abbas M A, Christen R, Jahns T M (1984) Six-phase voltage source inverter driven induction motor, IEEE Trans Ind Appl 20 (5), pp. 1251–1259.
60 Haylock J, Mecrow B, Jack A, Atkinson D (1997) Operation of fault tolerant PM drive for an aerospace fuel pump application, IEE Proc Electr Power Appl 145 (5), pp. 441–448.
61 Parsa L, Toliyat H (2004) Fault-tolerant five-phase permanent-magnet motor drives, IEEE Ind Appl Soc Annu Meeting IAS, pp. 1048–1054.

62 Mitcham A J, Antonopoulos G, Cullen J A (2004) Favorable slot and pole number combinations for fault-tolerant PM machines, IEE Proc Electr Power Appl 151 (5), pp. 520–525.

63 Levi E (2008) Multiphase electric machines for variable-speed applications, IEEE Trans Ind Electron 55 (5), pp. 1893–1909.

64 Aslan B, Semail E, Korecki J, Legranger J (2011) Slot/pole combinations choice for concentrated multiphase machines dedicated to mild-hybrid applications, IEEE Industrial Electronics Society IECON, pp. 3698–3703.

65 Bojoi R, Cavagnino A, Odhano S A, Tenconi A, Vaschetto S (2015) Experimental fault assessment on multiphase PM generators with fractional slot concentrated windings, IEEE Industrial Electronics Society IECON, pp. 4941–4947.

66 Barcaro M, Bianchi N, Magnussen F (2010) Analysis and tests of a dual three-phase 12-slot 10-pole permanent-magnet motor, IEEE Trans Ind Appl 46 (6), pp. 2355–2362.

67 Burkhardt Y, Spagnolo Y, Lucas P, Zavesky M, Brockerhoff P (2014) Design and analysis of a highly integrated 9-phase drivetrain for EV applications, International Conference on Electrical Machines ICEM, pp. 450–456.

68 Bianchi N, Bolognani S, Dai Pre M (2008) Impact of stator winding of a five-phase permanent-magnet motor on postfault operations, IEEE Trans Ind Appl 55 (5), pp. 1978–1987.

69 Locment F, Semail E, Kestelyn X (2008) Vectorial approach-based control of a seven-phase axial flux machine designed for fault operation, IEEE Trans Ind Electron 55 (10), pp. 3682–3691.

70 Khalik A S, Ahmed Sh, Massoud A M (2015) Low space harmonics cancelation in double-layer fractional slot winding using dual multiphase winding, IEEE Trans Mag 51 (5), pp. 839–842.

71 Dajaku G (2020) Advanced multiphase fractional slot concentrated windings: characteristics and potentials, Elec Eng 103, Springer, pp. 397–406.

72 Dajaku G, Gerling D (2018) Opportunities of advanced multi-phase concentrated windings, International Conference on Electrical Machines ICEM, Alexandroupolis, Greece, pp. 325–331.

73 Bianchi N, Fornasiero E (2009), Index of rotor losses in three-phase fractional slot permanent magnet machines, IET Electr Pow Appl 3 (5), pp. 381–388.

74 Zhu Z Q, Xia Z P, Wu L J, Jewell G. W (2009) Influence of slot and pole number combination on radial force and vibration modes in fractional slot PM brushless machines having single- and double-layer windings, IEEE Energy Conversion Congress and Exposition, ECCE 2009, San Jose, California, USA, pp. 3443–3450.

75 Yamazaki K, Fukushima Y (2011) Effect of eddy-current loss reduction by magnet segmentation in synchronous motors with concentrated windings, IEEE Trans Ind Appl 47 (2), pp. 779–788.

76 Krotsch J, Piepenbreier B (2012) Radial forces in external rotor permanent magnet synchronous motors with non-overlapping windings, IEEE Trans Ind Electron 59 (5), pp. 2267–2276.

77 Dajaku G, Gerling D (2012) A novel 12-teeth/10-poles PM machine with flux barriers in stator yoke, International Conference on Electrical Machines ICEM, Marseille, France, pp. 36–40.

78 Dajaku G, Gerling D (2012) Low costs and high-efficiency electric machines, 2nd International Electric Drives Production Conference EDPC, Erlangen-Nurnberg, Germany, pp. 1–7.

79 Li G J, Zhu Z Q, Chu W Q, Foster M P, Stone D A. (2014) Influence of flux gaps on electromagnetic performance of novel modular pm machines, IEEE Trans Energy Convers 29 (3), pp. 716–726.

80 Dajaku G, Gerling D (2014) Analysis of different PM machines with concentrated windings and flux barriers in stator core, International Conference on Electrical Machines ICEM, Berlin, Germany, pp. 375–384.

81 Dajaku G (2023), Comparison study of different FSCWs with flux barrier stator, IEEE International Electric Machines and Drives Conference IEMDC, San Francisco, pp. 1–6.

82 Yokoi Y, Higuchi T (2016) Design analysis of slit stator motors, International Conference on Electrical Machines and Systems ICEMS, Chiba, Japan, pp. 1–4.

83 Yokoi Y, Higuchi T (2018), Stator slitting of 12-slot 10-pole concentrated winding motors, IEEE Trans Ind Appl 54 (5), pp. 4377–4385.

84 Gerold J W, Gerling D (2018) Analysis of different arrangements of flux barriers and different pole pairs in a stator with concentrated winding, International Conference on Electrical Machines ICEM, Alexandroupoli, Greece, pp. 58–64.

85 Gerold J W, Gerling D (2019) An equivalent winding factor larger than 1 by using flux barriers in the stator, IEEE International Electric Machines and Drives Conference IEMDC, San Diego, USA, pp. 1855–1862.

86 Gerold J W (2022) Elektrische Maschinen mit Flussbarrieren im Stator, Ph. D. Thesis, Universitaet der Bundeswehr Muenchen, Shaker, Aachen, Germany.

87 Alban D, Dajaku G, Gerling D (2022) Evaluation of flux-barrier stator in five-phase PMSMs for electric aircraft traction, International Conference on Electrical Machines and Systems ICEMS, Chiang Mai, Thailand, pp. 1–6.

88 Roth Ch, Dajaku G, Gerold J, Greifelt A, Gerling D (2022) Distributed windings with flux-barriers applied to PM wind generators, International Conference on Electrical Machines and Systems ICEMS, Chiang Mai, Thailand, pp. 1–5.

89 Nollau A, Gerling D (2015), A flux-barrier cooling for traction motors in hybrid drives, IEEE International Electric Machines and Drives Conference IEMDC, Coeur d'Alene, USA, pp. 1103–1108.

90 Zhou R, Li G J, Zhu Z Q, Foster M P, Stone D A, Jia C J, McKeever P (2021) Novel liquid cooling technology for modular consequent-pole PM machines, IEEE International Electric Machines and Drives Conference IEMDC, (Virtual) USA, pp. 1–7.

91 Li G J, Ren B, Zhu Z Q, Foster M, Stone D A (2018) Demagnetization withstand capability enhancement of surface mounted PM machines using stator modularity, IEEE Trans Ind Appl 54 (2), pp. 1302–1311.

92 Reddy P B, Hu K K, El-Refaie A M (2012) Effect of stator shifting on harmonic cancellation and flux weakening performance of interior PM machines equipped with fractional slot concentrated windings for hybrid traction application, IEEE Energy Conversion Congress and Exposition ECCE, pp. 525–533.

93 Reddy P B, Hu K K, El-Refaie A M (2014), Generalized approach of stator shifting in interior permanent-magnet machines equipped with fractional slot concentrated windings, IEEE Trans Ind Electron 61 (9), pp. 5035–5046.

94 Kum-Kang H, El-Refaie A M (2014) Generalized approach of stator shifting in interior permanent-magnet machines equipped with fractional slot concentrated windings, IEEE Trans Ind Electron 61 (9), pp. 5035–5046.

95 Wang J, Patel V I, Wang W (2014) Fractional slot permanent magnet brushless machines with low space harmonic contents, IEEE Trans Mag 50 (1), pp. 1–9.

96 Wang K, Zhu Z Q, and Ombach G (2014) Synthesis of high performance fractional slot permanent-magnet machines with coil-pitch of two slot-pitches, IEEE Trans Energy Convers 29 (3), pp. 758–770.

97 Patel V I, Wang J, Wang W, Chen X (2014) Six-phase fractional-slot-per-pole-per-phase permanent-magnet machines with low space harmonics for electric vehicle application, IEEE Trans Ind Appl 50 (4), pp. 2554–2563.

98 Patel V I, Wang J, Nugraha D T, Vuletić R, Tousen J (2016) Enhanced availability of drivetrain through novel multiphase permanent-magnet machine drive, IEEE Trans Ind Electron 63 (1), pp. 469–480.

99 Domann N, Henke M (2014) Design and build-up of a high performance six-phase machine for an automotive application, International Conference on Electrical Machines ICEM, Berlin, Germany, pp. 2080–2086.

100 Spas S, Dajaku G, Gerling G (2015) Eddy current loss reduction in PM traction machines using two-tooth winding, IEEE Vehicle Power and Propulsion Conference VPPC, Montreal (Quebec), Canada, pp. 1–5.

101 Xie S, Chen H, Cao L Zou Y, Yuan X, Han B S, Hoang C C, Lee C H T (2021) Harmonic reduction for two-slot pitch winding permanent magnet vernier machines with stator shifting technique, IEEE Annual Conference of Industrial Electronics Society IECON, Toronto, Canada, pp. 1–6.

102 Rudden I A, Li G J, Zhu Z Q, Duke A, Azar Z, Thomas A (2024) Stator shifting in dual m-phase SPM machines with single-layer windings for space harmonic cancellation, IEEE Trans Ind Appl 60 (5), pp. 6867–6876.

103 Korthals-Altes W (1918) Motor winding, US Patent 1,267,232.

104 Hughes A (1970) New 3-phase winding of low m.m.f.-harmonic content, Electr Eng 117 (8), pp. 1657–1666.

105 Gjota R (1987) Winding arrangement of a stator and/or rotor of a three phase generator or electromotor with improved performances, US Patent 4,710,661.

106 Chen Y, Chen Z (1998) Investigation of a new AC electrical machine winding, IEE Proc Electr Power Appl 145 (2), pp. 125–132.

107 Fogarty J M (2004) Combined delta-wye armature winding for synchronous generators and method, US Patent 6, 704, 993.

108 Cistelecan M V, Ferreira F J T E, Cosan H B (2008) Generalized MMF space harmonics and performance analysis of combined multiple-step, star-delta, three-phase windings applied on induction motors, International Conference on Electrical Machines ICEM, Vilamoura, Portugal, pp. 1–5.

109 Cistelecan M V, Ferreira F, Popescu M (2009), Adjustable flux three phase a.c. machines with combined multiple-step star-delta winding connections, IEEE International Electric Machines and Drives Conference IEMDC, Miami, USA, pp. 219–224.

110 Lei Y, Zhao Z, Wang S, Dorrell D G, and Xu W (2011) Design and analysis of star–delta hybrid windings for high voltage induction motors, IEEE Trans Ind Electron 58 (9), pp. 3758–3767.

111 Misir O and Ponick B (2014) Analysis of three-phase induction machines with combined star-delta windings, IEEE Int Symp Ind Electron ISIE, Istanbul, Turkey, pp. 756–761.

112 Ferreira F J T E, (2015) On the star, delta and star-delta stator winding connections tolerance to voltage unbalance, IEEE International Electric Machines and Drives Conference IEMDC, Coeur d'Alene, USA, pp. 1888–1894.

113 Dotz B (2018) Windings, Design and Optimization of Electric Machines for Hybrid Traction Applications, Ph. D. Thesis, Universitaet der Bundeswehr Muenchen, Shaker, Aachen, Germany.

114 Raziee S M, Misir O, Ponick B (2018) Combined star-delta winding analysis, IEEE Trans Energy Convers 33 (1), pp. 383–394.

115 Vansompel H, Sergeant P, Dupré L and Bossche A V (2010) A combined wye-delta connection to increase the performance of axial-flux PM machines with concentrated windings, IEEE Trans Energy Convers 27 (2), pp. 403–410.

116 Dajaku G, Gerling D (2013) A novel tooth concentrated winding with low space harmonic contents, IEEE International Electric Machines and Drives Conference IEMDC, Chicago, USA, pp. 755–760.

117 Zhao W, Zheng J, Ji J, Zhu Sh, Kang M (2018) Star and delta hybrid connection of a FSCW PM machine for low space Harmonics, IEEE Trans Ind Electron 65 (12), pp. 9266–9279.

118 Dajaku G, Spas S, Dajaku Xh, Gerling D (2016) An improved fractional slot concentrated winding for low-pole induction machines, International Conference on Electrical Machines ICEM, Lausanne, Switzerland, pp. 114–119.

119 Chan C C (1987) Axial-field electrical machines – design and application. IEEE Trans Energy Convers 2 (2), pp. 294–300.

120 Patterson D, Spée R (1995) The design and development of an axial flux permanent magnet brushless DC motor for wheel drive in a solar powered vehicle. IEEE Trans Ind Appl 31 (5), pp. 1054–1061.

121 Huang S, Luo J, Leonardi F, Lipo T A (1999) A comparison of power density for axial-flux machines based on general purpose sizing equations. IEEE Trans Energy Convers 14 (2), pp. 185–191.

122 Cavagnino A, Lazzari M, Profumo F, Tenconi A (2002) A comparison between the axial flux and the radial flux structures for PM synchronous motors, IEEE Trans Ind Appl 38 (6), pp. 1517–1524.

123 Eastham J F, Profumo F, Tenconi A, Hill-Cottingham R, Coles P, Gianolio G (2002) Novel axial flux machine for aircraft drive: design and modelling. IEEE Trans Mag 38 (5), pp. 3003–3005.

124 Marignetti F, Colli D, Stefano D, Cavagnino A (2007) Design issues of a fractional-slot windings axial flux PM machine with soft magnetic compound stator. IEEE Annual Conference of Industrial Electronics Society IECON, Taipei, Taiwan, pp. 187–192.

125 Di Gerlando A, Foglia G M, Iacchetti M F, Perini R (2011) Axial flux PM machines with concentrated armature windings: design analysis and test validation of wind energy generators. IEEE Trans Ind Electron 58 (9), pp. 3795–3805.

126 Woolmer T J, McCulloch M D (2007) Analysis of the yokeless and segmented and segmented armature machine, IEEE International Electric Machines and Drives Conference IEMDC, Antalya, Turkey, pp. 704–708.

127 Gieras J F, Wang R J, Kamper M (2004) Axial Flux Permanent Magnet Brushless Machines. Kluwer Academic Publishers, The Netherlands.

Index

Note: Page numbers in *italics* and **bold** refers to figures and tables respectively.

a

asymmetric coils
  FSCWs with 87–95
  two-tooth FSWs with 204–205, *204*
asymmetric nine-phase FSCW 114–118, *115*, *117*, **118**
asymmetric six-phase FSCW 112–114, *113–114*, **114**
asymmetric six-phase winding 110
asymmetric star-delta FSCWs 219–221
  one-zone **220**
  two-zone **220**
  winding layout *220*, *221*
asymmetric stator shifting, AFMs with 261–264, *262–264*
asymmetric winding *217*
asymmetric zone distribution 58–66
  DL-FSCW with 63–66
  and SL-FSCW 59–63
  and unequal zone groups 59–63
axial flux machines (AFMs) 249, *250*
  asymmetric stator shifting 261–264, *262–264*
  concentrated windings 254–261
  construction 249–251, *250*, *251*
  double-layer FSCWs 255–259
  flux-barrier stator 264–265, *265*
  fractional slot concentrated windings (FSCWs) for 249–266
  realization 249–251, *250*, *251*
  single-layer FSCWs 259–261
  stator-shifting technique for 254–261
  stator winding MMF harmonics 251–254, *252*, *253*, **254**
  topologies 265–266, *266*, **266**
  unequal turns per coil 264

b

balanced star-delta FSCWs 222–228
  asymmetric FSCWs 225–228, *227*, **227**, **228**
  symmetric FSCWs 223–225, *224*, *225*
base winding 6, *6*, 8–9, 18–19, 22, 24, 27, 47, 252

c

coil configurations 87–106
  exemplary windings with special 104–106
coil MMF function 17–18
coil-pitch 8
coil-pitch effect 103–104, *103–104*
coil polarity 4–5
concentrated windings 9–10, *10*, *134*, 254–261
conventional stator core 133, *134*

d

distribution factors *vs.* shift angle 208, *208*
double-layer FSCW (DL-FSCW) 24–28, *24–25*, *27*, 154
  with asymmetric zone distribution 63–66
  with asymmetric zone-groups 84–85
  opposite polarity adjacent coils 26
  same polarity adjacent coils 26–27
  and stator-yoke flux-barriers 149–153, *150–153*, **152**, **153**
  with two unequal zone-groups 55–58
  with two unequal zones 53–55
  with unequal zone-groups 63–66, 83–84
  with unequal zones 53–58, 81–82
  winding factors for different 31, **32–33**

double-layer FSCW (DL-FSCW) (*Cont.*)
with $z = 1$, 39–41
with $z = 2$, 41–43
double-layer multiphase two-tooth FSWs 199–202, *199–202*, **200**
double-layer two-tooth/three-phase FSWs 188–194
double-layer FSWs with one zone/phase 188–191, *188–190*, **189**
double-layer FSWs with two zone/phase 191–192, *191–192*
double-layer FSWs with unequal-zones 193–194, *193*, **194**
double-layer winding 4, *4*
layout 122
dual slot-layer stator, star-delta FSCWs with 242–244, *242*, *244*
dual three-phase winding 110

e

eight-phase SL-FSCW 174, *174–175*
11-phase SL-FSCW 130
equivalent MMF function 141–143, *143*
exemplary star-delta FSCW machines 244–247, *245–247*
exemplary windings with special coil configurations 104–106, *105–106*

f

five-phase DL-FSCW 126
five-phase SL-FSCW 127
flux-barrier configurations 136, *137*
flux-barrier effect 136, *140*, *143*
flux-barrier stator 133, *154*, *168*, **169**
AFMs with 264–265, *265*
fractional slot concentrated windings with 133–177
five- and seven phase SL-FSCW with *171*
MMF distribution of DL-FSCW *150*
MMF harmonic winding factors *146*, *147*, *148*, *153*
with single coil *157*
flux density distribution 14, *138*, 154
flux transformation effect 154–155, *154*, **155**
flyer winding 9–10
fractional slot concentrated windings (FSCWs) 13–44, 133–177
with an arbitrary number of phases 118–125
with asymmetric coils 87–95
asymmetric nine-phase 114–118
asymmetric six-phase 112–114
with asymmetric zone distribution 58–66
for axial flux machines (AFMs) 249–266
coil MMF function 88–91
conclusions 241
examples 239–241
exemplary windings with special coil configurations 104–106
and flux-barrier stator 133–177
key observations 241
MMF analysis 47–66
MMF analysis of standard FSCWs 18–28
MMF characteristics of three-phase FSCWs with unequal zones 79–86, *80–85*, **86**
MMF characteristics of three-phase standard 35–44
multiphase 165–177
one-zone FSCWs 239–240, *239*, *240*
single-coil MMF formulation 13–18
with specific coil configuration 87–106
and star-delta connection 211–247
stator flux-barrier techniques 135–138, *135*
stator-teeth flux-barriers 155–165, 169–177, **170**, *171–172*, **172**
stator-yoke flux-barriers 138–155, *138–140*, 165–169, *166–169*, **167**, **169**
three-phase MMF distribution 33–35
two-zone FSCWs 240–241, **241**, *241*
with unequal coil-pitch 101–104
with unequal turns per coil 95–101
with unequal zones 45–86
with variable pole-pair number 122–123
winding configurations 45–47
winding factors 28–33, **29**, *30*, **31–33**, 66–79, **66–73**, *74*, **75–79**
fractional slot winding 7
full yoke flux-barrier 139, *139*
fundamental harmonic 9
fundamental winding specifications 7–9, *7*

g

greatest common divisor (GCD) 8

h

harmonics 30–31

i
insertion winding 9–10

l
linear winding 9–10, *10*

m
magnetomotive force (MMF)
- analysis 47–66
- analysis of standard FSCWs 18–28
- approach 102
- coil MMF function 17–18
- function analysis 120–122
- harmonic definition 9
- parameters 14–17
- three-phase FSCWs with unequal zones 79–86, *80–85*, **86**
- three-phase standard FSCWs 35–44
- winding function 236–238, *237*, *238*
- working harmonics 78–79, **79**

maximum winding factors 29, *30*, 74–78
multiphase double-layer FSCWs 126–127
multiphase fractional slot concentrated windings 107–131
- FSCWs with an arbitrary number of phases 118–125
- multi-three-phase FSCWs 112–118
- representative multiphase FSCWs 125–131
- with slot/pole combination 124–125
- topologies 107–111

multiphase single-layer FSCWs 127–130
multi-pole operation mode 123–124
multi-three-phase FSCWs 112–118
- asymmetric six-phase FSCW 112–113, *113–114*, **114**
- asymmetric nine-phase FSCW 114–118, *115*, *117*, **118**

n
needle winding 9–10, *10*

o
one-zone FSCWs 239–240, *239*, *240*
one-zone star-delta FSCWs 229–231, *230*, *231*, **231**
operating harmonic 9
operating wave 9
operation modes 122–124

p
partial yoke flux-barrier 139, *139*
permanent magnet synchronous motor (PMSM) 133, 211
phase zone 5–6, *5–6*

r
radial flux machine (RFMs) 249, *250*
representative multiphase FSCWs 125–131
multiphase double-layer FSCWs 126–127, *126–128*
multiphase single-layer FSCWs 127–130, *129–130*

s
sequential star-delta coil arrangements 234–241, *235*, *236*
- FSCW examples 239–241
- MMF winding function 236–238, *237*, *238*

seven-phase DL-FSCW 126
seven-phase SL-FSCW 128
single-coil MMF formulation 13–18
single-layer FSCW (SL-FSCW) 18–24, *18–20*, *22–23*, 154
- with asymmetric zone distribution 59–63
- FSCW with $z = 2$, 23–24, *23*
- FSCW with $z = 2$ and $n_c = 1$ 22–23
- MMF characteristics for seven-phase 173–174, *173*, **174**
- MMF characteristics for eight-phase 174–175, *174*, **175**
- MMF characteristics for 11-phase 175–177, *175*, *176*, **177**
- with $n_c = 1$, 22
- opposite polarity adjacent coils 21–22
- same polarity adjacent coils 20–21
- and stator-teeth flux-barriers 158–163, *159–160*, 163–165, **163**, *163–165*, **165**
- and stator-yoke flux-barriers 143–149, *144–148*, **149**
- with two unequal zone-groups 50–53
- with two unequal zones 48–50, *48–49*
- with unequal zone groups 59–63
- with unequal zones 48–53, 79–80, 163–165, *164–165*, **165**
- winding factors for different 31, **31**
- with $z = 1$, 35–37

single-layer FSCW (SL-FSCW) (*Cont.*)
with $z = 1$ and $n_c = 1$, 18–19, *18*
with $z = 2$, 22, 37–39
single-layer multiphase two-tooth FSWs 195–199, *195–198*, **197**, **198**
single-layer two-tooth/three-phase FSWs 184–188
single-layer FSWs with one zone/phase 184–186, *184–186*, **185**
single-layer FSWs with two zone/phase 186–188, *187*, **187**
single-layer winding 4, *4*
layout 121
sinusoidal MMF distribution, star-delta FSCWs with 233–234, **234**
six-phase winding 108–110
specific coil configurations 203–210
and two-tooth fractional slot winding (FSWs) 203–210
two-tooth FSWs employing unequal tooth-width stator core 207–210, *207–210*
two-tooth FSWs with asymmetric coils 204–205, *204*
two-tooth FSWs with unequal coils per zone 205–207, *205*, *206*
standard FSCWs 13–44
double-layer FSCW (DL-FSCW) 24–28, *24–25*, *27*
MMF analysis of 18–28
single-layer FSCW (SL-FSCW) 18–24, *18–20*, *22–23*
star-delta connection
exemplary star-delta FSCW machines 244–247, *245–247*
and FSCWs 211–247
FSCWs with sequential star-delta coil arrangements 234–241
star-delta FSCWs with dual slot-layer stator 242–244, *242*, *244*
star-delta FSCW topologies 216–221
star-delta windings 212–215
winding analysis 221–234
asymmetric star-delta FSCWs 219–221, **220**, *220*, *221*
with dual slot-layer stator 242–244, *242*, *244*
with sinusoidal MMF distribution 233–234, **234**
symmetric star-delta FSCWs 216–219, *217*, *218*, *219*
topologies 216–221
star-delta parallel connection 212, *213*
star-delta phase currents 213–214, *214*
star-delta series connection 212, *213*
star-delta windings 212–215, *213*
star-delta phase currents 213–214, *214*
winding MMF function 214–215
stator flux-barrier techniques 135–138, *135*
flux-barrier effect 136
realization 136–138, *136*, *137*
stator-shifting technique 254–261
stator-teeth flux-barriers 155–165, *156*, *157*, 169–177, **170**, *171–172*, **172**
MMF characteristics for seven-phase SL-FSCW 173–174, *173*, **174**
MMF characteristics for eight-phase SL-FSCW 174–175, *174*, **175**
MMF characteristics for 11-phase SL-FSCW 175–177, *175*, *176*, **177**
and SL-FSCWs 158–163, *159*, *160*, 163–165, *163–165*, **163**, **165**
and unequal zones 163–165, *164–165*, **165**
and winding factors 160–161
stator winding MMF harmonics 251–254, *252*, *253*, **254**
stator winding types 2–3, *3*
stator-yoke flux-barriers 138–155, *138–140*, 165–169, *166–169*, **167**, **169**
and DL-FSCWs 149–153, *150–153*, **152**, **153**
equivalent MMF function 141–143, *143*
flux transformation effect 154–155, *154*, **155**
key features 153–155
potentials 153–155
and SL-FSCWs 143–149, *144–148*, **149**
subharmonics 9
symmetric six-phase winding 110
symmetric star-delta FSCWs 216–219, *217–219*
symmetric winding *217*

t

10-phase double-layer winding 126–127
three-phase FSCWs
conversion 108–110, *108–109*
MMF characteristics of 79–86, *80–85*, **86**
with unequal zones 79–86, *80–85*, **86**

three-phase MMF distribution 33–35
three-phase standard FSCWs
  MMF characteristics of various 35–44
two-tooth fractional slot winding (FSWs) 179–210
  double-layer multiphase two-tooth FSWs 199–202, *199–202*, **200**, **203**
  double-layer two-tooth/three-phase FSWs 188–194
  implementation 180–183, *180–183*
  multiphase 194–203
  operating principles 180–183, *180–183*
  single-layer multiphase two-tooth FSWs 195–199, *195–198*, **197**, **198**
  single-layer two-tooth/three-phase FSWs 184–188
  and specific coil configurations 203–210, *203*
two-zone FSCWs 240–241, *241*, **241**
two-zone star-delta FSCWs 231–233, *232*, *233*

u
unbalanced star-delta FSCWs 228–233
  one-zone star-delta FSCWs 229–231, *230*, **231**, *231*
  two-zone star-delta FSCWs 231–233, *232*, *233*
unequal coil-pitch, FSCWs with 101–104
unequal coils per zone, two-tooth FSWs with 205–207, *205*, *206*
unequal tooth-width stator core, two-tooth FSWs 207–210, *207–210*
unequal turns per coil, FSCWs with 95–101
unequal zone groups
  and asymmetric zone distribution 59–63
  DL-FSCW with 63–66
  SL-FSCW with 59–63
unequal zones 163–165, *164*, **165**, *165*
  single-layer FSCW (SL-FSCW) with 48–53
  and SL-FSCWs 163–165, *164*, **165**, *165*
  and stator-teeth flux-barriers 163–165, *164*, **165**, *165*
  three-phase FSCWs with 79–86, *80–85*, **86**

v
variable pole-pair number 122–123

w
winding
  application 10–11
  comparison of winding performances 43, **44**
  concentrated 9–10, *10*
  configurations 45–47
  dual three-phase 110
  feasibility 8
  manufacturing methods 9–10, *10*
  six-phase 108–110
winding analysis 221–234
  balanced star-delta FSCWs 222–228
  star-delta FSCWs with sinusoidal MMF distribution 233–234, **234**
  unbalanced star-delta FSCWs 228–233
winding factors 28–33, *30*, 66–79, *74*, **75–79**, *152*, *155*, 160–161, **163**, *163*, **165**, *167*, **198**
  for different balanced star-delta symmetric FSCWs **227**, **228**
  for different DL-FSCWs 31, **32–33**, **66–67**, **254**
  for different sequential star-delta coils one-zone FSCWs **239**
  for different sequential star-delta coils two-zone FSCWs **241**
  for different SL-FSCWs 31, **31**, **67**
  double-layer winding with two unequal zone-groups **194**
  eight-phase SL-FSCW **175**
  11-phase SL-FSCW **170**, **177**
  formulations for standard FSCWs **29**
  five-phase double-layer winding **200**
  one-zone double-layer windings **189**
  MMF harmonic winding factors **68–73**
  seven-phase double-layer winding **203**
  seven-phase SL-FSCW **174**
  theory 1
  two-zone double-layer windings **192**
  two-zone single-layer windings **187**
winding MMF function 214–215
winding periodicity 8
winding realization 4–6
  base winding 6
  coil polarity 4–5
  double-layer winding 4, *4*
  number of layers 4
  phase zone 5–6, *5–6*
  single-layer winding 4, *4*
winding specifications
  coil-pitch 8
  fundamental 7–9, *7*

winding specifications (*Cont.*)
  number of slots 8
  slots per pole and phase 7
  turns per coil and phase 8–9
  winding feasibility 8
  winding periodicity 8
winding topologies 6
working harmonic 9

Printed and bound by CPI Group (UK) Ltd, Croydon, CR0 4YY
17/09/2026
14964239-0001